Nanonetworks

IEEE Press
445 Hoes Lane
Piscataway, NJ 08854

Nanonetworks

The Future of Communication and Computation

Florian-Lennert A. Lau
Universität zu Lübeck
Institute of Telematics
Lübeck, Germany

IEEE Press
WILEY

Published by John Wiley & Sons, Inc., Hoboken, New Jersey.
Published simultaneously in Canada.

Library of Congress Cataloging-in-Publication Data

Names: Lau, Florian-Lennert A., author.
Title: Nanonetworks : the future of communication and computation / Florian-Lennert A. Lau.
Description: Hoboken, New Jersey : Wiley, [2024] | Includes bibliographical references and index.
Identifiers: LCCN 2024009796 (print) | LCCN 2024009797 (ebook) | ISBN 9781394213108 (hardback) | ISBN 9781394213115 (adobe pdf) | ISBN 9781394213122 (epub)
Subjects: LCSH: Nanonetworks.
Classification: LCC TK7874.845 .L38 2024 (print) | LCC TK7874.845 (ebook) | DDC 621.39/81–dc23/eng/20240314
LC record available at https://lccn.loc.gov/2024009796
LC ebook record available at https://lccn.loc.gov/2024009797

Cover Design: Wiley
Cover Images: © Andriy Onufriyenko/Getty Images

Set in 9.5/12.5pt STIXTwoText by Straive, Chennai, India

Contents

List of Figures

List of Tables

About the Author

Florian-Lennert A. Lau attained his bachelor's degree in computer science in 2013, followed by his master's degree with a focus on computational complexity theory in 2016. Since 2016, he has been actively engaged as a PhD student at the German University of Lübeck. Distinguished as the head of the pioneering Nano Group Lübeck, established in 2016, he has significantly influenced the field of nanonetworks by introducing DNA as a multi-purpose material.

In 2020, Florian was awarded the title of best PhD thesis by the special interest group "Communication and Distributed Systems" (KuVS). Subsequently, he transitioned to the role of postdoctoral researcher at the University of Lübeck.

Next to several private publications, Florian authored his inaugural scientific book, *Nanonetworks – The Future of Communication and Computation*, in 2024. His research is intricately focused on DNA-based self-assembly systems and their pioneering applications in nanorobots and nanonetworks, marking him as a leading authority in this field.

Preface

At the time of writing this book, I was unknown in the world of nanonetworks. However, people usually judge the credibility of new information by the reputation of the person giving it to them. Since I have no reputation, I would encourage you to come and see for yourself if the information in this book is any good. I have tried my very best to explain everything as pragmatically and directly applicable as possible. As Richard Feynman once said, "If you cannot explain something in simple terms, you don't understand it."

In order to enable the largest possible audience to read a scientific work, it should be written as simply and unconditionally as possible. The understanding of a text can be accelerated, for example, by using repetitions in appropriate places. This may sometimes irritate an expert, but the additional explanations benefit inexperienced but interested readers. For a thorough understanding of the following work, however, it is assumed that the reader has the basics of complexity theory, as well as logic and general computer science, or is willing to acquire them using the referenced secondary literature. The repetition of all the basics is far beyond the scope of a single book. Mathematical procedures are always chosen in the course of the work in such a way that they provide practical use. For example, theoretical proofs of concepts are nice but often of little practical relevance. By means of a constructive approach or proof, however, it is possible to use results effectively or to derive procedures that offer real added value.

Why This Book

Before we begin with anything else, it is good to understand the reasons and driving forces behind this book. The field of nanonetworks is, in several regards, special and cannot be easily compared to other areas of research.

First and foremost, nanonetworks is a young discipline that has only been introduced as recently as 2008. That said, that is only true for the area itself – other, much older sciences have been and are still contributing to this area for a very long time. Like any other discipline, nanonetworks are also standing on the shoulders of giants.

Due to the interdisciplinary nature of the field, nanonetworks are standing on a surprising number of giants. Unlike other areas, nanonetworks incorporate results from biology, chemistry, physics, medicine, computer science, mathematics, and engineering, to name just the major scientific fields. As a result, researchers have to be well-educated in several of those domains. Concerning all other domains, contributors should at least be aware of all the major ideas and problems. Consequently, especially new researchers can have a difficult time entering this vast area of research, and it can be quite intimidating to even start.

Additionally, the many different backgrounds also lead to a number of problems concerning wording and definitions. It is not rare that identical words mean very different things to the contributors. As a result, communication can be surprisingly difficult, and it is necessary to find a common ground and vocabulary to effectively communicate and avoid doing research twice.

Further, nanonetworks still sound extremely futuristic to the average person, and it is surprisingly difficult to tell science fiction and actual science apart from each other. This is only aggravated by the general tendency of the media to unnecessarily blow up headlines. Headlines are so commonly exaggerated that it is more and more difficult to tell the hype and the actual science apart from each other.

All of those aspects and many more led me to the conclusion that it is time for a somewhat general book that summarizes all the major ideas and problems in this area. We need a common language and a good entry point for new researchers if we want to achieve lasting progress as a community. Interdisciplinary research can be challenging, but given the right prior education, it can be done effectively and efficiently.

Important Remarks

Before we start with any tutorial on reading order or anything else, a few words on the peculiarities of the book.

While I avoid opinions in a book such as this whenever possible, there will be some occasions where I make predictions about the future. In those and a few other cases, the use of opinions is signified by the use of "I," "me," or "mine." Thus, it should be clear at any point what is an opinion and what is following the scientific publications.

If you are completely new to the topic of nanonetworks, I would advise you to read all the presented content in the displayed order. If you are just interested in a single topic, you can jump to the respective chapter by clicking on the elements in the table of contents. All of the chapters can be read on their own and possess the necessary introductions and connecting elements that allow for an enjoyable and informative read. It could be beneficial to also read the preface and the introduction in any case, as that helps convey the idea behind this book.

As this is an interdisciplinary book, some topics will not be covered in the depth they might deserve. This book is mainly written for computer scientists and engineers; it should offer a good entry point for scientists from other areas too. Computer science is a tool science where it is often necessary to learn a new field in a short period of time. During that time, it is often necessary to isolate and understand the big questions and problems in the field. Thus all the relevant aspects of a field will likely be covered, but some details will likely be missing.

Acknowledgments

At this point, I would like to thank a few people who have actively or passively contributed to the success of this work – some of the people mentioned might not even be aware of it. My thanks go to these people:

I thank my parents, who always gave me the necessary freedom to do what I wanted. They taught me as much autonomy as needed to allow me to make advanced progress and develop a certain pride in my own achievements – this is the basis of my ambition and drive.

I thank my grandmother for being a kind-hearted person.

I would like to thank my former high school teachers, Peter Koerting and Wolfgang Malm, who taught me a deep need for formal correctness and an affection for the natural sciences.

I would like to thank Rüdiger Reischuk for taking the time and patience to put the knowledge I gained during my studies to the test.

I would like to thank Bennet Gerlach for the fact that he is available to me almost every day for discussion and reflection and thus closed many gaps in my school knowledge that would otherwise still exist today. Without him, my model of the world would be fundamentally different – and, I think, worse.

I would like to thank my girlfriend, Christin Grill, for always being by my side with discussions, advice, and action.

I thank my brothers Finn and Frederic for the free time activities when the pressure of study and work became overwhelming.

And last but not least, I would like to thank my doctoral supervisor Stefan Fischer for the many years of supervision and the introduction to the academic world with all its subtleties, as well as the financial support that I benefited from at his institute.

The rest of my thanks go to all members of the Nano Group and the Institute for Telematics, who have walked part of my life with me over the last few years and have always supported me when I asked for it.

1

Introduction

This chapter introduces the new reader to the topic of nanonetworks and nanotechnologies in general. As this is a scientific book that goes into some level of depth, some scientific knowledge is presupposed. We first have a look at the designation "nano" and explain where the entire idea came from. Then, we have a look at science fiction, which often serves as a basis for future developments. Afterward, we try to establish an intuitive understanding of nanotechnologies and present several example applications that can either be conceptualized or even realized in wet-lab experiments. Then we discuss those parameters and environmental constraints that distinguish nanotechnologies from the early days of computer science, where resources have been very constrained as well. This includes various challenges and "new" laws of nature that system engineers have to deal with. Lastly, we summarize the chapter and give a chapter-wise overview of the entire book.

1.1 Etymology

Before we begin with anything else, it is beneficial to understand the meaning of the word "nano" and its etymology. The word *nano* was derived from the Greek word "nanos," meaning "dwarf." Although the Greek concept of atoms might be much closer in meaning to what we call today as "nano," the word was already overloaded with meaning in physics.

Technically, the prefix "nano-" is used to designate anything that is a billion times smaller than a meter (1–4000^{-9} m). In numerical terms, everything between 1 and 999 nm lies in the nanorange. While that definition suits the physical perspective well, the nanonetworks community decided that anything between 1 and 4000 nm in size could be considered "nanoscale." The smallest human blood vessels, also called capillaries, are about 4000 nm in diameter. Thus, any "nanobot" that is designed to operate in the human body must be smaller than that.

A more relatable example can be seen in Figure 1.1. On the left, a soccer ball is compared to the size of the planet Earth. The same ratio applies to the same soccer ball compared to a carbon nanotube.

To put things into perspective, a single hair is between 20 000 and 200 000 nm in diameter. A typical human cell has a diameter of about 10 000 nm, while DNA only has a diameter of about 2 nm. Hence, the nanoscale is extremely small.

Nanonetworks: The Future of Communication and Computation, First Edition. Florian-Lennert A. Lau.

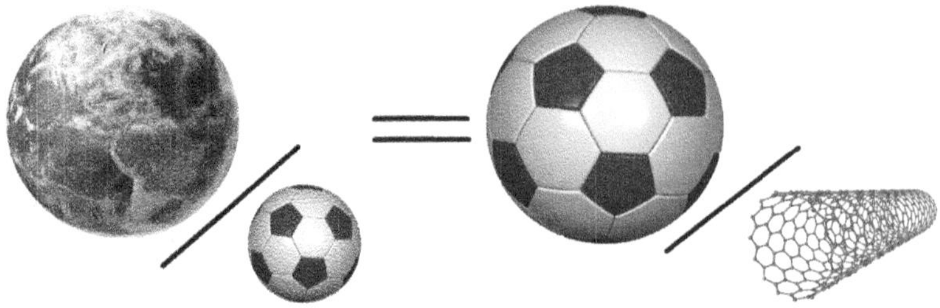

Figure 1.1 A comparison between a soccer ball and the planet Earth made by Marc Stelzner at a university seminar. *Source*: ITM (Stelzner).

1.2 Science Fiction

While the entire idea of "nano" already surfaced more than 2500 years ago in ancient Greece and India, it only picked up speed in the last 200 years. While the majority of the historical development is discussed in Chapter 2, the popular science fiction part is discussed here. Science fiction is technically not part of history, yet many ideas in modern physics have been inspired by even older science fiction literature. One could even go as far as claiming that such early literature supplies modern science with visions and ideas to strive for. While this has been true for many decades, there seems to be a current shortage of new ideas that might motivate future research.

The science fiction history that explicitly names nanotechnologies began in the 1950s with the famous author Arthur C. Clarke. In 1956, he wrote the story "The Next Tenants," where he writes about machines that are only micrometers in size (Clarke, 1957). It is a story about crazy scientists where biological and technological elements have been combined. The protagonist of the story discovers this mad scientist on a Pacific island, where he trains termites in the use of technology as he is convinced that humanity is doomed to self-destruct. This overarching theme describes the Cold War mentality quite well and combines it with the first taste of nanotechnologies as we envision them today.

While many of the elements of the story seem foreign to us, nanoscale technology is by no means limited to science fiction literature.

Another famous author who picked up the idea of nanotechnology is Stanislaw Lem, who is well-known for his "Star Diaries." In 1964, he wrote the book *The Invincible*, about a spacecraft from which the name of the book derived (Lem, 1964). The crew of the ship is sent out to investigate the fate of its sister ship the "Condor". During the investigations, the crew of the spacecraft discovers traces of quasi-life based on an evolutionary selection process of self-replicating machines. This might be one of the, if not the first, dystopian stories in which nanotechnology has "gone wrong."

In 1984, Stanislaw Lem wrote another book titled *Peace on Earth* (Lem, 1984). This book combines visionary aspects of artificial intelligence and nanorobotics. Lem envisions a peaceful society where most of the arm production is outsourced to factories on the moon. In the process, the nations on Earth agree to demilitarize the Earth completely. While a lot of autonomy is programmed into the stationary technology on the moon, many mysterious events begin to happen. A similar development as in *The Invincible* starts, and the result is a strange "metallic moon dust" that turns out to be a swarm of nanobots. During an expedition to the moon, some of that dust is brought to the earth, where the nanobots cause a near complete devastation of technology infrastructure.

Possibly motivated by this story, the famous TV show *Futurama* adopted a similar plot. Here, tech genius Professor Farnsworth is dissatisfied with the political conditions on Earth and decides to relocate to another location. To make the comet he chose more habitable, he deploys a swarm of nanorobots to purify the water. Due to unknown circumstances, the nanorobots get caught in a process of rapidly accelerated robotic evolution, which ultimately leads to the emergence of intelligent robots.

In the 1990s, even Mickey Mouse comic books contained ideas about nanorobots and the potentially devastating consequences they might bring. All of this might create a rather bleak outlook on the potential of nanotechnology to improve living conditions for humans. We will analyze the immense medical potential in Chapter 3 and have a look at the potential dangers in Chapter 13.

1.3 Nanotechnology Intuition

While technology might not have reached the stage described in science fiction, there is still an ever-decreasing trend in size. There are manifold driving forces behind this development. For example, the transistor density on memory chips is roughly doubling every 12–18 months (Moore, 1965). Gordon Moore observed this while working at Intel in 1965. A similar rule also existed for the computational power of CPUs but that law has come to a halt. Since 2010, the number of basic operations performed by a single commercial processor has remained more or less stagnant.

That said, improvements are happening, just not in the simple number of operations. For example, instead of limiting ourselves to a single processor, many devices have been introduced with multiple cores. Thus, a new dimension of parallel/concurrent computing is technically keeping the trend of accelerated computation alive. Yet, the "new" natural laws that apply at the nanoscale create a host of other challenges for system designers.

To give a better intuition of the different size scales, a comparison between different objects is shown in Figure 1.2. While transistors of the current generation might not be at the very left end of the scale, they are somewhere between the size of DNA and viruses.

Five years before Moore discovered this trend, famous scientist Feynman gave a revolutionary talk on the topic "There is plenty of room at the bottom." In this trend, he predicted the massive trend of miniaturization we have experienced over the last 60 years. While this might not seem like much to some readers, it is surprisingly difficult to predict future developments like that. Feynman more or less started the current "race to the bottom" when it comes to the limits of miniaturization.

While Feynman mainly limited himself to the areas that were available at the time, nanotechnologies turned out to be an interdisciplinary field. Over the last 50–100 years, an enormous body of knowledge has been gathered by humanity as a whole. For example, a big industrial interest

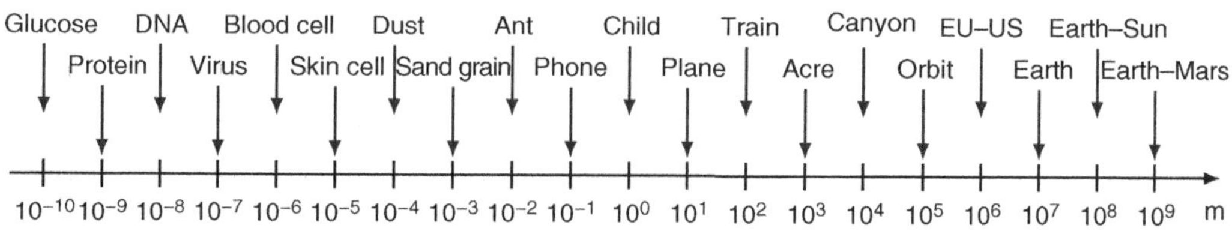

Figure 1.2 A comparison between different structures and their size. The interval between 1 and 100 nm (10^{-9} to 10^{-7}) contains many interesting, naturally occurring protein-based structures. *Source*: Florian-Lennert A. Lau.

stems from the automotive industry, medicine, and chip manufacturers. The automotive industry is mainly interested in new technologies like terahertz antennas to prevent collisions by detecting close objects in advance. Chip designers want to increase the efficiency of their products to be ahead of the competition. Medicine offers a variety of possible applications for nanotechnology, like cancer treatments on the basis of metallic nanoparticles. We analyze many of these medical applications in Chapter 3. The area draws a lot of interest as health might be the greatest concern for humans. Even the Buddha said 2600 years ago that "health is the greatest worldly good" (Gotama, 600–520 BCE approximately 500 BCE, n.d.).

The general interest in nanotechnology exceeds the industrial aspects by far. Vast areas of basic science in STEM concern themselves with aspects of nanotechnologies. Examples are physics, chemistry, biology, computer science, and many other areas. Among other things, phenomena, tools, or other technologies that either measure or manipulate matter at the nanoscale are researched.

A small overview of the relevant areas to the explicit area of nanonetworks that is discussed in this book is shown in Figure 1.3. As nanonetworks are a subdiscipline of computer science, the hierarchical mind map starts in the middle with communication and computation at the nanoscale. The map can also serve as an overview of the table of contents of the entire book.

It covers the major areas of computation, communication, and construction at the nanoscale in great detail as computer science and computer network science have most expertise to offer in those areas. It can be surprisingly difficult to construct simple structures at the nanoscale as we have few available tools and many external influences hinder precise construction processes. Computation at the nanoscale faces similar challenges. While transistors and circuits themselves are only a few nanometers in size, the voltage supply to power them is not. Overall, the constraints in memory, energy, and computational power are comparable to the early days of computer science.

Communication faces similar challenges. Energy is a scarce resource at the nanoscale, especially when trying to use the same principles as in modern computers. Wireless sensor networks use about 95% of their energy for communication, and the trend will likely only be worse in nanonetworks. As a result, it might be necessary to switch to alternative means of communication based on molecules. These have been tested by nature and are proven to work – maybe it is possible to adapt them for nanonetworking purposes.

In addition, the book progresses along the topics of locomotion and localization. Both problems can be surprisingly difficult to solve using artificial, nanoscale devices. Locomotion requires energy and is, at least in the human body, always subject to several other influences. Every small enough structure is subject to Brownian motion, diffusion, and blood flow. Finding the exact location of a structure or device on this basis can be surprisingly challenging. It is even more difficult as the memory of devices might be so limited that they cannot even store a unique identifier.

Other very important areas are actuators and sensors. Both are necessary to either capture the state of the environment or manipulate it. Actuators and sensors form the basis for communication and locomotion. While nanoscale sensors and actuators exist at least as concepts, it is surprisingly difficult to combine them with other devices. For example, the process of creating carbon nanotubes involves many stochastic processes, and the final position of a finished carbon nanotube is more or less random. Precise placement of components in 3D is a huge unsolved problem.

For networking purposes and many algorithms, it is necessary to have at least a primitive understanding of time and randomness. Without time, it is impossible to gather data in a meaningful way for medical applications. A diagnosis is only relevant for a time, and many algorithms rely on random numbers. This is especially true for heuristic methods that might save energy while not producing optimal results.

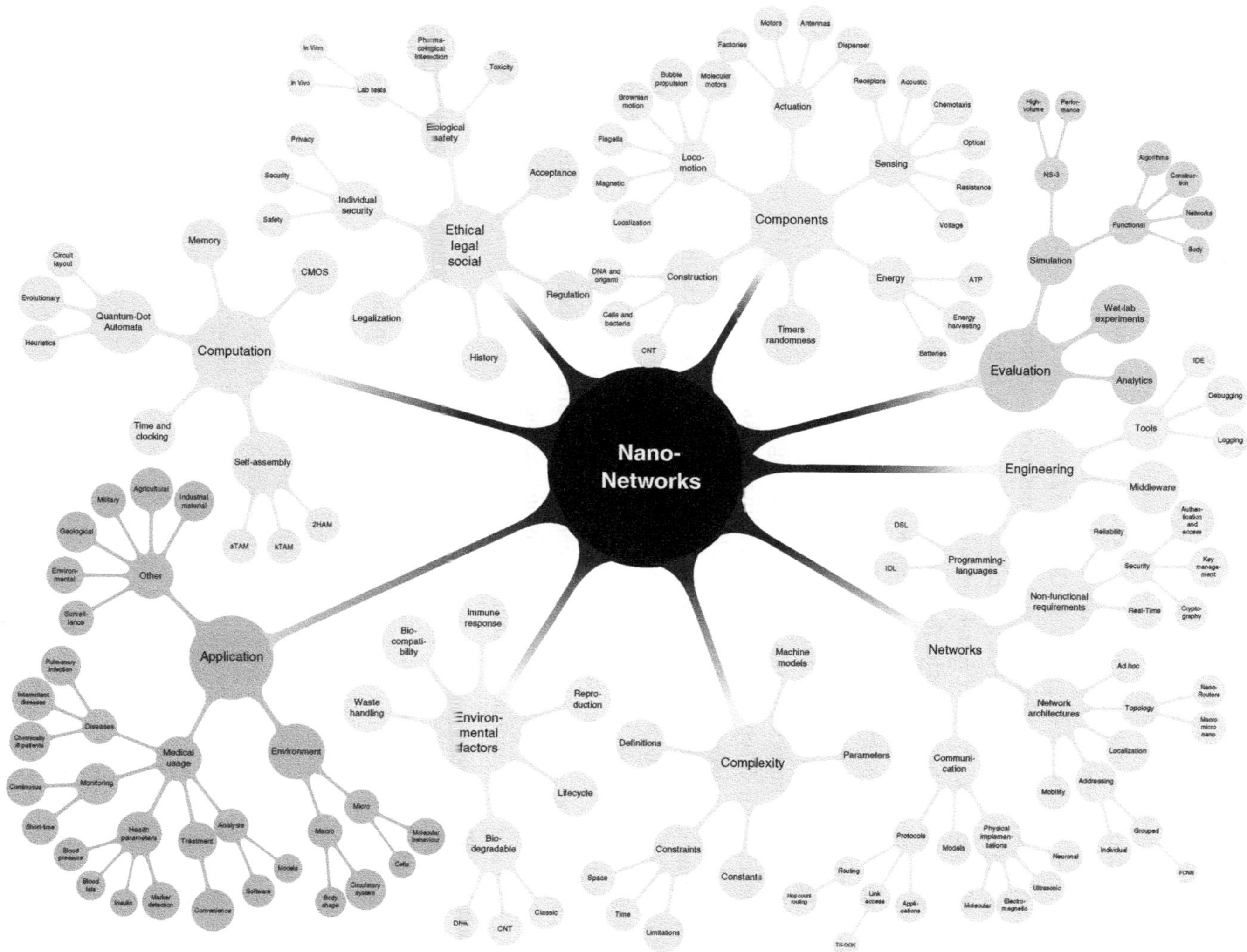

Figure 1.3 A mind map of nanonetworks. The different branches represent different areas of research and each node is related to its neighbors. *Source*: ITM (Florian-Lennert A. Lau).

The last analyzed area in the book is ethical, legal, and social issues (ELSI). ELSI is concerned itself with the ethical, legal, and social aspects of other disciplines. While it might be possible to create certain systems, it might not be possible to get a permit to use them or they might be ill-received by the society as a whole. For example, DNA research and gene manipulation have a pretty bad reputation that can be avoided given better management beforehand.

Arguably the biggest challenge is the combination of the individual solutions into a fully functional nanonetwork. As single devices are too constrained to solve difficult problems, nanodevices are expected to collaborate (Akyildiz et al., 2008). Yet, it is challenging to even assemble the different components of nanorobots into one fully functional device. The challenge of creating an entire network of such devices is even greater. In Chapter 12, we analyze the possible architecture in detail.

Yet, one especially promising approach for the described problem of assembly and the component combination is self-assembly. The general idea of the work is to grow nanoscale devices and networks of devices like crystals. Crystals are to be understood as a lattice structure made up of simple building blocks. At the nanoscale, there are few other ways to manipulate matter on a large scale.

Crystal formation is based on the principle of self-assembly. Tiny and often simply structured components assemble themselves into larger and often complex structures according to local rules. These complex structures could be nanodevices, nanosensors, packets of messages, entire nanonetworks, or even computing machines.

Although most known crystals follow the paradigm of self-assembly, the process can hardly be influenced. An exception is certain building blocks made of DNA. These also follow the self-assembly paradigm and can be treated like crystals. However, these building blocks have the property that some DNA strands can only meaningfully form a bond with their inverse.

It is possible to create arbitrary strands of DNA in the laboratory. Consequently, one can create building blocks of DNA that behave like pieces of a puzzle or even arbitrary shapes from a single DNA strand. As will be shown later, basically all the problems can be solved using just DNA as a building block. Unlike artificial approaches, DNA already works at the nanoscale and can assemble with the necessary degree of precision/accuracy. Thus, the biggest problems of precise placement and the component combination might come "for free" when focusing on DNA as an atomic building block.

In addition to DNA as a possible building block, there are other ideas about how nanodevices can be created. For example, there are many approaches that are based on naturally occurring cells or bacteria. They suggest using existing resources in nature and, if necessary, adapting them to one's own needs. For example, biological nanodevices.

These are explained in detail in Chapter 4.

Artificial approaches are also possible. An example of a nanorobot based on classic electronic components is shown in Figure 1.4. The approach is based on wireless sensor networks as used in research and parts of the economy. Novel materials, such as carbon nanotubes, are used, which are a promising building block for sensors, actuators, and storage technologies at the nanoscale (Baughman et al., 2002).

1.4 Example Applications

While the state of the art in nanotechnologies has not yet reached what science fiction literature envisioned, there are still hosts of innovations available. Today, nanoparticles are used regularly (Yao et al., 2017; Wang et al., 2015; Li et al., 2018), for example, in cancer research (Singhal et al., 2010). Some nanoparticles are structured in such a way that they only bind to

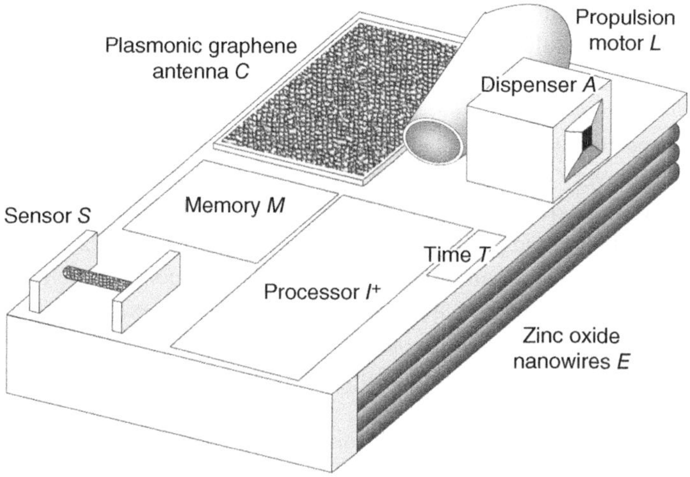

Figure 1.4 Schematic representation of an electric nanorobot. The individual components of a nanorobot are implemented using electronic components (Büther et al., 2017). *Source*: Buther et al. (2017)/with permission of Springer Nature.

cancerous tissue and are heated there by external influences. This can be done, for example, by laser. The affected tissue is destroyed and the surrounding, healthy cells can recover. Visionary approaches envision the use of nanodevices in the human body to better label or directly destroy cancer cells.

This approach can easily be generalized. In general, nanodevices can be a useful, supportive measure in the detection and treatment of diseases. This includes, among other things, the treatment of diabetes (Gu et al., 2013), the detection and local treatment of inflammatory diseases (Staples et al., 2006; Amato et al., 2010; Stelzner et al., 2016b), and many other diagnostic and therapeutic options (Benenson et al., 2004). It is often suggested that nanodevices may have to collaborate because of their size limitations (Akyildiz et al., 2008).

Complex scenarios describe the use of molecular automata, which are already able to recognize several markers for disease in the laboratory and start producing specific RNA sequences as soon as all markers are present (Anderson et al., 2006). The sequences produced do not serve any purpose at the moment, but show in principle that one can react to environmental parameters – also constructively.

In Ghavami et al. (2012), the detection of diseases is further abstracted. Among other things, applications such as intracellular operations are proposed. Such technologies require a high degree of precision (Freitas, 2005).

Another popular use case for nanodevices is to support the human immune system. It often happens that people's self-healing powers are no longer sufficient for certain clinical pictures. If a patient has reached old age, regeneration processes often take significantly longer than that in young people. A general improvement in the immune system is also conceivable, regardless of the initial situation (Freitas, 2005; Akyildiz et al., 2008; Gu et al., 2013).

1.5 Unique Problems and Challenges

People who are new to the field of nanonetworking often intuitively object that the resource constraint imposed at the nanoscale resembles early-day computer science. This criticism is indeed legit as engineers and programmers only worked with a few bits of available memory at the beginning. Thus, there is a multitude of already available knowledge concerning resource-constrained computation and “tricks” that may be used to save scarce memory or computational power.

Yet, the constraints and unique properties at the nanoscale far exceed a mere reduction in available resources. One could argue that the very laws of nature change at the nanoscale and system designers sometimes have to completely switch paradigms to make things work. Indeed, it is very likely that there is a multitude of yet undiscovered problems and challenges waiting for us.

For example, it is fairly easy to conceptualize a nanonetwork that might transport and deliver drugs to a target area. Yet, it is often overlooked that basically *every* structure that enters the human body is immediately “coated” by proteins and fats that are part of the surrounding environment. Thus, every task of engineering requires a decent amount of trial-and-error methodology until the envisioned system roughly fulfills the desired task.

Yet another unique aspect of the nanoscale is the transition from classical physics to quantum mechanics. While many systems do follow the intuitive behavior as predicted by classical physics, that type of consistency breaks down if the size of a system is small enough. For example, energy suddenly no longer appears to be continuous but comes in discreet energy packages called “quants.” If the energy in a system is low enough, there is suddenly a moment when the level of a single electron is reached. That electron can either be there or not, leading to a breakdown of the classical behavior of circuits.

Relatedly, there is the so-called tunnel effect where electrons can suddenly change positions through a barrier. This is especially relevant for circuit design, where wires cannot be placed too close together due to electrons spontaneously switching wires. This can lead to bit-switches that corrupt the state of a memory or the result of a computation. This can be especially bad for decision problems or circuits that have a single yes/no or “1”/“0” as a result.

Yet another major paradigm shift stems from the strange behavior of measurement instruments like electron microscopes. Famous scientist Heisenberg observed that every act of measurement at the quantum level also influences the system. Most optical or laser-based measurement instruments work on the basis of a particle beam that is reflected by the object that is about to be magnified and later captured. In the process, the particle beam obviously influences the object it tries to amplify. The intensity of this effect obviously increases, the smaller the objects of interest become. You could compare the process by pointing a flashlight at a soccer ball. The photons that are emitted by the flashlight will not have a visible effect on the ball or move it in any way. However, a flashlight that emits photons the size of a soccer ball will definitely influence the state of the soccer ball itself.

In addition there is no clear distinction between waves and particles at the nanoscale anymore. Sometimes particles behave like waves and sometimes waves behave like particles. Interestingly, the simple act of measuring a wave function can lead to it “collapsing” and becoming a discrete particle. This has been proven in the famous double-slit experiment. Relatedly, there is also the phenomenon of quantum entanglement, where the states of two particles can be tied together so that finding out the state of one particle automatically determines the state of the other particle. It might even be possible to use this “spooky action in the distance” for communication purposes. There are many theories that try to explain this behavior, yet have no really satisfying answer. Interestingly, the simple “knowability” of a state is sufficient for a wave function to collapse, which might indicate that the entire effect might be a property of the observer, not of the system.

One last major peculiarity is the so-called Schrödinger effect. When isolating particles from their environment, it is possible to have isolated particles assume two states at once. This leads to the famous Schrödinger's cat thought experiment, where a switch that releases poison onto a cat is tied to the state of the isolated particle. As soon as the particle is measured and taken out of isolation, it will assume one out of two possible states, and, as a result, the cat will either be alive or dead. Earlier, the cat was technically in a third state of dead and alive at once.

Hence, the nanoscale is full of strange and often unintuitive effects that start emerging at the quantum level. Based on this, at least eight overarching, yet unsolved problems emerge that justify the existence of fields like nanonetworking or nanorobotics. They are:

- construction of nanoscale functional components for nanodevices or nanonetworks
- communication and communication protocols at the nanoscale
- computations taking into account nanoscale constraints
- compatibility of all partial solutions for holistic, nanoscale functional units or computing networks
- using the third space dimension effectively
- precise placement of building blocks
- coating of nanodevices by protein and fats in the human body and
- uncontrolled replication or degradation of nanostructures

For these problems, there are solutions based on a wide variety of principles. However, these are largely examined independently of one another and rarely checked for compatibility.

So far, for example, it has been possible to produce primitive nanodevices under laboratory conditions. However, these can at best be understood as proof of the feasibility of the technology in general. It is completely unclear how these can communicate with each other or perform computations (Huang et al., 2016).

Conversely, there are approaches for communication between devices, which, however, do not appear to be feasible with the currently available nanostructures as the necessary components for information processing are not yet compatible (Akyildiz et al., 2008).

Although computations can be implemented at the nanoscale, there are no real concepts of how these can take place autonomously within nanodevices (Windeck, 2019).

So far, after an extensive search, no concept has been found that combines construction, communication, and calculations in a holistic model.

Many problems can be explained by the current inability to precisely place objects at the nanoscale, especially in the third dimension. This might be one of the biggest unsolved problems where great industrial interest exists. Interestingly, chip design is very much limited by our inability to use the third dimension effectively. Currently, all chip manufacturing happens via the removal of parts from a pre-prepared 2D piece of silicone. We will learn more about these methods in Chapter 2.

1.6 Summary

We have acquired a good intuition for nanonetworks and nanotechnology. We started by analyzing the etymology of the word "nano" and tried to understand the philosophy that first came up with the idea of indivisible particles. Those serve as a foundation for the concept of objects that are way below the size of what humans can perceive with any sensory organ – at least unassisted by precise measuring instruments.

Another very important source for new visions and ideas bordering what is realistically possible is science fiction literature. Science fiction could rightfully be called the cradle of visions and ideas for future technologies, and generations of young scientists grew up with those books. Famous authors like Arthur C. Clarke or Stanislaw Lem shaped the visionary mindset of generations and thus contributed to current developments in many areas. One of the most difficult problems in science is surely the generation of new ideas, to which we have no formal approach. The exploration of Plato's world of "potential ideas" and thereby "clawing ideas into existence" is not something many people can do.

Then, we had a look at the current state of nanotechnologies and their interdisciplinary nature. Unlike many other fields of research, basically, all natural sciences have something to contribute to that area. This creates a host of new challenges in knowledge communication as it is often the case that many different designations for identical concepts exist. For example, it is not rare that some people use the words "nanomachine," "nanodevice," and "nanorobot" interchangeably, while others assign different meanings to all these words (Büther et al., 2017). We then gave a short overview of all the relevant areas in the discipline of nanonetworks as a mind map. All the parts of the mind map will be explained in the following chapters so that the reader has a solid foundation of knowledge to start their own contributions toward yet unsolved problems.

Then, we analyzed several unique properties of the nanoscale and the quantum level and listed several possible applications where nanotechnologies can prove to be beneficial. The topic will be discussed in Chapter 3 in much more depth as improvements in medicine due to nanotechnologies are one of the biggest areas of interest and likely the main monetary motivation for advancements in nanotechnologies. While applications are not limited to medical use cases, they make up the majority of the scientific interest. Apart from that, military applications, improvements in agriculture, or simply IT technology are the possible areas that might also benefit from nanonetworks.

In summary, there already exists a multitude of nanotechnologies in the industrial sector and an astonishing amount of new concepts and ideas surface every year. It is to be expected that both the number of applications and the number of ideas and visions continue to grow as the entire discipline of "nano" is not even 70 years old. In terms of nanonetworks, the term was only coined in 2008 (Akyildiz et al., 2008). Nanotechnologies will undoubtedly shape the technological landscape of the 21st century and likely beyond.

1.7 Chapter Overview

This section briefly explains the structure of the book. Each chapter is written to be readable on its own. However, the presented sequence is not chosen randomly. It is best to have a solid foundation of knowledge in the preceding chapter before starting with the next one. Furthermore, it is important to note that none of the chapters is optional. However, they can seem optional if you already have a solid foundation of knowledge.

The remainder of the book is structured as follows: Chapter 2 explains the historical developments in the area of nanotechnologies and nanonetworks in detail. We explore the major ideas of the past and try to find out the first ideas that ultimately led to the rise of nanotechnology, in general, and nanonetworks, specifically. In that context, the possible manufacturing accuracy is a very important parameter. Finally, we explore the current state of the art to give the interested reader an intuitive understanding of what is and is not possible.

Chapter 3 introduces different areas of our modern life where nanodevices or nanonetworks might provide benefit. In that context, the areas of medicine and pharmaceuticals are of special interest as many nanonetwork applications are of a medical nature. However, there are also applications in farming, the military, geology, and even early warning systems for natural catastrophes that might be an option.

Chapter 4 creates the necessary foundation of knowledge to follow, along with the remainder of the books. Here, we discuss many of the possible materials future nanodevices or nanonetworks might consist of. Furthermore, we discuss several possible construction mechanisms and try to define many of the desired nano-objects formally. Finally, we discuss the state of the art in nanoconstruction and briefly dive into the topic of simulation.

Chapter 5 discusses the topic of nanocomputation, which is of special interest to engineers and computer scientists alike. Without the ability to compute, complex applications are unthinkable. To estimate how difficult it is to solve a given problem, we introduce the basics of complexity theory and the concept of reductions. We then introduce several potential computational models that might be applicable at the nanoscale and try to determine how realistically they might be physically implemented in the near future. Finally, we briefly dive into the topic of trying to programmatically generate sufficiently good programs for swarms of nanodevices given the expected resource constraints at the nanoscale.

Chapter 6 introduces a new layer of complexity by introducing the ability of devices to communicate into the equation. We start with a brief history of digital communication and provide formal definitions for many of the involved components. We then analyze the different types of communication that might be applicable at the nanoscale. Among those are electromagnetic communication, molecular communication, acoustic communication, and DNA-based communication. We then compare the different types and try to find out what aspects of nanonetworks might be useful or applicable. Finally, we briefly get into the area of multihop communication and introduce the necessary models we need in Chapter 12.

In Chapter 7, we explore the area of locomotion and movement at the nanoscale. We divide the chapter into two broad areas: passive and active movement. At the nanoscale, every nano-object is subject to a number of different effects that change its position in space. Among those are the blood flow and Brownian motion. It is further important to keep in mind that ambient temperature and movement are very related. We then discuss possible active means of locomotion and try to find out how to localize nanodevices if their position in space is subject to continuous change. Finally, we discuss the organ-on-chip technology as a potential candidate to realistically test previously generated hypotheses about nanoscale movement.

Chapter 8 discusses the sensor and actuator components of nanodevices. Without sensors and actuators, communication, locomotion, and many other tasks would be impossible. Both sensors and actors allow nanodevices to interact with their environment in a targeted manner.

Chapter 9 explores the topic of energy harvesting and energy storage at the nanoscale. We analyze ultracapacitors, batteries, and various different technologies for the generation or harvesting of energy at the nanoscale from a number of different sources. We further discuss the possible means to efficiently use or save energy in nanonetworks.

Chapter 10 analyzes the topics of time and randomness that are important for a number of applications and algorithms. Without a basic understanding of time, it would be impossible to use nanonetworks for many medical surveillance applications and medical measurements are only relevant for a certain time. Additionally, synchronization among nanodevices or in a nanonetwork is only possible if all participants have at least a logical understanding of time. Furthermore, randomness is required for a number of important cryptographic, routing, and even synchronization algorithms. As soon as the number of possible options is too big to be explored in total, approximate procedures are necessary that often rely on restarting after random time intervals.

In Chapter 11, we explore the area of security at the nanoscale. Given the resource constraints, special environments, and unique applications in the area of personalized medicine, it is important to come up with satisfying and feasible solutions to ensure data security. Due to their small size, nanodevices are likely not able to perform state-of-the-art cryptographic algorithms and different low-weight solutions are necessary. Especially when it comes to molecular and DNA-based communication, there are several unsolved problems. Furthermore, there are a number of new network components that might offer new attack possibilities.

Chapter 12 takes all the previously generated knowledge and combines them into different reference architecture for nanonetworks that function based on different underlying principles. We further analyze a number of role models for nanonetworks and explain in what ways they differ from the nanoscale context. We discuss DNA-based nanonetworks in some detail as this type of nanonetwork is likely the least well known. Finally, we analyze how to verify or falsify a given nanonetwork concept on different levels of realism.

Chapter 13 explains the ethical, legal, and social aspects of nanonetworks. We try to understand the possible influences on the environment, explore the process from idea to finished medical product, and highlight the unsolved problems of waste disposal. In the process, there are several problems that medical nanonetworks might face in heavily regulated regions like the EU. Finally, we analyze the topic of acceptance of medical nanotechnologies and explore the possible dangers and fears the general public might have when it comes to possible disruptive novel technologies.

Finally, Chapter 14 summarizes the entire book and offers a list of yet unsolved problems, exciting future possibilities, and attempts to predict the immediate, near, and distant future of nanonetworks to some extent. There is also a list of key messages that an interested reader may memorize.

2

History

While history is not a hard science in itself, it offers us valuable insights into the prevalent problems that motivated and enabled certain ideas and developments. Without a solid basis in historical trends, it is difficult to fully understand and predict possible future developments. This chapter first discusses the philosophical basis for modern nanotechnology both from a Western and an Eastern point of view. After that, we analyze one of the most important parameters concerning nanotechnology: manufacturing accuracy. The manufacturing accuracy can be used as a parameter to define the advancement of culture as a whole. Based on that, we analyze the modern tools that we can use to measure, analyze and manipulate matter at the nanoscale. Lastly, we analyze the state of the art in nanostructures that can be produced with currently available tools and methods.

2.1 Philosophy

Philosophy is the discipline that concerns itself with the love (philos) for wisdom and knowledge (sophia) (Plato, 375 BCE, n.d.). Philosophy is the closest we can get to the discipline modern science is derived from. Long before the structures that we know today emerged, people were already thinking and pondering about how the world worked. Basically, any method or model modern science is based upon has its roots in ancient philosophy. From a Western point of view, Aristotle might very well be regarded as the founding father of structured scientific methods. If we include Eastern philosophy, Gotama the Buddha might be the first person who possessed the necessary formal expertise and logical thinking as well as the means of teaching and communication to be called the original source of science. In Section 2.1, we go through the philosophical lines of thought that ultimately lead to advancements in nanotechnology today.

2.1.1 Ancient India

In ancient India, there existed a rich tradition of wandering ascetics, thinkers, and philosophers. Around the time Gotama the Buddha was born (563 BCE), the Indian society was transitioning from tribalism to civilization (Thera, 1972). In many places, big kingdoms were already established. Yet, the first metaphysical views were already well established and might have been the forerunner for many modern physical theories. In the lectures of the Buddha in Digha-Nikaya 2, the Buddha mentions that "many episodes of world-expansion, followed by world-contraction" that beings have transmigrated through (Gotama The Buddha, 1987c). One could argue that this view might be the basis for the modern theory of the big bang, as well as the "big crunch" and some others.

Nanonetworks: The Future of Communication and Computation, First Edition. Florian-Lennert A. Lau.

The Buddha further argues that "there is nothing external to experience as a whole" that is accessible to us. He even said that the "entire universe lies in this fathom-long body" to illustrate that we are ultimately participants in a larger system and not observers from afar (Gotama The Buddha, 1997). This is indeed very similar to Heisenberg's observation that any measurement influences a system as a whole. In fact, Heisenberg was very much inspired by Eastern philosophy in his thinking.

Yet another aspect of interest in nanotechnology might be the concept of kalapas in Buddhism (Lerner, 1995). It is a term from Theravada Buddhist phenomenology and describes the smallest unit of physical matter, or as it is called in Buddhism: rupa/form. A kalapa is said to be 1/46 656th the size of a particle of dust from the wheel of a chariot (Lerner, 1995). While this theory does likely not date back to the Buddha himself, as it was first mentioned in written texts in the 11th century, it still shows that people were concerned with such questions. Yet, even without modern machinery and measuring instruments, people still tried to solve the modern "big questions" of physics. They might not have used the term "nano," yet the concern was the same.

2.1.2 Greece

A full analysis of the Eastern and Western origins of natural philosophy concerning nanotechnologies is beyond the scope of this book. Yet, the interested reader is referred to Kenny (2012) for a much more in-depth analysis of Western philosophy and much more.

The idea of a world that consists of tiny particles that "cannot be further divided down" originated in Greek philosophy as early as 500 BCE. The modern word "atom" was derived from the ancient Greek word "atomon," meaning "uncuttable" or "indivisible." While the ancient Greeks did not possess the necessary equipment to answer any of their questions and speculations empirically, they still formulated the hypotheses that were tested much later by modern scientists. Western science started to accept the proposed model somewhere between 1600 and 1900. Yet, the ultimate empirical test that proved the existence of atoms came much later.

To the best of our knowledge, the first Greek philosopher who proposed the idea of the atom was Leucippus of Miletus around 500 BCE. While he never used the specific name "atom," he clearly described the idea of the "smallest" particles. His disciple Democritus later introduced the commonly known designation.

According to Leucippus, atoms come in various shapes and sizes from which the macroscopic properties of matter emerge. For example, some atoms were needle-like, some hexagonal, and some yet other shapes. Each of them, so his theory, was responsible for a different property of matter we can detect with our senses.

The famous philosopher Epicurus picked up the same idea and argued for a more materialistic view of the entire universe that is much more akin to the current scientific worldview. He was among the first philosophers who rejected the idea of almighty gods and other divine beings. According to him, even gods had to play by the same laws of nature. This idea is very much like the reasoning Gotama the Buddha used 300 years earlier. Whenever he used gods or divine beings in his lectures, he mainly used them as examples of how even the divine status will not protect you from the ultimate laws of nature.

Yet, the most important influence on modern science that dates back to ancient Greece is Aristotle. His methods in many fields like biology remained unrivaled for close to 2000 years. In his time, it was common to write poems about important topics. On the topic of atomism, he wrote the famous poem "De rerum natura" (On the Nature of Things). Again, Aristotle might be the one person on whom most modern science is based.

While all the presented ideas relate to the modern scientific view, the ancient philosophers did not follow a rigorous scientific system as we do today. The ancient philosophers were much more interested in holistic ideas that incorporate everything into one explanation for the "all." Whenever the idea of atoms was challenged, the contemporary philosophers did so on the basis of thought and not empirics. For example, Plato is most famously known for his "Theory of ideas," which had its own set of problems and inconsistencies (Kenny, 2012). The work of the ancient philosophers contributed to the level of ideas and theories that ultimately motivated the following generations of scientists. In that sense, their contribution is of a similar kind to that of science fiction authors like Arthur C. Clarke. The main difference would be the degree of "seriousness."

2.1.3 Modern Era

The modern era introduced many paradigm shifts to classical natural philosophies and science as we know it today was born. Due to the industrial revolution and fast communication networks, big improvements in measuring technologies were made. One could argue that the ever-present prospect of war was also a catalyst for technological growth.

No matter the reason, several innovations made it possible to test the hypotheses that were untestable before. The development of ever-improving microscopes allowed humanity a deeper and deeper look into the structure of matter.

It is difficult to pinpoint the exact event that sparked the quest that ultimately led to the emergence of modern nanotechnologies. Depending on the point of view of the discipline of physics, Max Planck and Albert Einstein, in particular, are possible candidates who majorly contributed to the field unknowingly. Many of the relevant (quantum-)effects that make the nanoscale special have been researched or hypothesized by Einstein and his colleagues. For example, the "spooky action in the distance" through quantum entanglement was "named" by Einstein, who doubted the possibility of such an effect to happen. Yet, it turns out that it is indeed possible to recreate that effect over relatively large distances.

In 1900, Max Planck developed a formula that first described the measured frequencies of a black body (Planck, 1900). He assumed that the "strange" quantization of energy was a property of matter and not a property of light. In 1905, Einstein refined the quantum theory by suggesting that the photoelectric effect can be explained by a quantization of the energy of light itself (Einstein, 1905). This is one of the major shifts from classical to quantum physics – events no longer appear continuous.

Eight years later, Niels Bohr used Einstein's model to explain the spectral lines of hydrogen atoms. One of the most famous models of the atom (the Bohr model) dates back to that time period. All of those and many more discoveries gradually unveiled the unique properties and natural laws that matter at the nanoscale is affected by.

In 1953, Watson and Crick famously discovered the DNA, which is only 2 nm in diameter (Watson and Crick, 1953). This discovery likely inspired and influenced later advancements and ideas in many other fields. Today, DNA is one of the most promising and well-researched nanoscale structures and a promising building block for many possible future technologies.

The real development of nanotechnologies started in 1959 with Richard Feynman's famous topic "There's Plenty of Room at the Bottom." He vividly described the future possibility of manipulating matter at the atomic scale. His prediction later came true, and the trend of gradual miniaturization was discovered and named in 1965 by Moore Moore (1965).

The term "nanotechnology" itself was likely first used by Norio Taniguchi in 1974. The discipline of nanotechnology itself only really started emerging in the 1980s with the major discovery of the scanning tunnel microscope. For the first time, it was possible to visualize matter at the atomic scale. The microscope itself will be analyzed in some detail later.

Shortly after, many major discoveries have been made. These include the discovery of fullerenes, nanoclusters, carbon nanotubes, and many other structures. All of this was possible thanks to the advancements in microscopy. Without the prior work on the quantum tunneling effect, this would not have been possible.

2.1.4 Since 2008

With the modern developments in communication and the emergence of the Internet, a new era in research began. Now it was possible to communicate and share research results at an astonishing speed. Information exchanges that took several weeks before required mere minutes once the first communication networks between universities were established. Based on that trend, the Internet as we know it today accelerated the development of basically all scientific areas, especially telecommunication.

As early as 2002, Ian F. Akyildiz Akyildiz et al. (2002) already gathered available knowledge on research in the then-emerging area of wireless sensor networks. This discipline served as a direct role model for the developments in nanoscale sensor networks. Once research in the area of wireless sensor networks attracted more and more industrial interest, researchers branched out and further miniaturization was the logical next step.

In 2008, Akyildiz published the paper "Nanonetworks: A New Communication Paradigm" (Akyildiz et al., 2008), which sparked worldwide interest. The paper first introduced the paradigm of nanonetworks. Apart from the differences in size, nanonetworks face numerous new challenges, such as those mentioned in Section 1.5. At the time of writing this book, the research discipline was already close to 15 years old. A lot of knowledge from different areas is accumulating, and it is hard for new researchers to contribute to the area. There is simply too much interdisciplinary knowledge available and no single person can be an expert on all of them. This book attempts to close the aforementioned gap.

While nanonetworks appear to be a realistic idea nowadays, that has not always been the case. While there has been considerable progress in manufacturing structures at the nanoscale, the process is by no means efficient or accurate enough. In fact, manufacturing accuracy might be the single most important parameter in assessing the "progress" of a civilization as a whole.

2.2 Manufacturing Accuracy

This section gives a short overview of the history of manufacturing accuracy. We start with the earliest known tools that humans have used and gradually move toward the nanoscale.

2.2.1 Antiquity

The history of human tool usage started about 1.6 million years ago with simple chopping tools. Several primitive examples are shown in Figure 2.1.

While many of them are barely distinguishable from naturally occurring stones, archaeologists are certain that they serve a purpose. They have been used to crack open nuts, seeds, and bones. Other versions were viable tools for cutting. The manufacturing accuracy, however, was not very

Figure 2.1 The earliest known tools that humans used. These chopping tools are around 1.6 million years old and have been used to crack nuts or bones. The manufacturing accuracy is in the range of a few centimeters (Descouens, 2010). *Source*: Descouens (2010)/Wikimedia Commons/CC BY-SA 4.0.

advanced and was in the range of a few centimeters. The ability to use tools was roughly on the level of today's primates, and it took a long time until new and more refined tools emerged.

The first pieces of art date back to around 30 000 years ago. An example is shown in Figure 2.2 (b). The Venus vom Galgenberg is a comparably refined piece of art that is already recognizable as the shape of a female body. It is hard to tell how precisely it has been manufactured as 30 000 years of exposure to the elements led to considerable wear and tear. However, it is safe to say that the statue requires even more precision than a hand ax and is estimated to be around 3 mm. Furthermore, art is only really of interest when the most basic needs are taken care of. Thus, the statue marks a period of further cultural development where survival has not been the only purpose.

Another major milestone has been unearthed in the Acheuléen region about 200 000 years before the common era. The hand ax in Figure 2.2 (a) is clearly much more refined. The manufacturing accuracy is in the range of 5 mm and likely required many cultural developments to keep the knowledge of how to create such tools alive.

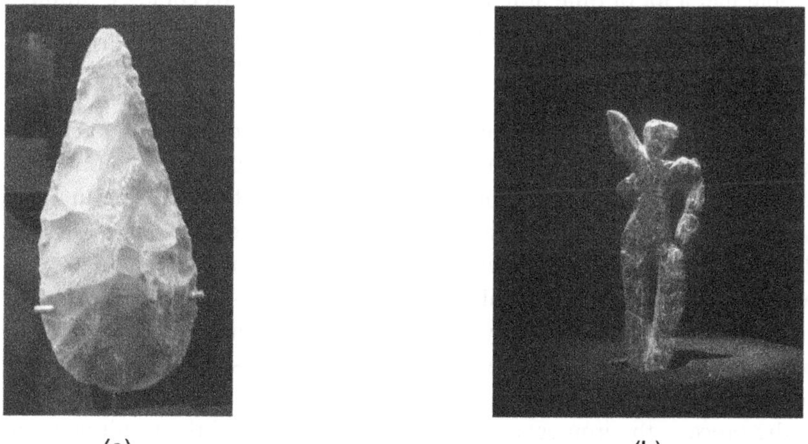

(a) (b)

Figure 2.2 A 200 000-year-old hand ax (a) and a 30 000-year-old statue (b). *Source*: (a) Luis García/Wikimedia Commons/CC BY-SA 3.0; (b) Aiwok/Wikimedia Commons/CC BY-SA 3.0.

Figure 2.3 One of the first historical images of a potter's wheel that allows for extremely precise manufacturing down to 0.1 mm (Didia, 2014). *Source*: Didia (2014)/Wikimedia Commons/CC BY-SA 3.0.

The next major milestone was achieved in ancient Egypt around 3100 BCE. While most prior technologies served a direct purpose, the pottery wheel in Figure 2.3 was among the first tools that enabled the more accurate production of *other tools*. Due to the clever use of rotational forces, new levels of accuracy were made possible. While it is comparably difficult to accurately move an object in three dimensions, it is much easier to stably move tools or even your hands in just one or two dimensions. This advancement allowed for smooth surfaces and near perfectly rounded surfaces. As a result, the number of pots that a single potter could produce in an hour increased drastically. Logically, people had a lot of time at hand once many basic needs could be fulfilled more efficiently, which allowed for advancements in other areas. The same principle was applied centuries later when people started utilizing water wheels to power machines that allowed for the accurate processing of wood or iron.

The history of human civilization is often categorized and named by major discoveries that influenced entire eras. For example, the Stone Age (4000–2000 BCE) is named after the predominant material of tools and ends with the emergence of copper- and bronze-based technologies. In Asia, the Bronze Age ended around 1200 BCE, while the same trend started around 500 BCE in Europe. During that time, the dominant manufacturing material was iron, followed by steel.

One especially noteworthy technology is the so-called damascene steel displayed in Figure 2.4. While many of the other technologies and tools only indirectly influenced modern nanotechnology, damascene steel implicitly used nanotechnologies. Unlike iron, steel is created by adding a set amount of carbon to the melting process of iron. The resulting steel is much sturdier and can no longer be forged. In the process, the iron gets saturated with carbon until the resulting steel has about 1.7% carbon content. Interestingly, the resulting steel is strengthened by carbon nanotubes (Reibold et al., 2006). People thus used nanotechnologies way before the official discovery

Figure 2.4 A blade made from damascene steel. *Source*: Inazakira/Wikimedia Commons/CC BY-SA 2.0.

of carbon nanotubes in 1991 (Iijima, 1991). While the resulting steel contained carbon nanotubes, the manufacturing accuracy remained at around 0.1 mm.

2.2.2 Middle Ages

The knowledge to create steel was also available in Northern Europe in the early Middle Ages. About 170 medieval swords called "Ulfberth" date back to 800–1100 (Figure 2.5). The process is similar to that of creating damascene steel, but people during that time used animal bones as a source of carbon. They believed that the strength of the animal entered the sword and made it especially sturdy.

While the entire period is known for a "gap" in knowledge and advancements, a few influences on modern nanotechnologies occurred. Among those are ambitions by alchemists to change the chemical composition of materials, e.g., turning lead into gold. While alchemists never succeeded in their ambitions, some of their knowledge remained. One of their discoveries was the so-called "colloidal gold." While the substance had already been used in the fourth century to color glass, Francis Anthony popularized it in 1618 (Anthony, 1618).

Apart from various claims concerning medical benefits, colloidal gold is interesting for nanotechnologies as it contains tiny gold particles the size of 2–100 nm. Colloidal gold was mainly interesting

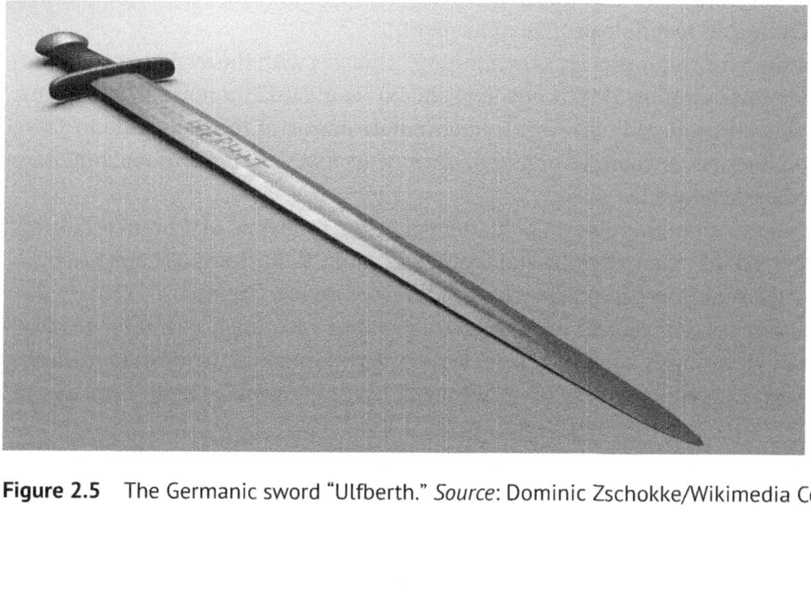

Figure 2.5 The Germanic sword "Ulfberth." *Source*: Dominic Zschokke/Wikimedia Commons/CC BY-SA 4.0.

for its coloring effects on glass surfaces. Yet, by modern definitions, it would be regarded as nanotechnology, just like carbon nanotube-reinforced metals.

2.2.3 Modernity

The modern period brought the biggest changes to civilization thus far. 1800–1900 granted us the widespread use of trains, printing press, telephones, cameras, automobiles, and efficient agriculture. 1900–2000 granted us planes, plastic, radio, TV, antibiotics, the Internet, and so on. Basically, the entirety of modern civilization has been created in a little over 100 years. Earlier, things were more or less the same for several thousand years.

As a small anecdote, I am 33 years of age and my father was born in 1943. When he was a child, chariots were still more common than cars in parts of Germany and there was basically no TV, radio, or most of the other luxuries we have today. The last 100 years truly led to a remarkable shift compared to thousands of years before.

2.2.3.1 Manufacturing Methods

The progress toward that immense increase in the standard of living was enabled through the use of machines that allowed for more and more accurate manufacturing. One could argue that an ongoing exponential trend in miniaturization that started with the Industrial Revolution is the driving force of progress. Around 1850, the first machines were created that allowed for more accurate manufacturing than the 0.1 mm that are the limit of what the human eye can even perceive without tools like magnifying glasses. From then on, numerous advancements allowed for gradually more refined machines, and a positive feedback loop started.

Once the limits of hand-operated machinery had been reached, computers emerged and accelerated the process yet again. The demand for more accurate components was driven by the military, nuclear energy, space travel, and computer parts themselves. In the 1960s, the true race to the nanoscale in commercial electronics began.

A variety of different methods for manipulating matter at a very small scale are shown in Figure 2.6. The y-axis shows the level of precision and the x-axis the time since 4000 BCE in years (nonlinearly). On the right, we see different methods of manufacturing or varying degrees of precision. The curved lines indicate that many methods can be refined and thus span several levels of precision, depending on technological advancement.

Furthermore, the definition of "high-precision machining" changes with the available technologies. The best achievable precision from 1930 is now considered "standard" for normal machining. In a few (expensive) cases, it was already possible to manipulate matter at the nanoscale in 1960. While the trends indicate that we can achieve higher degrees of accuracy, all types of machines are frequently used and further developed.

The first method that allows for manipulation of matter below the threshold of 0.1 mm is cutting. A rotating head of a cutter usually removes material layer by layer until the desired shape has been reached. Today's cutting technologies can reach accuracy levels of up to 0.01 mm.

The next level of accuracy can be reached with ordinary grinding machines. Grinding is a process where either through vibrating or rotating a level and hard grinding head is lowered toward a material to even its surface. Depending on the materials used, accuracy levels of up to 1 μm can be reached.

To exceed the accuracy of grinding, even finer materials have to be used. Naturally, it is only possible to achieve the same level of accuracy on any object that the surface of the grinding or polishing head has. Lapping is a process in which two surfaces are rubbed together using an abrasive

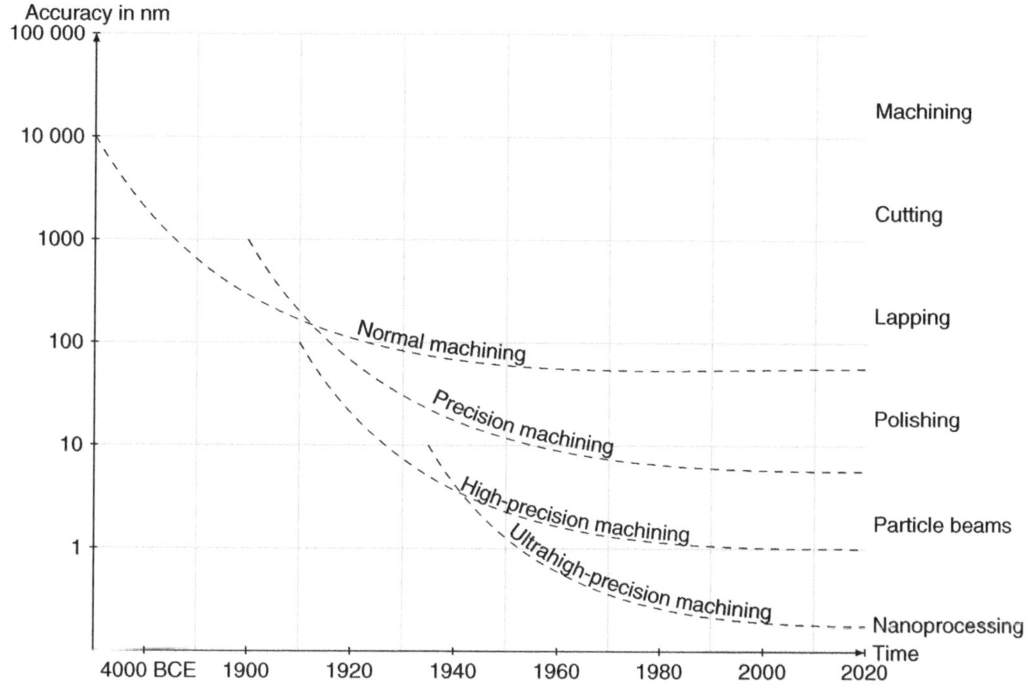

Figure 2.6 Various methods for manipulating matter at a gradually smaller scale (Venkatesh and Izman, 2007). *Source*: Venkatesh et al. (2007)/Tata McGraw-Hill Publishing Company Limited.

between them. That abrasive could be any material that fulfills the necessary conditions of hardness and size. Typical examples are glass particles in a milky fluid. Through lapping, the accuracy can be lowered down to 0.1 μm for conventionally available technologies and down to 30 nm in extreme cases. Following the same principles, polishing allows for even more accurate processing down to 0.01 μm.

To achieve levels of accuracy that are higher than that, new technologies are necessary. Those are usually based on particle beams and the most widespread manufacturing technology in that range are wafer steppers. Wafer steppers are an essential part of the photolithography process that is basically the only known mass manufacturing method for creating nanoscale circuit elements on the surface of silicon wafers. In the process, an entire array of ultra-precise optical lenses is necessary to compensate for any manufacturing inaccuracies. The wafer-stepper projects (using UV lasers) a circuit blueprint onto a wafer of silicon to transfer a geometric design onto a light-sensitive chemical that has been applied to the silicon beforehand. That UV laser either hardens or weakens the chemical and thus prepares the wafer for the next step, where the remaining material is etched off. Then, only the desired circuit structure remains. This technique allows for the mass production of computer chips and is responsible for the computer revolution that is ongoing. It is possible to achieve levels of accuracy down to 20 nm or less. Modern technologies even mention processors that use structures down to 7.7 nm (Windeck, 2019).

While the overall trend looks exponential, there are limitations to the achievable levels of accuracy and precision for each method. Many methods are limited by either systemic or random errors that might occur due to vibrations that are caused by the machine itself. If a piece of material in a metal cutter vibrates by 50 nm, then it is not possible to achieve higher degrees of accuracy. While such systemic errors might be compensated to some degree, random errors will always occur and thereby limit the overall achievable accuracy. Furthermore, the size of atoms is a hard limit on further miniaturization; thus, we cannot manufacture more accurately than 0.3 nm. It is to be expected that the curve will flatten somewhere between 2020 and 2050 (Haron and Hamdioui, 2008).

2.2.3.2 Microscopes and Imaging

One of the biggest challenges in achieving higher degrees of manufacturing accuracy is the limited resolution of the human eye. We can only influence what we see in a targeted manner and that requires tools that enhance our ability to see at a small scale. Thus, microscopes of any kind are a necessary precondition for targeted research of nanoscale materials. Without these, it is nearly impossible to create novel materials.

Among the first such devices are magnifying glasses. As early as 424 BCE, Aristotle already wrote jokingly about magnifying lenses to ignite tinder. While the lenses were capable of some magnification, it was typically limited to a magnifying power of two. Some rare archaeological findings that exceeded the limits of human eyesight likely used such tools for more accurate manufacturing.

The earliest microscopes that combine multiple lenses to achieve greater degrees of magnification date back to 1620 (Murphy, 2002). While the actual inventor remains unknown, several other optical discoveries have been made at that time. For example, in 1608, the first patent for a telescope that works based on the same principle was filed by Galileo Galilei. The magnifying power of those devices is estimated to be around 3–9.

The first significant success of a magnifying power of 270 was achieved by Antonie van Leeuwenhoek using only a single lens. Other microscopes between 1600 and 1700 only achieved powers of around 50 at best. Today, the most refined versions of optical microscopes can achieve resolutions that are only limited by the wavelength of light (400–800 nm). Using optical microscopes, it is possible to see cells, but smaller structures like viruses or DNA remain undetectable. Beyond that, only a very few improvements are possible and other technologies are necessary.

The next step that went beyond the optical microscopes was the invention of the electron microscope around 1931 by German physicist Ernst Ruska. He directed a beam of electrons at an object of usually less than 100 nm thickness. The beam passes through the object and creates an image on an imaging device below. This technique is called "transmission electron microscopy" and marks the beginning of a new era. Using electrons instead of light allows for much higher resolutions.

There are many different versions of the electron microscope that can achieve different levels of magnification. The "cheaper" versions can detect objects of 100 or 200 nm. The high-end versions require special environments that thermally stabilize the microscope to avoid vibrations. It is thus possible to achieve resolutions of around 1 nm.

Another class of microscopes that achieves even higher resolutions is the scanning probe microscope. They can roughly be categorized into two subclasses: the scanning tunneling microscope and the atomic force microscope. For the tunneling microscope, a physical probe is placed close to an object, a voltage is applied, and the resulting tunnel current is measured. This technique requires a conductive sample; otherwise, the quantum tunnel effect does not apply.

The atomic force microscope utilizes a very fine cantilever that touches a surface and bends in the process. A laser is directed at the cantilever and the reflection is measured by a photodiode. The amount of reflected laser light varies with the bending of the cantilever. It is thus possible to visualize even non-conductive materials.

The electron microscope and the scanning probe microscope allow us to visualize nanoscale structures. Without such tools, it is impossible to verify the results of wet-lab experiments involving nanonetworks. That said, it is not possible to visualize, e.g., living organisms as the sample is usually destroyed in the process, which is yet another limitation of imaging techniques.

Several other imaging techniques allow for three-dimensional images:

- *positron emission tomography* (PET): Michel Ter-Pogossian and Michael E. Phelps
- *computed tomography* (CT): Hounsfield, 1973
- *magnetic resonance imaging* (MRI): Paul C. Lauterbur, 1973
- *optical coherence tomography* (OCT): David Huang, 1991

While all these techniques allow for a varying degree of three-dimensional imaging, the resolution is nowhere near the necessary nanoscale values. For example, typical PET techniques allow for a resolution of around 10 mm, while the MRI can achieve spatial resolutions of around 1 mm. Especially the OCT technique offers a standard resolution of around 10 μm. In theory, most of these devices are not limited to the resolutions offered by standard medical equipment. However, no suitable system for visualizing nanoscale systems exists as of today.

2.2.4 From 2D to 3D

All the methods and techniques for construction that we analyzed have been limited to two dimensions. The top-down approaches we use to mass-manufacture chips only manipulate the surface of materials. In fact, we currently have no approach that utilizes the third dimension in a meaningful way when it comes to the top-down manufacturing of nanoscale structures. Utilizing the third room dimension to, for example, stack chips is one of the biggest unsolved problems. An ability to do so would massively enhance our storage capacity per unit of space. At present, photolithography is limited to modifying the surface of materials just like microscopes usually only visualize surfaces as well.

While some modern approaches actually utilize the third dimension in some way, nearly all of them are based on modifying or using naturally occurring structures. Nearly all nanostructures that emerged through evolutionary processes utilize all three room dimensions. It is only natural to take naturally occurring components to modify them until they fulfill the desired function. One could even argue that nearly all knowledge is ultimately inspired by observing naturally occurring processes.

2.2.5 Placement Accuracy

However, even if we were able to create three-dimensional structures in a meaningful way, the problem of placement accuracy would still remain. It is surprisingly difficult to move nanostructures to specific places. Connecting several components with the necessary accuracy of a few nanometers is a near-impossible task with current technologies.

For example, we can efficiently produce isolated carbon nanotubes. The building block for carbon nanotubes is graphene, which is a single-atom-thick sheet of carbon. The edges of those graphene sheets have a tendency to form bindings with other carbon atoms, which results in an energetically more efficient state.

Currently, there are three known methods for rolling graphene sheets into carbon nanotubes: arc discharge, laser ablation of graphite, and chemical vapor deposition. The first two techniques start with graphite, which is a compound material of many layers of graphene. Combusting the graphene produces a gas from which the carbon nanotubes are formed. A laser removes layers of graphene, which produces a gas that is blown into a suitable environment where carbon nanotubes can form. However, the process involves no accurate placement of carbon nanotubes at all. The final product is more or less randomly scattered and researchers have to "search" for the material with microscopes. Given these difficulties, it seems to be a difficult task to actually create nanoelectronics by combining multiple carbon nanotubes into circuits.

2.3 State of the Art

This section gives a brief overview of the state of the art in nanoscale manufacturing. In various areas, it is possible to create an array of different nanoobjects that exceed mere theoretical interest. We briefly analyze advancements in artificial, natural, and hybrid approaches.

2.3.1 Artificial Materials

The simplest artificial nanostructure are nanoparticles. These have a variety of uses and are promising candidates for treating diseases like cancer. An example of healthy (left) and cancerous cells (right) is shown in Figure 2.7. The idea behind this treatment method is rather simple. Metallic nanoparticles are coated with specific antibodies that allow them to bind to specific areas in the human body. In the case of cancer, the now-bound nanoparticles can then be heated through special lasers, which leads to the death of cancerous cells, while the surrounding tissue is left unharmed.

Similar treatment ideas also exist for several parasitic diseases like leishmaniasis. While this kind of therapy is possible today, it is far from maturity or practical use.

Another impressive proof of the ability to manipulate matter at a very small scale is shown in Figure 2.8. The figure displays a simple snake pattern that has been 3D-printed and may someday serve as a basis for new battery technologies.

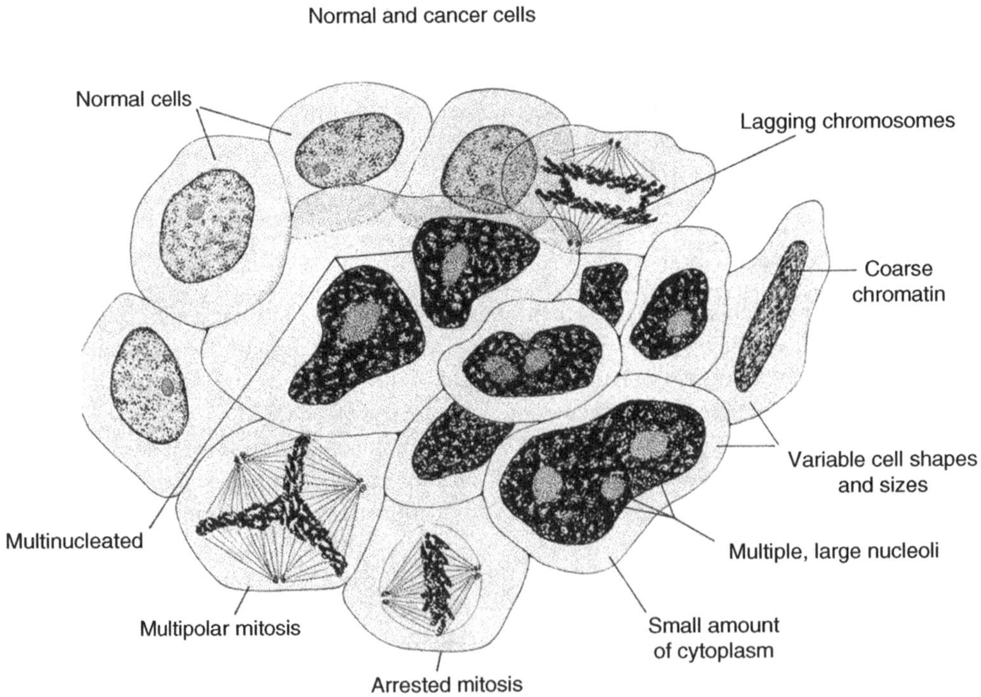

Figure 2.7 A comparison of healthy and cancerous cells. *Source*: Pat Kenny / National Cancer Institute / Public Domain.

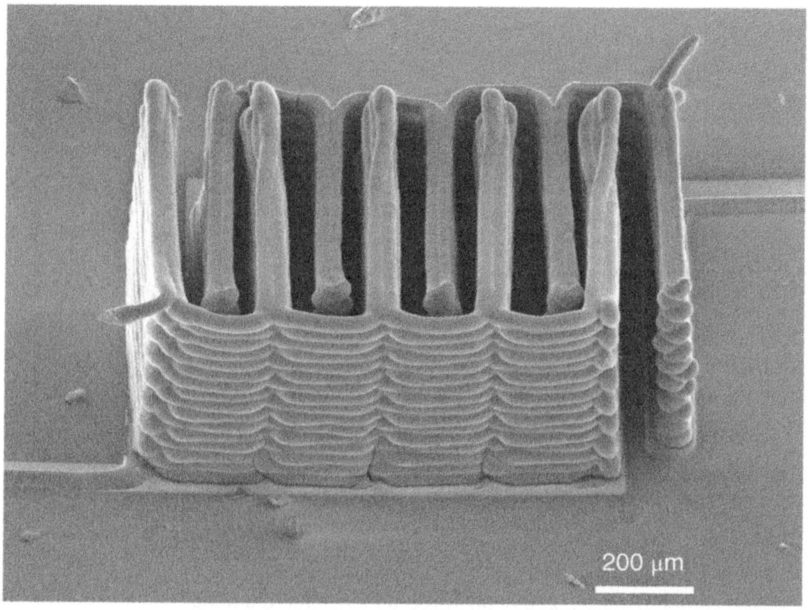

Figure 2.8 A proof of concept for microscale manipulation of matter (Grybauskaite, 2018). *Source*: With permission of John Wiley & Sons.

One element spans just 200 μm, and the entire structure was created as early as 2013 (Sun et al., 2013).

Yet another advancement in nanoscale manufacturing is shown in Figure 2.9. The image shows several different molecular configurations of carbon atoms. From top left to bottom right, there are diamond structures, 3D graphite, 2D graphite, a carbon nanotube, a one-dimensional nanoribbon, and a fullerene. While there are many more possible molecules, it is still impressive to see how many different structures can be created from a single atom. Many of these structures could be the basic material for future (medical) nanoscale applications.

For example, carbon nanotubes have been proposed for a variety of novel materials or technologies. These range from nanoscale three-dimensional circuits over sensor types to possible rope technologies that could someday enable space elevator technologies.

Another interesting step toward nanoscale robotics is shown in Figure 2.10. Harvard created a fully functional coin-sized robot. While such robots cannot be created at the micro- or nanoscale, some pioneering work created an array of primitive microbots of roughly 100 μm size. While it was already possible to create circuits of that size, there was no compatible actuator that could move such circuits around. The designed robots can be operated by just 200 μV and require less than 10 nW. It is even possible to mass-manufacture this robot and create millions of such devices on just a 4-inch silicon wafer. While these microbots fulfill no real function, they mark an important step toward achieving a degree of compatibility between very small components. As is known, the compatibility of different technologies is among the biggest unsolved problems at a very small scale.

2.3.2 Programmable Matter

After microscopic robots, some researchers focus more on the interaction between devices. Both size and collaboration are crucial factors to realize applications at the nanoscale.

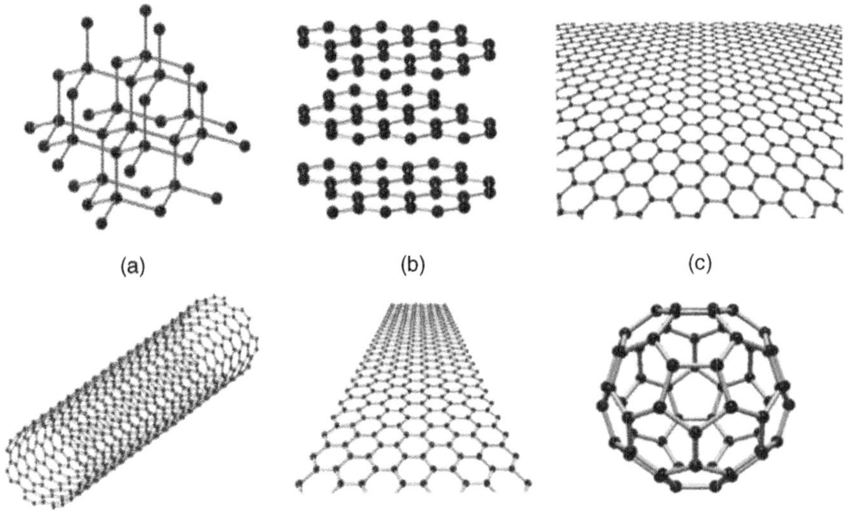

Figure 2.9 Several allotropes of carbon (Li et al., 2009). (a) Three-dimensional diamond. (b) Three-dimensional graphite. (c) Two-dimensional graphene. (d) One-dimensional nanotube. (e) One-dimensional nanoribbon. (f) Zero-dimensional fullerenes. *Source*: Li et al. (2009)/With permission of IEEE.

Figure 2.10 A very small "microbot" that can move on its own (Miskin et al., 2020). *Source*: With permission of IEEE.

While current modular robots are not even close to the nanoscale, the overarching paradigms used in their creation can just as much be applied to smaller devices. For example, there are reconfigurable sliding cubes of about 1 cm^3 in size (Romanishin et al., 2013). They can assume basically any voxel shape given sufficient modular robots (Kawano, 2015).

An even more advanced technology is *3D Catoms* (Kirby et al., 2005). They are near-spherical devices of only 3.6 mm in diameter that can perform a variety of actions. They have a unique identifier and can communicate with other nearby Catoms. In doing so, they can execute distributed programs and assume basically any imaginable shape. Thus, they can serve as a role model or at least an inspiration for construction at the nanoscale.

2.3.3 Biology

While researchers struggle to create fully functional devices using top-down approaches, bottom-up approaches already delivered promising results. Various researchers analyze existing nanoscale structures in terms of their viability for nanoscale applications like targeted drug delivery. As shown, drug delivery is among the biggest unsolved problems in modern medicine.

The idea behind most of those approaches is identical. We take a naturally occurring structure and modify it to fulfill the specific needs of a given application. For example, there are so-called liposomes that are essentially small spheres of fat with empty space in the middle. We analyze the structure in more detail in Chapter 4. The empty space in the middle can be used to encapsulate medication to deliver it to a specific location. The protective layer can also shield the medication from being metabolized too early or allow it to reach certain areas in the body where access is restricted to molecules with a specific surface structure.

The same principle can also be applied to dead bacteria, viruses, or cells. A slightly different approach works based on boxes that have been created from modified DNA molecules. While DNA also naturally occurs in all living organisms, the box has been created from entirely artificial DNA molecules. These were designed such that they self-assemble into a box that can open and close when specific molecules bind to it as shown in Figure 2.11. On the left, the box is closed, and on the right, the box is opened due to the presence of a specific marker sequence. In 2018, it was demonstrated that it is also possible to open such boxes using molecules other than DNA (Tang et al., 2018).

This type of structure is of repeated interest throughout this book due to its unique and useful properties.

Apart from the ability to form boxes, DNA has a variety of additional use cases that might make it the single most useful molecule for nanonetworking purposes. Figures 2.12 and 2.13 show a

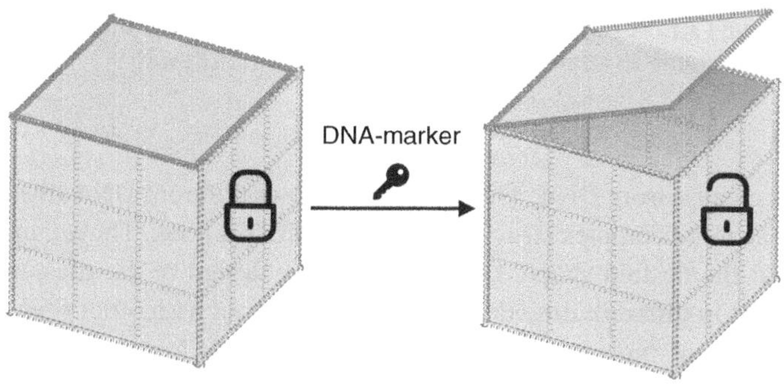

Figure 2.11 A box from DNA-origami (Andersen et al., 2009; Tang et al., 2018). *Source*: Florian-Lennert A. Lau.

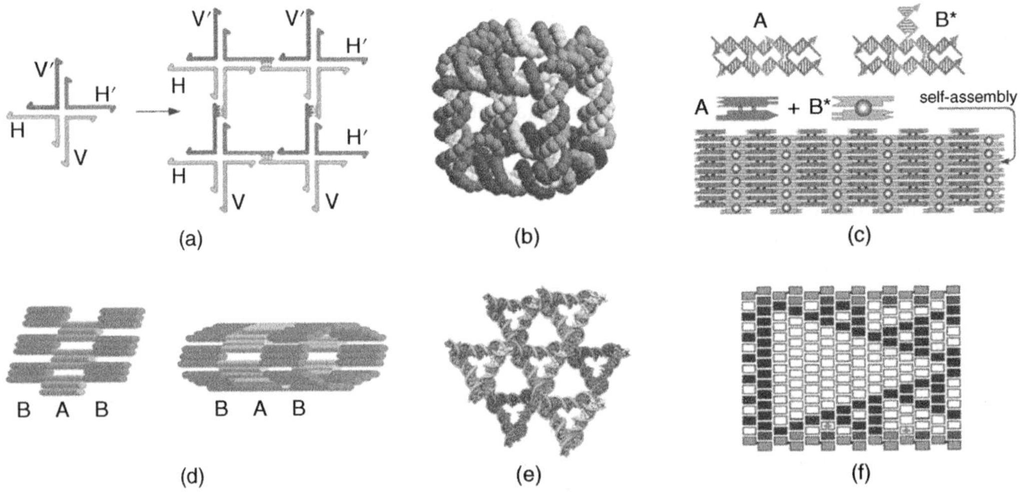

Figure 2.12 Various tile-based nanostructures created until 2015 (Nummelin et al., 2018). (a) Holiday junction + sticky-ends, (b) DNA cube, (c) DX-tile: 2D lattices, (d) TX-tile: 2D lattices, nanotubes, (e) 3D tensegrity triangles and crystals, and (f) Algorithm self-assembly (Sierpinski triangles). *Source*: Nummelin et al. (2018)/with permission of John Wiley & Sons.

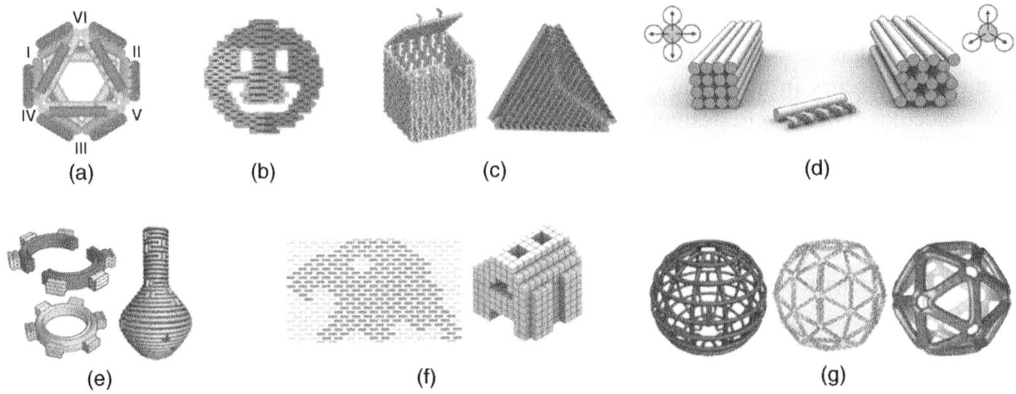

Figure 2.13 Various nanostructures created until 2015 (Nummelin et al., 2018). (a) DNA octahedron, (b) 2D DNA origami, (c) Hollow 3D DNA origami, (d) Lattices-based 3D DNA origami, (e) curved 3D DNA origami, (f) single-stranded tile assembly, and (g) wireframe structure. *Source*: Nummelin et al. (2018)/with permission of John Wiley & Sons.

selection of artificial nanostructures created from DNA. Figure 2.12 (a) shows different DNA tiles that are the building blocks for various more complex structures presented in this book. These can be programmed to self-assemble into the desired shape. Depending on the design of the tiles, a large number of DNA strands can point in different directions. Those can interact with other tiles.

Figure 2.12 (c) shows different kinds of molecules/tiles that can be used to form grid-like structures. Regular grids are especially interesting for chip designers as DNA might be a suitable technology to allow for the accurate placement of components. This DNA templating approach

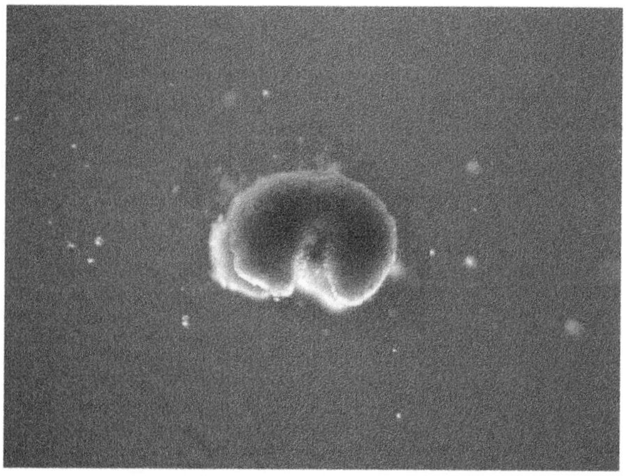

Figure 2.14 An artificial living organism created from frog cells (Wikipedia, 2023). *Source*: Artem.G/Wikimedia Commons/CC BY 4.0.

uses the self-assembly properties of DNA to circumvent the current limitations of top-down approaches.

Figure 2.13 (b)–(g) shows various structures created using the DNA-origami method. This technique utilizes a single strand of DNA that is stapled together with smaller pieces of DNA to create an almost arbitrary shape. The aforementioned DNA box was also created using DNA origami.

Figure 2.13 (d) shows, among more two-dimensional shapes, structures that have been created from DNA bricks. DNA bricks behave in a similar way compared to DNA tiles but can utilize the third-room dimension for self-assembly. DNA bricks are among the first technology that allows researchers to artificially create three-dimensional structures from scratch.

2.3.4 Hybrid

Apart from strictly artificial and strictly biological approaches, there are also possible hybrid nanotechnologies. The simplest version of such a hybrid technology might be the aforementioned templating of miniature components using the self-assembly properties of DNA (Ma et al., 2010).

More interesting approaches have recently surfaced featuring entirely new living organisms, as shown in Figure 2.14. The figure displays a multicell artificial organism created from frog cells. The so-called "xenobot" can move around like a scallop and self-regenerate in case of damage. While this organism serves no real purpose, it illustrates future possibilities and serves as a proof of concept.

Many other possible hybrid approaches are also possible that include bacteria, viruses, or cells that get enhanced with artificially created components.

2.4 Summary

Humanity has come a long way since the first emergence of tools nearly 1.6 million years ago. We started with simple tools to crack open nuts and bones and slowly developed them into more and more beneficial technologies. Tools that allowed for the creation of other tools enabled humanity to create food and other products more and more efficiently. On the basis, civilization emerged as we know it today.

Once survival was not the most pressing concern anymore, different versions of natural philosophies started to emerge in the East and West alike. These still form the foundation of modern science and thinking and provided the first ideas and visions for nanotechnologies as we know them today.

With the Industrial Revolution, the ongoing trend of a lot of accurate manufacturing picked up exponential speed. In a little less than 200 years, humanity went from the steam engine to modern computers and medicine. Earlier, things were more or less the same for thousands of years.

Nowadays it is possible to create a variety of useful nanostructures and the discipline in itself is rapidly growing. Many scientists suggest nanotechnologies as a solution for yet unsolved problems in medicine and other areas. In Chapter 3, we will go through the biggest proposed applications that are currently being researched.

3

Current and Future Applications

This chapter analyzes and explains many of the current and possible future applications of nanotechnologies. First, we discuss the surprising amount of nanomaterials that already serve a purpose in a variety of industrial sectors. Then, we examine the possible future scenarios in which nanonetworks explicitly serve a purpose. These include military, material science, agricultural, and many medical purposes. Then, we highlight a few key challenges for nanonetworks and try to motivate the mid-term future improvements of especially medical applications of nanotechnology. Finally, we summarize the chapter by listing the most important aspects.

3.1 Nanotechnology in Materials and Industry

Before we begin with possible future uses of novel nanotechnologies, we analyze the areas in which nanotechnologies already contribute. Unlike what most people believe, nanotechnology is heavily used in variety of areas. As discussed later, the trend is only really accelerating. The following brief list of current applications is based upon (Nau, 2016). While the less "strict" political definition of "nano" is used, it is still fitting to call most of what follows real nanotechnology.

- *Motor vehicles*: A surprising number of nanotechnology applications exist in the motor vehicle sector, ranging from titanium dioxide, for their anti-fog effect on the windshield, to silicon dioxide, to reduce the friction of tires on wet surfaces. Furthermore, there are uses for cerium dioxide, copper, and platinum are used to reduce the wear and tear on the motor and reduce exhaust emissions.
- *Food*: In the food industry, there is a surprising amount of use cases for nanotechnologies too. While pretty much everything we eat gets broken down into nanoscale building blocks, some supplements contain zeolite or silicon dioxide to improve the mineral uptake of the body or to serve as a filler material in capsules. Zeolite is also used as a filter to remove alcohol from beer.
- *Agriculture*: Zeolite is used to improve mineral storage in soil for farming.
- *Pharmaceuticals*: Nanotechnologies or nanoparticles are used to create implants, contrast media, tooth replacements, vaccines, and all kinds of pills, suppositories, creams, and gels. For these purposes, titanium nitride, cellulose, zirconium dioxide, iron, silver, gold, and even quantum dots are used.
- *Medicine*: Silver, cellulose, and gold are used in vaccines, band-aids, and bandages. In addition, a number of potential applications of nanotechnologies are currently being researched.
- *Cosmetics*: Titanium dioxide and zinc oxide are used as sunscreen. Silicon dioxide is used in all types of creams, lotions, and make-up. Some fullerenes are used as antiaging creams and even carbon is used to create mascara.

Nanonetworks: The Future of Communication and Computation, First Edition. Florian-Lennert A. Lau.

- *Textiles*: Nanomaterials are mainly used to coat the surface of clothes. For example, titanium, silver, and zinc have antimicrobial properties, sometimes offering additional protection against UV light or help repelling water and dirt.
- *Construction materials*: In the construction sector, there are several creative use cases for nanoparticles. For example, titanium oxide is used to create cobblestones with self-cleaning properties. Asphalt and cement are much sturdier when zeolite is added to the mixture. A similar effect can be achieved by adding carbon nanotubes to certain construction materials in lightweight construction.
- *Paints and varnishes*: Paints and varnishes are one of the areas in which nanotechnologies are used. They serve a variety of purposes, including protecting wood from weather and sunlight, paints, clear coats, varnishes, coloring glass, coatings, and printer ink. For example, zirconium oxide can increase the scratch resistance of surfaces, while aluminum makes colors appear more vivid. Titanium oxide is used in dirt protection; copper is used against fungi, insects, and other pests; and silicone oxide improves the transparency of some varnishes as well as scratch resistance. Carbon can simply be used in printer inks to improve the contrast of black ink. Overall, nanotechnologies have many uses in the painting industry, which cannot be discussed here due to space constraints.
- *Electronics*: (Consumer) electronics are another sector that heavily uses certain nanomaterials. Barium sulfate is used for stabilizing the surfaces of circuit boards, indium tin oxide is used to achieve better transparency of touch screens, and gold is used for better conductivity of certain printer ink. In organic light-emitting diodes (OLEDs) and modern monitors, quantum dots are used for reducing energy needs and for better illumination. For certain contacts, diamonds are necessary, and for smoother surfaces, for example, smartphones, a variety of nanoparticles like silicon, aluminum, or cerium dioxide are used.
- *Exercise equipment*: Carbon is often used to increase the stability of fishing rods, tennis bats, or golf clubs. These materials are much lighter and provide better stability. The use of carbon composite materials is by no means limited to these areas. Wherever light and stable materials are required, the material innovations originating in space travel might be useful.
- *Plastics*: Nanomaterials are frequently used in plastic industry. Nanoscale clay tiles are commonly used in the production of plastic bottles to keep the carbon dioxide inside the bottle. In contrast to people's belief, plastics allow certain gases to pass through, which is why beer and other lemonades were predominantly sold in glass bottles or tin cans. Indium tin oxide can be used in plastics to prevent the electrostatic charging of the material. Titanium oxide and silicon oxide can be used to protect materials against UV light and thus increase their durability. There is also a variety of additives that allow the improvement of certain material properties of plastics. These include carbon to increase stability, (thermal) conductivity, coloring, and UV protection. There are countless additional uses that would also exceed the scope of a short overview.
- *Renewable energies*: In the renewable energy sector, nanomaterials like titanium oxide are used in semiconductors. Zirconium dioxide, platinum, gold, and cerium dioxide are used for creating fuel cells to improve their efficiency. These cells have a variety of purposes as discussed in Chapter 9.
- *Toolmaking*: Tungsten carbide, diamonds, and zirconium dioxide are used to improve the resilience of machine parts. This especially applies to drills, cutters, and grinders.
- *Water purification*: Zeolite is used for water softening, binding radioactive substances, and for all-purpose filtering. Diamonds, iron, and gold can be used for removing mercury and other pollutants from water.

- *Chemistry*: Nanotechnologies are mainly used as filters or catalysts. For these purposes, zeolite, gold, platinum, and aluminum are used. Hence, chemistry is further involved in the production of nanomaterials that are used in other areas.

3.2 Medicine

Next main area of interest where nanotechnologies are used is medicine. Health is the most important aspect of human life. As early as 2600 years ago, the Buddha already declared health to be the "greatest worldly good" (Gotama The Buddha, 1987a). He was even frequently called the "great surgeon" for his ability to teach people the way out of their mental health problems.

Currently, things have not changed a bit and the healthcare industry is among the biggest sectors in every developed country. Health is desired by all people, yet the emergence of the 2020 pandemic showed us yet again that sickness and diseases constantly change. As a result, our treatments must change too and nanotechnologies offer promising solutions to yet unsolved problems in modern medicine.

Three overarching types of problems draw a lot of attention from the nanonetwork community. All these cannot be solved with currently available knowledge and technologies but nanonetworks might offer promising research directions that might offer future solutions. Even though the topic might still sound futuristic and likely will not be implemented in the future, the area attracts a lot of funding due to the universal interest in good health. The three problems are as follows:

1. The affordable long-term monitoring of health parameters in high-risk groups while maintaining quality of life: For example, medical nanorobots can be helpful in monitoring health parameters of unhealthy patients. This will help the patients in daily life without being depended on inpatient monitoring.
2. Fighting diseases as they arise without causing symptoms: Many diseases can be prevented much more efficiently in this way. If, for example, the exponential growth of bacteria can be detected and stopped in just a few specimens, hardly any serious interventions are necessary.
3. The local treatment of disease without systemic delivery of drugs through targeted, local administration: One of the biggest unsolved problems in modern medicine is the increasing resistance of pathogens to antibiotics. These can be reduced by only using medication where it is really needed. In case of bacterial pneumonia, this would correspond to local treatment of the lungs by nanorobots.

In what follows, we analyze the most frequently named unsolved problems in medicine and analyze the potential benefits of nanonetworks.

3.2.1 Convenient Permanent Health Monitoring

The use case most frequently mentioned in the technical literature is the constant monitoring of human health (Akyildiz et al., 2011). At present, such monitoring is hardly possible or involves an enormous effort and deprivation. Blood tests, for example, are only carried out on a random basis – usually only as soon as there is a reasonable suspicion of disease. However, many diseases can hardly be detected in this way or are detected much later (Freitas, 2005).

Figure 3.1 Storyboard representation of continuous medical surveillance using nanodevices (Lau, 2020). *Source*: Florian-Lennert A. Lau.

Figure 3.1 illustrates the process using an example. During treatment, nanodevices are used, which measure health parameters and continuously transmit them to an external software. This helps patients to follow daily routine without being confined to the hospital. Furthermore, the 2020 pandemic has shown that the healthcare system in most countries is underfunded and it is very difficult to provide equipment to constantly monitor the health of many patients during times of emergency. Nanonetworks promise a convenient solution to this problem.

3.2.2 Targeted Drug Delivery

Another extremely popular use case for nanonetworks is targeted drug delivery. This topic has drawn attention from various other disciplines where researchers examine possible nanostructures to use them as potential transport vessels for medication. Figure 3.2 shows the various examples of currently researched technologies. As mentioned before, researchers mainly use existing structures and try to modify them to their needs.

For example, liposomes are frequently used in dietary and nutritional supplements to prevent the body from breaking down active substances. Liposomes are lipid-based nanoparticles that have a cavity in the center where other substances can be stored. They can also be coated with other substances to incentivize movements to specific parts of the human body (Wang et al., 2015). It is believed that liposomes might be a suitable candidate for future nanonetworks that assist in targeted drug delivery.

The same problems also exist with regular medication (Arts and Hollman, 2005). Usually, only a portion of medication survives contact with the gastric system and the liver. Storing medication inside liposomes can circumvent this problem and thus lead to much better absorption by the body.

Many other nanoparticles are usually based on the same or a similar principle. An active substance is stored away in a transport vessel and is later released by a mechanism that is natural to the human body or through an external stimulus. In doing so, it is possible to avoid the systemic effects of drugs and only apply them to areas where they are needed.

A more extensive overview of possible nanoparticles in drug delivery is shown in Figure 3.2. In general, there are two broad classes of nanoparticles: organic and inorganic. The organic nanoparticles can be further categorized into polymeric, lipid-based, DNA-based, and "others."

Inorganic nanoparticles are often better researched than organic nanoparticles and have lower complexity. They are usual candidates for the top-down creation of nanodevices. While the lower complexity is appealing to basic research, it is by no means clear how to transfer the prevailing principles to the nanoscale. Furthermore, inorganic nanoparticles usually do not naturally occur inside living organisms and might cause yet unknown health problems.

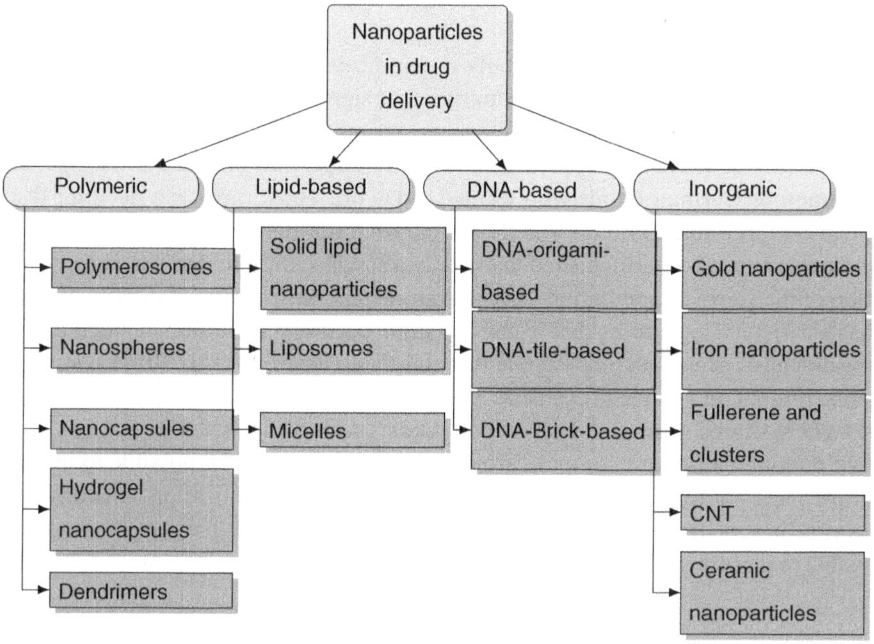

Figure 3.2 Extended list of nanoparticles for potential use in drug delivery systems based on Aghebati-Maleki et al. (2020). *Source*: Adapted from Aghebati-Maleki et al. (2020).

Organic nanoparticles have a different set of problems. While their building blocks often do occur in living organisms, the modified versions do not and thus might pose health risks. Especially with bacteria and viruses, special care is necessary to ensure that the human immune system does not react in a negative way. Another problem with DNA-based technologies is the potential for DNA to cause unwanted side effects. It must be clear that any DNA used for creating nanodevices cannot replicate on its own and cause an unwanted chain reaction in a living organism.

3.2.3 Immediate (Local) Treatment

For many diseases, it is important to act as quickly as possible once symptoms start surfacing to prevent possibly incurable consequences. An example of such a disease is cancer, where metastases may occur and make it difficult to cure the disease. While it is unclear how long it will take until negative effects start accumulating, it is desirable to detect and react to a disease as soon as possible, preferably before it starts showing any major symptoms. Medical nanonetworks offer a possible solution to this problem by continuously monitoring the health parameters of susceptible people.

A typical problem where immediate and even local treatments would be beneficial are bacterial diseases that are typically treated with antibiotics (Stelzner et al., 2016b). Bacterial pneumonia is a major cause of morbidity and mortality worldwide. Especially hospital-acquired pneumonia is associated with high mortality (Bassetti et al., 2015). The current treatments of bacterial pneumonia are becoming increasingly difficult to apply as bacteria develop resistance to antibiotics (Organization et al., 2014).

In such a nanonetwork, nanodevices may aid in detecting a bacterial lung infection by detecting proinflammatory cytokines released by the immune systems' effector cells. The nanodevices could then communicate the measured concentration levels to an external entity for further analysis.

3.2.4 Smart Medicine

Many diseases cannot reliably be detected using only one indicator. In fact, most existing diagnostic tools work based on the presence of a single marker that signifies the presence of a disease. However, many of those tests only work based on presupposed context knowledge. For example, the SARS-CoV-2 test can easily be tricked by applying a droplet of Capri Sun onto the test stripe, as demonstrated by numerous German pupils who wanted a day off. Thus, there is a need for test technologies that can compute smart choices and thus avoid such unintended problems.

In fact, many diseases cannot be distinguished using only a single symptom. For example, a relatively huge portion of the German population suffers from *chronic obstructive pulmonary disease* (COPD). COPD has a variety of symptoms that it shares with nosocomial pneumonia, and it is not easy to find a fitting diagnosis even with modern macroscopic tools (Restrepo et al., 2018). A variety of symptoms must be present all at once to correctly identify and distinguish both diseases. This is one of the diseases where shared symptoms make a correct diagnosis difficult.

Since DNA-based nanonetworks can compute arbitrary functions, it is even possible to base a diagnosis on multiple pieces of information. DNA-based molecular communication can consider an arbitrary amount of inputs. Since certain proteins only occur in specific parts of the body, it could even be possible to identify their location in the human body.

3.2.5 PCR Alternative

The 2020 pandemic showed us once again how important fast and reliable diagnosis tools are during times of crisis (Lau et al., 2021a). To accurately identify virus DNA, the *polymerase chain reaction* (PCR) is the gold standard. PCR tests need to be conducted in a laboratory, and the provision of results therefore takes quite a while. It has been shown that quick and frequent testing is a key component to contain the spread of a newly emerged virus.

A faster testing method is hence needed, like the antigen-detection diagnostic tests that were adopted during the ongoing pandemic to directly detect SARS-CoV-2 proteins. The sensitivity of those rapid diagnostic tests that are based on a chromatographic immunoassay is orders of magnitude lower than that of the DNA-based PCR tests. Another concern in current testing strategies is that mutations in the virus could suppress diagnostic detection (Ascoli, 2021). So there is still a demand for tests with the sensitivity of PCR tests, but for near-patient use providing quick results and with the option for fast adaptation to virus mutants. With DNA-based nanonetworks, one can even imagine rapid diagnostic tests that can distinguish between critical mutants of viruses.

DNA-based nanonetworks can be used as an alternative approach to detect even small amounts of DNA, thereby rendering the PCR amplification of DNA segments unnecessary. The error rate of the process can be fine-tuned, and it can finish relatively fast. This approach works based on a DNA box that can be opened in the presence of specific marker proteins or DNA sequences. If one of the boxes detects a specific DNA sequence, it releases its payload and other similar DNA sequences to propagate the finding through the medium. The payload could be a fluorescence marker to indicate a successful diagnosis.

3.2.6 Personalized Medicine

Since the field of medicine emerged in antiquity, the overall approach remained more or less identical. Medical professionals observe sick people and try to find a solution to the given problem. Individual differences are usually ignored in the process and treatments are designed for an "idealized patient" right in the middle of a Gaussian distribution of possibilities. With the current potential medical nanonetworks, this might change in the future.

Medical nanonetworks might create a paradigm shift in medicine where the previously mentioned general approaches are further individualized (Duffy, 2016; Mousa et al., 2020). Instead of prescribing one medication for all people if a single predefined indicator is present, the shift is toward diagnosis, prevention, and treatment for every single person individually. For example, DNA-based nanonetworks illustrate that nanonetworks can help with detecting health parameter anomalies from a personal norm instead of a global average, taking precision medicine to the most personalized level possible (Lau et al., 2022b).

While humans overall are similar, the normal ranges of various health parameters differ from person to person. For example, the blood pressure of humans who lift heavy weights is naturally elevated. However, that is usually not critical and just a bodily adaptation to weight training. For other people, the same value could be dangerously high, and blood pressure is just one of the many examples. Generally, many health parameters follow Gaussian distribution.

In classical medicine, medication is usually determined by comparison to a deviation from the population average, and the treatment follows a one-drug-fits-all model. While this can be a useful tool for the population average, it fails to account for individual variation. Especially people who deviate from the average cannot be reliably diagnosed. This observation is most evident in women, where it is difficult to gather empirical information about new medication as the hormonal cycle introduces a number of difficult-to-account-for parameters. As a result, it is cheaper to test medication on men alone. The generated data are then extrapolated for women, hoping that the medication has similar effects.

3.2.7 Vaccines

Yet another aspect of human healthcare that might benefit from nanonetworks is vaccines (Patrick, 2022). Currently, there are several distinct types that work based on various effects. Among the most modern kinds are *messenger RNA* (mRNA) vaccines. Those vaccines work based on the communication channels inside human cells. Messenger RNA is used to transport gene information from the DNA inside the core of a cell to the place where the desired protein is created from the DNA blueprint. By supplying special mRNA sequences, the cell can be encouraged to produce harmless parts of a virus, like spike proteins, that the body can use to learn about a new disease.

While mRNA vaccines are not new, nanonetworks can assist in the process of creating new active substances. Similar to the actual function of the human immune system, we envision a nanoscale "factory" that can immediately react to a new virus and create a fitting mRNA vaccine. Since mRNA has a remarkably simple structure, it is possible to manufacture desired RNA sequences in the human body. In addition, it is possible to communicate the findings to external research facilities or directly into the nano factories in other bodies to immunize them as well. Thus, the process of vaccination might be completely automated. However, this will likely take a long time to become reality.

3.2.8 Immune Enhancement

Another idea that is related to vaccines is the enhancement of the human immune system (Patrick, 2022). Creating any type of vaccine works based on the principle of "directing the attention" of the natural immune system capacities to a specific problem.

This approach is by no means limited to viral or bacterial diseases. It might be possible to artificially aggravate signals in the human body to draw the focus of the immune system to or away from a problem. In doing so, certain over- or underreactions of the immune system might be compensated.

For example, many autoimmune diseases like rheumatism have the effect that the immune system attacks the body. Weakening such a response in a targeted manner might alleviate the symptoms of many patients who suffer from excruciating pains.

A similar effect can be achieved by increasing the immune response in places where the signal is too weak or increased too gradually for the immune system to react. By artificially creating a substance the body naturally reacts to, the focus of the immune system might be drawn to such an area. As a result, it might be possible to treat chronic inflammations for which no good cure exists today. Many other applications can be envisioned based on the same principle.

3.3 Military

Historically, the military was always one of the driving forces of technology (Akyildiz et al., 2008). There is no better motivation than matters of vital concern and the military budget of most countries on earth reflects that. While the military already has plenty of options when it comes to killing, there is an ongoing interest in acquiring information. In fact, it is not far-fetched to assume that the majority of future wars will take place online. Getting information on the course of action of competitors or enemies is crucial in such scenarios.

Nanonetworks are a prime candidate for surveillance tasks as collecting medical data of patients and collecting behavioral data of people is not too different. While the ethical aspect of such applications is highly questionable, it is nonetheless to be expected that the military will test the waters and evaluate the benefits and shortcomings of such approaches. If proven beneficial, nanobots might be part of the military espionage arsenal until suitable countermeasures have been discovered. Nanotechnologies might also play a relevant role in counteracting such technologies.

3.4 Agriculture and Geology

The agricultural industry is always interested in further increasing efficiency. A whole book was recently devoted to the subject (Axelos and Van de Voorde, 2017). Given the unchecked population growth in Africa and parts of Asia, it is becoming increasingly important that the available arable land is used as efficiently as possible so that everyone can be fed.

But there are also luxury problems. The richer a country becomes, the higher the consumption of meat. The production of animal biomass is a factor less efficient than the production of plant biomass. Currently, research is mainly on genetically modified foods – in the hope that this technology can bring about a sufficient increase in efficiency.

In addition, nanodevices could also be used in a supportive manner. A suitable example would be in pest control or as a catalyst for the release of nutrients bound in the topsoil. In addition, it is conceivable that nanodevices can be used to deliver the required nutrients to the roots of crops or wherever they are needed. This can possibly prevent joint use by other plants or weeds. Finally, it is conceivable that nanotechnology can be used to have a positive impact on sensory and communication channels, such as rhizome and fungal networks (Almpanis et al., 2019).

Geological use cases are also of interest. The most interest in nanonetworks comes from early warning systems, for example, for natural disasters. Nanotechnologies could reach places that are hardly accessible for macroscopic technologies. This principle can also be applied to the exploration and surveying of underground oil fields. The use of nanodevices could help evaluate the profitability of natural resource reservoirs, which cannot be assessed using conventional methods (Jin et al., 2019).

3.5 Future Developments

So far, it has not been possible to synthesize nanodevices with a significant range of functions. However, as biological technologies continue to evolve, DNA-based nanodevice implementations as well as bacterial nanodevices may be soon available (Seeman and Sleiman, 2018).

In this chapter forecasts the future of nanotechnology, which is discussed in-depth in Chapter 13. Two options are chosen to approach the question:

- extrapolation based on already researched technologies
- the tracking of industrial interests and driving forces.

The first point deals with the theoretical feasibility and the second with the bundled social interest in nanotechnologies. In combination, it is possible to infer why the nano research area will likely continue to advance.

The starting point is formed by the technologies that have already been implemented. These are analyzed in detail in the following chapters. In summary, primitive approximations of nanodevices are already manufacturable. However, these are usually just proofs of concept without use cases in reality.

Driving forces and incentives, on the other hand, form the basis for the speed at which innovations continue to develop. The starting point for an analysis on this basis is the capitalist social system, in which a prompt response to a demand is provided in the form of a product, which gradually improves in the course of competitive processes.

By analyzing and comparing the financial value of different industries, it is possible to infer which products are of great social interest. Table 3.1 shows an example of the US economy in 2021 (The Biggest Industries in the United States, 1994).

It can be seen that the parts of the government spending with a low layer of Maslow's hierarchy of needs (Maslow, 1943) take a large financial share. Figure 3.3 shows the hierarchy for those who are unfamiliar with the concept. The hierarchy shows how important certain aspects of life are given that there is a lack of these. They are ordered by personal significance. If there is no food to eat, all the higher layers are of little to no immediate interest.

Applying Maslow's hierarchy of needs to economic statistics results in rough categories that arouse particularly strong social interest. A particularly large interest offers resources for intensive research on novel technologies, which tackle currently unsolvable problems.

It can be seen that healthcare is of particular interest as it addresses level 1 of Maslow's hierarchy of needs and a large proportion of the government budget is invested. Only social security attracts more funding. However, one could argue that social security supplies all the other basic needs of the Maslow hierarchy. As a result, one could argue that Medicare and health are the single largest positions when it comes to the government budget.

In addition, medicine is often suggested as the potentially biggest area of application for nanotechnology. In the following chapters, the focus is therefore on medical applications for nanotechnology.

Table 3.1 US total government spending fiscal year 2021 divided by sectors and compared with Maslow's hierarchy of needs.

Sector	Total $	%	Maslow
Social security	2.81 tn.	39.63	1
Medicare and health	1.61 tn.	22.67	1
Military	773. 28 bil.	10.91	2
Education	439.98 bil.	6.21	4
Interest on debt	303.03 bil.	4.27	5
Veterans' benefits	256.04 bil.	3.61	1–3
Food and agriculture	240.61 bil.	3.39	1
Housing and community	227.01 bil.	3.20	1–4
Transportation	209.08 bil.	2.95	4
Government	65.31 bil.	0.92	1–5
International affairs	65.2 bil.	0.92	5
Energy and environment	54.76 bil.	0.77	2
Science	38.64 bil.	0.54	5

In total, 1.6 trillion dollars are distributed every year (Street and MA, 2022).
Source: Florian-Lennert A. Lau.

Figure 3.3 Maslow's hierarchy of needs. The further down the hierarchy an entry is situated, the more important it is when taken away. Health is at the very bottom of the hierarchy, together with food, water, shelter, and rest. *Source*: Florian-Lennert A. Lau.

3.6 Summary

We can see that nanotechnologies already have countless applications and industrial uses. These range from simple chemical ingredients for manufacturing other products to computer chips that require nanoscale precision for the final product to work as intended. Many of these technologies are slowly approaching the physical limit around 0.3 nm or 260 pm (cesium). It is not possible to create technologies with a smaller resolution than that as that is roughly the size of single atoms.

In addition to currently available nanotechnologies, there is an entire range of possible future applications that attract considerable funding worldwide. Among those are military espionage applications, increases in efficiency in agriculture, and early warning systems for natural catastrophes.

However, the biggest area for potential application of future nanotechnologies is medicine. Health is extremely important. In fact, Medicare and healthcare are among the biggest government expenses worldwide. Possible applications range from targeted drug delivery over localized, immediate treatment of diseases to entire automatically generated and administered vaccines.

It is safe to say that nanotechnologies will shape the scientific landscape of the current century and likely beyond.

4

Construction

This chapter analyzes the most fundamental problem of all nanoscale systems in detail: construction. The best ideas and concepts are absolutely useless when it is not possible to actually manufacture the proposed systems in the near or even distant future. We first analyze the two major construction paradigms of bottom-up vs. top-down. Afterward, we enumerate the most promising construction materials for nanoscale systems. Based on that, we define suitable machine and system models as well as all their constituting components.

Given a suitable machine model, we can test how realistically implementable specific construction approaches are. We analyze both *carbon nanotubes* (CNT) and DNA in various forms as fundamental building blocks for nanonetworks. Next, we present a short overview of the state of the art in nanodevice and nanonetwork concepts. Finally, we list and explain the software for the simulation of construction purposes at the nanoscale. We conclude the chapter with several wet-lab results and a final comparison of the different approaches.

4.1 Construction Paradigms

There are several fundamental paradigms for construction at the nanoscale. They can roughly be categorized into top-down, bottom-up, and hybrid approaches.

The most fundamental difference exists between top-down and bottom-up manufacturing. Top-down approaches typically start with, for example, a silicon wafer and remove materials until the remaining structure fulfills the desired function. The process can be compared to working with a block of marble until it has the desired shape and form. Most chip manufacturing and basically all IT technologies are based on top-down mass manufacturing. Many aforementioned limitations of precise/accurate component placement and the inability to utilize the third room dimension apply to this method.

Bottom-up approaches start with certain building blocks and combine them into bigger and more complex structures. The process can be compared to building the wall of a house brick by brick. At the nanoscale a suitable example would be the self-assembly of DNA molecules where the simple bases adenine, thymine, guanine, and cytosine self-assemble into the DNA double helix. However, most bottom-up approaches use artificial molecules that fundamentally use the same self-assembling forces. The process is by no means limited to DNA alone, as basically all nanoscale structures naturally self-assemble in one way or another. Thus, it is an almost universal construction principle that emerged through an evolutionary process.

In contrast, it might be possible to combine both approaches into a hybrid version. By taking naturally occurring components like cells, bacteria, or viruses as a basis, it might be possible to enhance

Nanonetworks: The Future of Communication and Computation, First Edition. Florian-Lennert A. Lau.

them with artificially created components. In doing so, a lot of provably functioning components might be readily available. While it might be a challenge to combine the various necessary components of nanodevices at the nanoscale, repurposing organelles of a cell might be vastly easier.

While it is unclear which of the approaches is superior, some evidence points in the direction that bottom-up approaches are more realistic in the near future.

4.2 Materials

No matter if it is a top-down or bottom-up approach or even anything in-between, nanonetworks are created from certain materials. Those building blocks of nanodevices are the subject of this section. While most of those building blocks contain carbon, they do so in varying quantities and sometimes inside or outside living cells. A frequently mentioned inorganic material for the creation of nanodevices is CNT. They have a variety of interesting properties that might help overcome a variety of challenges at the nanoscale.

Other materials are based on molecules like DNA for self-assembly that we analyze in detail. Yet, other approaches focus on the surface manipulation of material via coating. Another special case of this would be so-called metasurfaces. The basics of all those materials are discussed in this section.

4.2.1 Inorganic Carbon

The most important building block of all nanodevices is carbon. Carbon and carbon-based nanoparticles have drawn a lot of research interest over the years, and a large array of different structures exist. This section just gives a brief overview, and the interested reader is encouraged to have a look at the work of Dresselhaus et al. (1996).

Due to its special properties, carbon is also called the "element of life." Carbon has the element symbol "C" and is the sixth element of the periodic table. Furthermore, it is the most important of the six elements from which the majority of living organisms are created. The elements are hydrogen, oxygen, nitrogen, phosphorus, sulfur, and most importantly carbon. An example carbon atom can be seen in Figure 4.1. Pure carbon is pitch black and often used as a coloring agent.

Carbon differs from the other elements due to its unique binding properties. They are based on the special electron configuration in the half filled outer layer. The maximum number of electrons

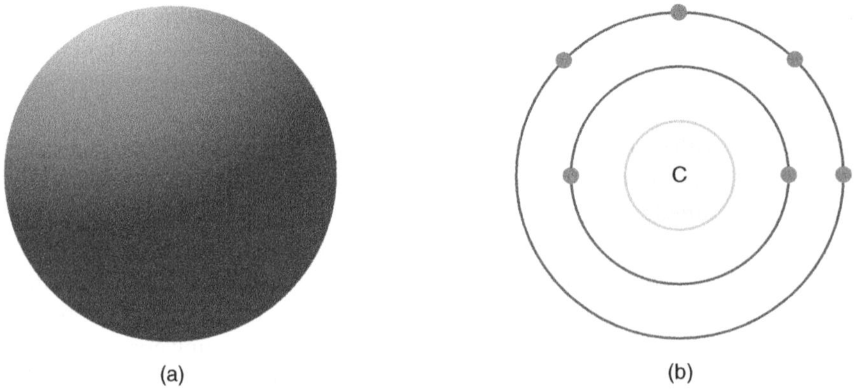

Figure 4.1 Single carbon atom. (a) The atom. (b) The atom with its electrons. *Source*: Florian-Lennert A. Lau.

per layer can be computed by the formula $2n^2$. As the outer layer is only half-filled, carbon has a tendency to form stable bindings with a variety of other atoms and even itself. At least 200 000 different inorganic and over 20 000 000 organic molecules exist. This extreme complexity serves as a basis for complex life forms.

In nature, carbon has two pure manifestations: diamonds and graphite. Both of them have different internal structures. Graphite consists of sheets of nanometer-thin layers of carbon atoms that form stable bindings with each other. The layers themselves only weakly bind via Van der Waals forces. A practical example and application of graphite can be seen in pencils. By dragging the tip of the pencil over a paper surface, the weakly bound graphite layers break loose and form a thin graphite line.

Diamonds have a different structure and are the hardest naturally occurring material. Each carbon atom has four neighbors. Due to the 109.5° angles of the connections, a very stable pyramid structure is formed, as illustrated in Figure 2.10(a).

4.2.2 Molecules

Up to a certain size, everything that consists of more than one atom and is held together by covalent bonds called a *molecule*. An example of a standard molecule representation can be seen in Figure 4.2. The colored spheres represent different atoms and the edge connections/bindings between atoms. As discussed before, especially molecules that contain carbon can produce an immense variety of complex structures.

In nature, all complex structures and life forms self-assemble from simple molecules. Given a suitable enviroent, they assemble like "crystals" without any additional external guidance. In the case of living beings, the "program" for the crystal formation is encoded in the DNA. Interestingly, the DNA also encodes parts of the environment in which the self-assembly takes place.

Based on those observations, it makes sense to assume that self-assembling molecules might be a suitable building block for nanonetworks. This is especially true as the process provably works in nature, and we would just have to adapt it to our needs. Nanoscale communication has taken a similar route where the initially proposed electromagnetic communication might be replaced by molecular communication that has been inspired by hormones and pheromones.

While the largest artificial molecule that has been created contains more than 17 million atoms, many structures that contain only 60 atoms are already called "nanoclusters" (Zhang et al., 2011). Some molecules that are created in living organisms can be even bigger. Nanoclusters are the topic of the Section 4.2.2.1.

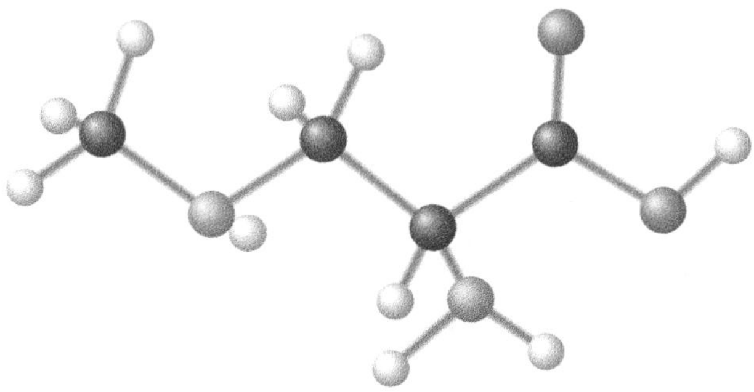

Figure 4.2 An example representation of a molecule. Atoms are represented by spheres and connections by thick lines. *Source*: Florian-Lennert A. Lau.

4.2.2.1 Carbon-Based Nanoclusters and Fullerenes

Lesser known, artificially created molecules from pure carbon are nanoclusters/fullerenes, as displayed in Figure 4.3(a) (Kroto et al., 1991). The figure displays the so-called Buckminsterfullerene, also called c60. The structure is named after famous architect Richard Buckminster Fuller who designed buildings that resemble the structure of the newly discovered carbon allotrope. In its structure, the c60 fullerene resembles a soccer ball and consists of a combination of pentagons and hexagons. Without the pentagons, the structure would not bend in the third dimension to form a sphere. In fact, all fullerenes contain exactly 12 pentagons to achieve a spherical shape.

Overall 60 carbon atoms are part of the structure, and due to its structural properties, the c60 fullerene is exceptionally stable. The structure has an overall diameter of 10 Angstrom, and the atoms are either 2.8 or 2.92 Angstrom apart from each other. The entire structure has a density of around 1.7 g/cm^3 and has similar electrical properties as graphite.

Apart from the first-discovered c60 cluster, a whole range of different fullerenes exists. The c540 fullerene can be seen in Figure 4.3(b). They mainly differ in the number of involved carbon atoms, as indicated by the name. Interestingly, the number of involved atoms is always even. If any more atom is added, it usually just lies on the surface of the remaining fullerene.

The smallest possible fullerene is c20 which only consists of pentagons and 20 carbon atoms in total. c32 is the smallest somewhat stable fullerene. In theory, the number of hexagons per fullerene is arbitrary but abides by the so-called "isolated pentagon rule." From c60 onward, the pentagons of the spherical surface do not touch anymore as that is energetically more efficient.

As a rule of thumb, if less than 1000 carbon atoms form a structure, it is called a fullerene or cluster. If more atoms are involved, it is called a nanoparticle. Until roughly 1000 atoms, each atom has a significant impact on the geometrical structure of the fullerene and thus on its emergent properties. As an example, diamonds and c60 nanoclusters only differ in their geometrical structure, yet have vastly different electrical properties. The c60 fullerene is a semiconductor, while diamonds are not conductive at all.

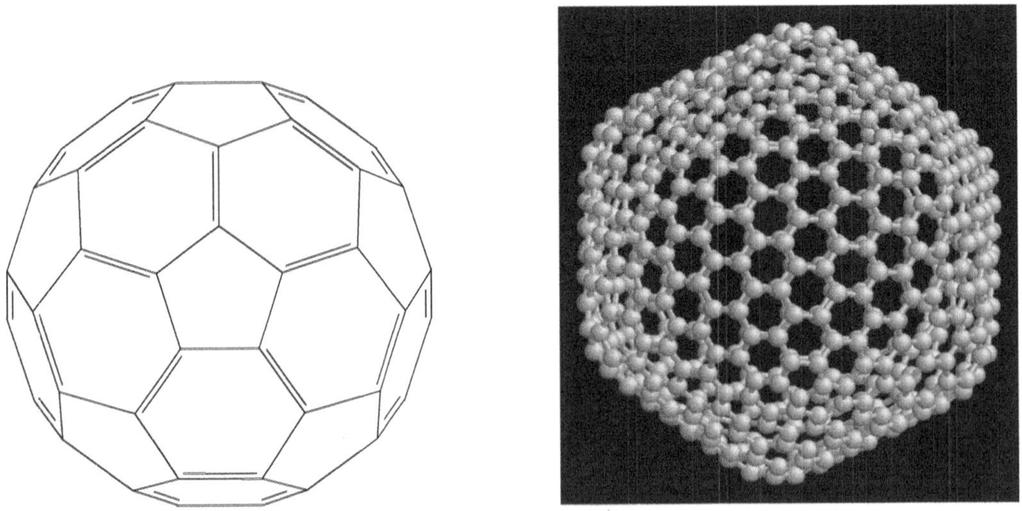

(a) (b)

Figure 4.3 c60 and c540 fullerenes. (a) c60 fullerene, also called "Buckminsterfullerene." (b) c540 fullerene. *Source*: Public Domain.

The process of creating clusters and fullerenes is fairly simple in theory but is also subject to the aforementioned limitations. While the clusters can be created, they can by no means be placed precisely or accurately to utilize their unique properties.

In essence, most carbon structures like CNT, clusters, or fullerenes are created following the same principle. The process can be compared with natural cloud formation. There is a gaseous medium where carbon monomers are added via vaporization using lasers. To turn, for example, graphite into gas, temperatures of around 3700 °C are necessary.

Oftentimes, a noble gas like helium is used to prevent unwanted chemical reactions. By either reducing the temperature or adding more monomers, a condensation process begins. Depending on the time, the basic building blocks/monomers, and the external conditions, different structures will form.

The resulting structures are of varying size and follow a size distribution. By fine-tuning the environmental conditions, it is possible to favor specific, very stable, fullerenes. If, for example, the temperature is very precisely chosen, only the most stable of clusters can form, while all other fullerenes will break down almost immediately. To get isolated clusters of a specific size, additional filtering processes are usually necessary.

4.2.2.2 Carbon Nanotubes

CNTs are yet another allotrope of carbon that was only recently discovered in 1991 (Iijima, 1991). Different examples can be seen in Figure 4.4. CNTs are usually created from thin sheets of graphite in a similar process as fullerenes. The graphite sheets form little rolls instead of spheres and depending on the angle, the resulting CNTs have different properties. The different ways of rolling the graphite sheets are illustrated by the coordinate system on the top left. The process can be compared to rolling a sheet of paper. If the edges do not perfectly overlap, a "twisted" paper roll is the result that has unique electrical properties.

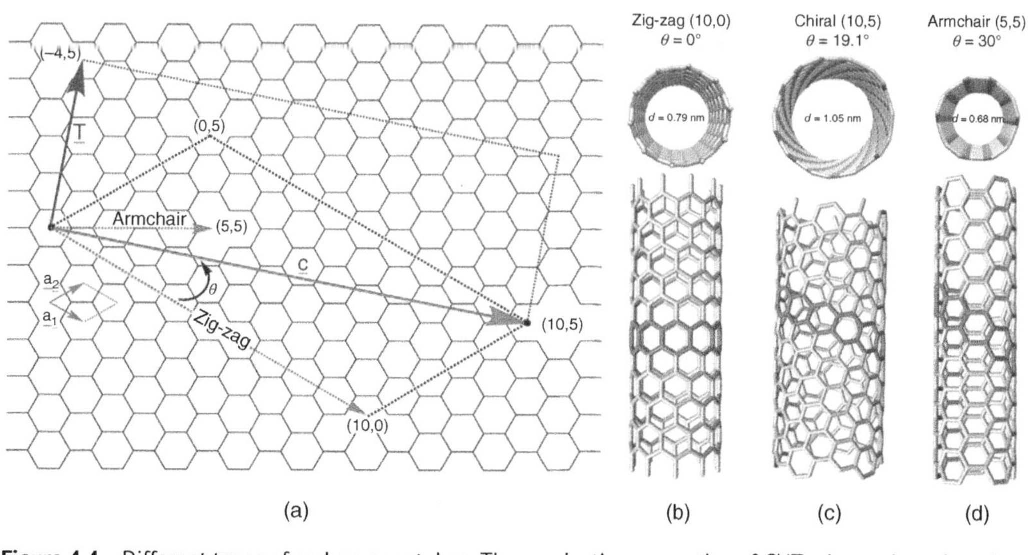

Figure 4.4 Different types of carbon nanotubes. The conductive properties of CNTs change based on the chirality of the graphite sheet attaching to itself. (a) Different possible angles of rolling a graphite sheet. (b) Zigzag rolling. (c) Chiral rolling. (d) Armchair rolling. *Source*: Barbero and Strank(2016)/with permission of John Wiley & Sons.

If the sheet of graphite is rolled "normally," it is called a "zig-zag carbon nanotube" as the border of the tube shows a regular zig-zag pattern. An example can be seen on the top right of Figure 4.4. If the graphite sheet is rolled diagonally in a specific/proportional way, the resulting CNTs are of the "armchair type," as can be seen at the bottom right. Any other rolling pattern is called "chiral," as illustrated at the bottom left. In this case, the border of the CNT is irregular. Depending on the type of rolling, the electrical properties of the CNTs change. It is thus possible to create CNTs with either metallic or semiconductor properties. It is even possible to create *y*-junctions or diodes from CNTs.

In addition to the rolling angle, several CNTs can also be nested. Furthermore, the diameter can vary from CNT to CNT. As you can see, there can be a huge variety of different CNTs and we have no reliable method for creating just a single type. However, specific applications might require exactly a huge amount of single-walled CNT with a diameter of 14 of the zig-zag type.

Many of the other properties of CNTs can be seen in Table 4.1. Small CNTs can have a diameter of as little as 1 nm, which is half the diameter of DNA. The density of CNTs is extremely small for its stability. They are even lighter than aluminum while having 2–20 times the tensile strength of steel ropes and being true to form.

The electrical properties of CNTs are equally extraordinary. CNTs have 1000 times the ampacity of copper and heat conductivity comparable with diamonds. While having all the above properties, CNTs have comparable heat stability to regular wires. All of this comes at a price; however, they are often 150 times as expensive as gold.

Due to all these extraordinary properties, CNTs have been suggested as a building material for a variety of things. They range from novel computer technologies over fishing lines to CNT ropes that could one day realize a space elevator. Figure 4.5 shows an example procedure of how ropes can be woven from CNTs. While a single CNT has 20 times the stability of steel, a rope created from multiple CNTs is internally only held together by Van der Waals forces. Thus, the stability is "only" two times that of steel.

Many scientists in the nanonetworking community view CNTs as the most promising material for the implementation of nanoscale systems. However, there is still no accurate/precise method available to manipulate CNTs at the nanoscale. Furthermore, a lot of the initial hype between 1991 and 2010 has died down by now as CNTs could not live up to the promises thus far.

Table 4.1 Properties of CNTs or composite ropes from CNTs.

Properties	CNT
Diameter	1 nm
Density	1.4 g/cm^2
Tensile strength	2–20 of a steel rope
Mechanical stability	True to form
Ampacity	1000 times of copper
Heat conductivity	Twice that of diamond
Heat stability	Comparable with wires
Costs	150 times the price of gold

Source: Florian-Lennert A. Lau.

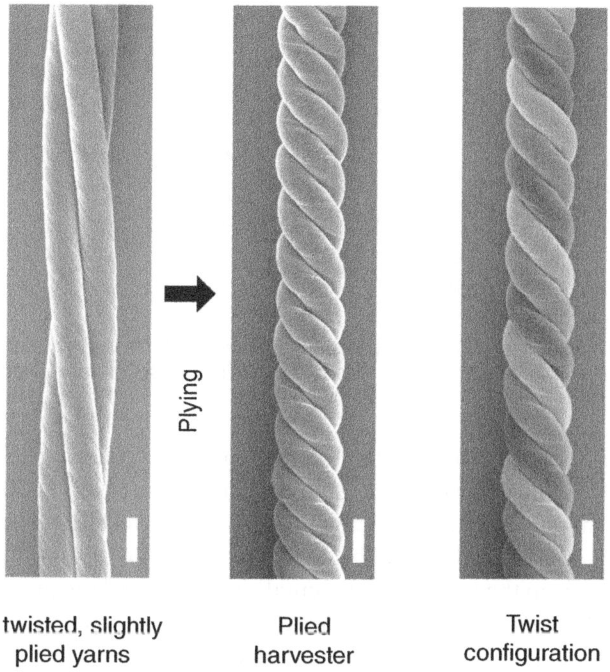

Figure 4.5 Rope made out of carbon nanotubes (Zhang et al., 2023). *Source*: With permission of The University of Texas at Dallas.

4.3 Nanoparticles

The next class of nanoscale structures is nanoparticles. Nanoparticles are often defined to be between 1 and 100 nm in diameter (Büther et al., 2017).

Nanoparticles are mainly distinguished from nanoclusters by their size and other emergent properties. If adding more atoms to a cluster does not change the emergent properties sufficiently, then the structure is called a nanoparticle and no longer a nanocluster. As a rule of thumb, a structure that contains more than 1000 atoms is usually a nanoparticle and is no longer regarded as a molecule or cluster.

There is a large variety of different nanoparticles that have a variety of different use cases. We have already discussed some of them in Chapter 3, like coloring glass or treating cancer. Whenever necessary, we introduce new nanoparticles on the fly instead of listing them here.

4.3.1 DNA

A special molecule that deserves more attention is DNA. As mentioned before, DNA has self-assembly properties and is part of all living organisms. DNA contains the assembly program and the necessary information to create the "machine" (assembly environment). DNA is a very "regular" and symmetric molecule that can be artificially manufactured, and the cost for this is getting cheaper and cheaper.

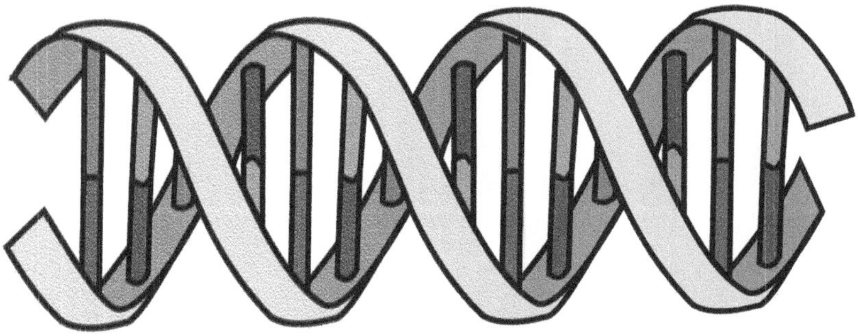

Figure 4.6 Example piece of DNA with its sugar–phosphate backbone (yellow) and the bases adenine (A), thymine (T), cytosine (C), and guanine (G) (colored connectors). *Source*: Leyo/Wikimedia Commons/Public Domain.

An example piece of DNA can be seen in Figure 4.6. The structure consists of a sugar-phosphate backbone and the bases adenine, thymine, cytosine, and guanine that connect both backbones. The interesting property of DNA is that only adenine and thymine as well as cytosine and guanine can form a stable binding with each other. Based on that behavior, half a double helix will always self-assemble into a full double helix given the right environmental conditions and the presence of the basic building blocks. Thus, DNA is *self-replicating*, which is the basis for procreation in all life forms. Without it, evolution would be impossible as evolution is based on slight *errors* in the replication process that might prove advantageous given certain changes in environment. The pieces of DNA that provide the resulting creatures with an advantage have a higher chance of passing their genes on to the next generation. The result of such long chains of "advantages" is complex life forms like humans.

DNA was only "recently" discovered in 1953 by Watson and Crick Watson and Crick (1953). However, the molecule has reshaped the scientific and medical landscape to an extreme degree. Figure 4.7 shows a timeline of important DNA-related discoveries and inventions. We start with the first discoveries that go beyond DNA as genetic code and explore its properties as a material for construction.

Between 1982 and 1983, the idea of using (artificial) branched DNA connections came up for the first time. Eight years later, a cube was synthesized from DNA. These steps represent the important basics for three-dimensional self-assembly systems and transport mechanisms. In 1996, the DX molecule was introduced, which is the basis for DNA templating and modern self-assembly systems. Tiles are the basic building blocks for self-assembly systems and are created on the basis of this molecule or variations of it.

In 1998, Winfree was the first to grow a two-dimensional lattice from DNA. Six years later, *ribonucleic acid* (RNA) was proposed as a building material for nanotechnology for the first time. Unlike DNA, RNA consists of a single strand with the base uracil instead of thymine. In 2006, DNA origami technology was introduced and Rothemund used it to assemble the first nano-smiley face. A year later, the DNA-guided assembly of gold nanoparticles was achieved. They play an important role in fighting cancer and other diseases. One more year later, the DNA origami process was generalized to three dimensions. In addition, a DNA box was created, which could be opened by a signal. These two milestones further motivate tile-based nanonetworks. In 2010, the DNA box was further refined. It was possible for the first time to load nanoparticles into a 3D DNA structure and later release them again. This idea forms the basis for the local application of drugs using nanodevices.

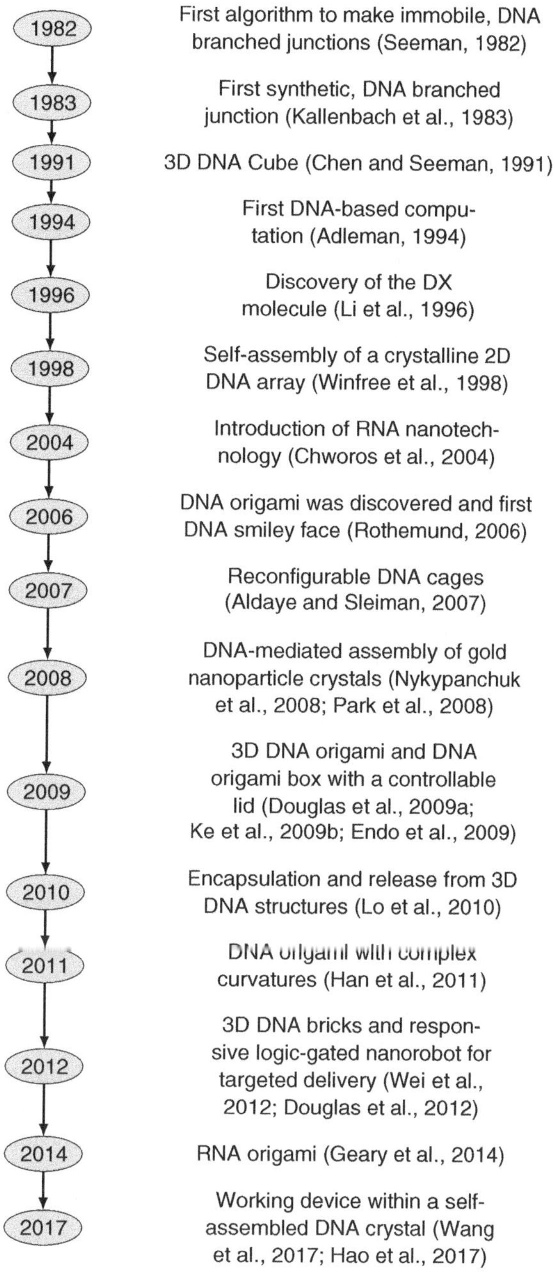

Figure 4.7 Timeline of important DNA-related discoveries with carryover into nanoscale systems and applications (Seeman and Sleiman, 2018). *Source*: Seeman and Sleiman (2018)/with permission of Springer Nature.

In 2011, the creation of DNA origami with complex curves succeeded. In 2012, DNA "bricks" were presented, which can serve as a building material for more complex structures. In addition, drug application has been further improved by nanodevices. Two years later, the origami approach was extended to RNA, and in 2017, complicated nanodevices were developed for the first time. Semiconductors and nanodevices have been embedded in three-dimensional DNA crystals. The latter result shows that it is possible to use DNA to transport other structures within it.

As DNA is one of the most promising materials for nanoscale manufacturing, the following sections and chapters explain many DNA-based structures in detail. Here, we just provided the absolute basics for the new but interested reader.

4.3.2 Metamaterials and Metasurfaces

Metamaterials and metasurfaces are artificial nanoscale structures with properties that usually do not occur in nature (Walia et al., 2015). The word "meta" is derived from ancient Greek and means "beyond" – metamaterials thus "go beyond conventional matter/materials." The same idea also applies to metasurfaces, which are a subclass of metamaterials.

The envisioned applications for metamaterials range from planar lenses to the possibility of turning objects invisible. For a detailed overview, interested readers refer to the work of Oliveri et al. (2015). While there are many proposed applications for metasurfaces, mainly the electromagnetic characteristics are of interest to the nanonetworking community. As an example, it is already possible to create antennas from metamaterials or to coat materials to change their electromagnetic properties (Cai et al., 2007).

The principal idea behind the creation of metamaterials is a repeating array of very small, often nanoscale elements. These repeating elements are sometimes called *unit cells*, and they are typically smaller than the wavelengths of the phenomena they influence. Those unit cells are typically "hard-coded" at the moment of manufacturing and cannot be changed later. While this is true for most classic metamaterials, a new class of *software-defined metamaterials* (SDMs) allows for real-time manipulation of a material's characteristics (Taibi et al., 2019). Due to the possibly high levels of achievable accuracy and precision, SDMs are a possible candidate for improvements in the area of *wireless sensor networks* (WSNs) (Abadal et al., 2020). As nanonetworks are an extension of WSNs, it is to be expected that metamaterials might also influence this new area of research and enable advancements that have been regarded as impossible until now.

4.4 Defining Complex Nanostructures

With all of the "simple nanostructures" explained, we can define more complex, artificial nanostructures. While most people still use various terms interchangeably, there have been efforts to streamline the vocabulary in the nanonetworking community (Büther et al., 2017). In this section, we enumerate and explain the currently available, most precise definitions for nanostructures and more complex nanodevices. Based on that, we derive an accurate machine model for nanonetworks and all of their constituting parts using Markov decision processes.

We now formally define nanoparticles, nano-objects, nanostructures, nanodevices, nanomachines, nanosensors, nanonodes, nanorobots, and nanonetworks. In addition to that, there is a brief overview of all the constituting components of the respective devices.

4.4.1 Component-based Approach

A brief overview of all the interesting structures and their relationship can be seen in Figure 4.8 (Büther et al., 2017; Lau et al., 2022a; Lau, 2020). The diagram is structured in various layers with several overlapping ellipses in the center. The further outside a structure is, the simpler it is. Nano-objects are the simplest kind of artificial structure, and nanorobots/nanobots are the most complex. From the nanonetworking point of view, molecules, clusters, fullerenes, and nanoparticles have a special place as many of them are not artificial constructs.

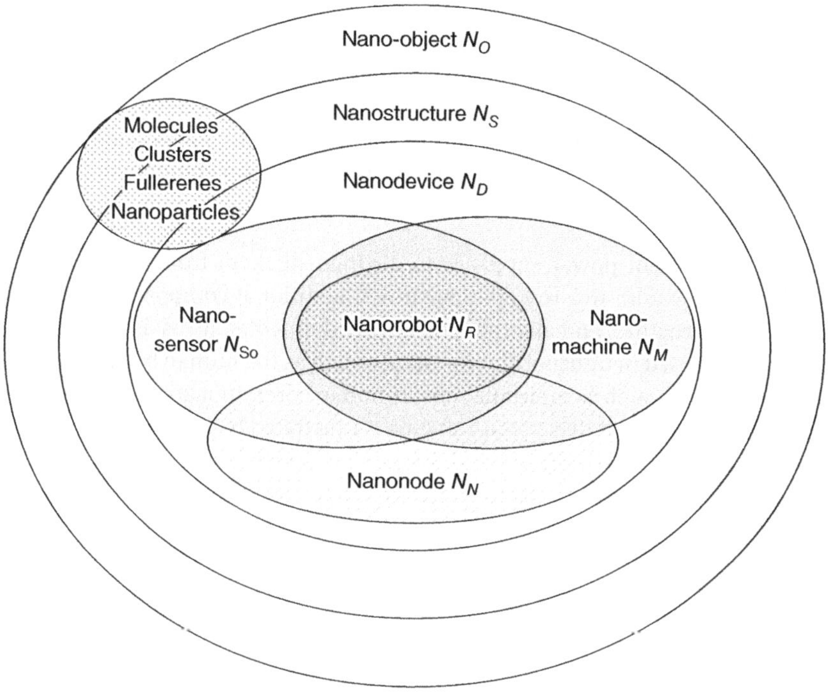

Figure 4.8 Venn diagram of the various nanostructures and objects. The overlap between circles represents a functional similarity in components. *Source*: Lau et al. (2022a)/with permission of Association for Computing Machinery.

We start with the definition of nano-objects/nanoparticles and only highlight the additional properties, constraints, or components. All of the following definitions follow the same pattern.

Definition 4.1 A *nano-object* or *Nanothing* $\mathcal{N}_O$ is an object of nanoscale size in at least one dimension in an environment S modeled as a random variable. A *nanoparticle* $\mathcal{N}_P$ is a nanoscale object of a maximum of 1–100 nm in all dimensions. A nano-object/particle consists of zero or more optional components $K_{opt} \subseteq \{A,C,E,I,L,M,O,T\}$. Nanoclusters or fullerenes are "proto" nanoparticles with special properties.

Examples of very small nanoparticles are grid-like sphere structures called clusters or fullerenes. An example of a nano-object would be a terahertz antenna as used in some cars for distance measurements. While paint does not seem like a nano-object, it does fulfill the necessary preconditions in some cases.

Definition 4.2 A *nanostructure* $\mathcal{N}_S = K_{opt}$ is an artificial nano-object between 1 and 4000 nm in all dimensions. It is designed to perform a specific function in an environment S modeled as a random variable. A nanostructure consists of zero or more optional components $K_{opt} \subseteq \{A,C,E,I,L,M,O,T\}$.

After particles and nano-objects, nanostructures are the most general, potentially complex structures that can be created for use at the nanoscale. The *nanoscale* corresponds to a size of 1 nm to a few micrometers in any dimension. Even though more than 1000 nm are technically no longer "nano-sized," the term has nevertheless been used in the literature because, for example, the human capillary system allows blood cells of 4 μm in size (Mohrehkesh et al., 2017; Pierobon and Akyildiz, 2010).

Definition 4.3 A *nanoscale device* or *nanodevice* $\mathcal{N}_D = K_{\text{mand}} \cup K_{\text{opt}}$ is a nanostructure that has a set of necessary components $K_{\text{mand}} = \{\text{E}\}$ and a set of zero or more optional components $K_{\text{opt}} \subseteq \{\text{A,C,I,L,M,O,T}\}$.

Nanodevices require a self-sufficient power supply E to distinguish them from fully passive nanostructures or particles. Just as with macroscopic machines, additional components may be present. A nanodevice is a nanostructure, and all other devices are defined on its basis.

Nanodevices are expected to be used in unusual environments, such as the human body. The size of nanodevices poses new challenges such as molecular absorption in communication or quantum effects (Bush et al., 2015). The existence of these requirements is illustrated by the environment S. The environment S designates a list of constraints or conditions. One possible constraint is the size of a device. For example, the human capillary system constrains some nanodevices to be less than 4 μm in size.

Definition 4.4 A *nanomachine* $\mathcal{N}_M = K_{\text{mand}} \cup K_{\text{opt}}$ is a nanodevice with necessary components $K_{\text{mand}} = \{\text{A,E}\}$ and a set of zero or more optional components $K_{\text{opt}} \subseteq \{\text{C,I,L,M,O,T}\}$ in an environment S.

This definition interprets nanomachines as very small machines. An artificial enzyme that folds proteins can be a nanomachine according to this definition, ignoring a direct energy supply. Similar to machines, the concept of a sensor node can be reformulated for the nanoscale (Xie, 2003).

Definition 4.5 A *nanosensor* $\mathcal{N}_{\text{Se}} = K_{\text{mand}} \cup K_{\text{opt}}$ is a nanodevice with the necessary components $K_{\text{mand}} = \{\text{E,O}\}$ and a set of zero or more optional components $K_{\text{opt}} \subseteq \{\text{A,C,I,L,M,T}\}$ in an environment S.

A simple nanomachine or nanosensor can endlessly repeat a simple task without ever evaluating information from the environment for interaction. However, to solve more complex tasks, it is necessary to autonomously create an abstraction of the environment S.

Definition 4.6 A *nanoagent* or *nano(ro)bot* $\mathcal{N}_R = K_{\text{mand}} \cup K_{\text{opt}}$ is an autonomous and programmable nanodevice in an environment S. It has a set of mandatory components $K_{\text{mand}} = \{\text{A,E,I,M,O}\}$ and a number of zero or more optional components $K_{\text{opt}} \subseteq \{\text{C,L,T}\}$.

Since both computational space and memory are limited at the nanoscale, it can be assumed that many nanodevices must be involved in the solution of a problem in a network (Akyildiz et al., 2008). A prerequisite to participating in a nanonetwork is the ability of the potential participants to communicate. This allows nanodevices to exchange information so that a common goal can be worked toward.

Definition 4.7 A *nanonetwork* is a directed ad hoc graph $G = (V, E)$, where V is a set of at least *nanonodes* $\mathcal{N}_N$ operating in an environment S. Nanonodes are nanodevices that additionally have the necessary communication component $C \in K_{\text{mand}}$. E is a set of edges $e \in V \times V$.

Analogous to macroscopic networks, nanonetworks can be modeled as directed graphs. Here, vertices describe nanonodes and edges correspond to the possibility of communication with other devices that are within range and not otherwise prevented from communicating. The environment S influences and constrains the graph significantly. The communication range and noise are conditioned by the environment. In an unstable or mobile environment, it is easy for both node positions and edges to be variable and thus harder to model.

In Section 4.4.2, the various components of nanostructures are explained in more detail and Section 4.4.3 proposes a holistic model based on Markov decision processes.

4.4.2 Components

In Section 4.4.1, a number of components were mentioned, which can be part of the various devices presented. In order to make the above definitions more precise, the individual components are explained in more detail here. All of the components shown here fall into the hardware category. They are physical constructs or parts of machines and robots. All behavior is a software part of the information processing I component. The individual components are explained in the following paragraphs.

4.4.2.1 Actuators A

Manipulation or actuation describes the ability of a nanodevice to influence its environment. This can happen via various environmental parameters, for example, physical, chemical, or biological. Actuators can be designed using complementarity. Molecular attraction and repulsion reactions are used. For example, an opening mechanism can be created by closing a container with a weak bond, as shown in Figure 2.12. This can be opened by bringing a stronger binding molecule close to the locking mechanism and thus opening (Andersen et al., 2009).

4.4.2.2 Communication C

Communication is an exchange of information between two or more entities that have the necessary capability. A suitable environment variable that is as reliable as possible is manipulated and then measured by another device. Both actuators A for manipulation and sensors O for the observations/measurement are required for this.

Communication must take place with the intention of exchanging information since otherwise natural processes could be understood as communication. Communication merely describes the "capability" to exchange information. Effects such as noise can permanently prevent a nanodevice from successfully communicating. The components used for communication do not differ technically from conventional sensors and/or actuators. In the case of biological nanodevices in particular, which communicate via the exchange of molecules, these are often identical. In order to be able to distinguish measurement and actuation from communication, a previously defined intention is required. A nanodevice, therefore, communicates when manipulation and measurement are done with the intention of communication, otherwise only an environmental parameter is manipulated or measured.

In some cases, a nanodevice will not be able to distinguish between an attempt to communicate and a random change in the observed parameter. Communication protocols can help here by applying error correction mechanisms or using checksums to detect errors. At least three important methods of communication at the nano-level are distinguished:

- acoustic communication,
- electromagnetic communication, and
- molecular communication.

Table 4.2 Comparison of different molecular communication methods from Farsad et al. (2016).

Parameter	Propagation	Invasiveness	Bits/particle
Tiles (DNA)	Diffusion/flow	Very low	$1/k$
Particles	Diffusion/flow	High	$1/k$
Concentration	Diffusion/flow	High	$1/k$
Type	Diffusion/flow	High	$1/\log_2 n$
Order	Diffusion/flow	High	$2/n$
Bacteria (DNA) (Balasubramaniam and Lio', 2013)	Chemotaxis	Low–high	Several
Viruses (DNA) (Walsh and Balasubramaniam, 2013)	Diffusion/flow	Low–high	Several

The variable k describes the particles per message, and n is the number of particle types (Lau, 2020).
Source: Florian-Lennert A. Lau.

Table 4.2 shows a list of molecular communication concepts and their associated properties. The types of communication are categorized based on their underlying concepts. The table makes it clear that most communication mechanisms spread information on the basis of diffusion. Only bacteria-based techniques deviate from this, they use chemotaxis (Balasubramaniam and Lio', 2013). *Chemotaxis* describes the forces of attraction and repulsion used by bacteria.

Particle-based methods mostly use molecules that are present in the environment. If the environment is the human organism, this can affect the body's own control circuits. Tiles and bacteria probably have little potential for conflict. Nevertheless, it is difficult to compare the different types of communication with each other.

Above all, tile-based communication poses a problem since message molecules are first assembled from tiles, which in turn can be used in communication like conventional molecules. The hierarchical process of message composition has hardly been studied in terms of networks.

Table 4.2 also shows that tile-based methods can be assessed as positive in comparison since they are less bioinvasive and a lot of information can be sent per particle. The low level of bio-invasiveness is due to the fact that DNA can occur in practically any place in the human body, and, therefore, there are hardly any antibodies. It can therefore be assumed that DNA tiles have less of an impact on existing communication mechanisms in the human body than other methods.

Concentration-based methods convey comparatively few bits per particle since only the total concentration of particles in a medium can be measured, which often has to be very high in order to be able to meaningfully distinguish between a logical "1" and a logical "0."

4.4.2.3 Information Processing I^+

The information processing component I allows a nanodevice to manipulate and transform data. I stands for a device that can be configured/hard-coded, while I^+ stands for a programmable nanodevice. In the simplest case, information processing can be represented by a Boolean operation, which is implemented by a single transistor. Information processing assumes that nanodevices can be programmed to control the behavior of an information transformation. Computational models such as circuits or self-assembly systems cannot be programmed in the classic sense. In this case, it must be possible to create a new circuit or a new self-assembly system that produces different outputs for the same inputs.

In principle, programmability can be interpreted as a form of device configuration. Programmability is thus a spectrum, starting with a few bit switches that control coarse machine behavior, up to Turing-complete interpreters for a machine language.

Possible models for information processing are classic computers based on transistor technologies such as CMOS. These can be organized into circuits to represent the Turing-complete logic. A logic of similar power can be generated based on quantum dot cellular automata, which are presented in Chapter 5.

Another computational model that can be used for information processing is self-assembly systems, which are also Turing-complete. These are explained in detail in Chapter 5 as well.

4.4.2.4 Locomotion L

The locomotion component L describes the ability of a nanodevice to change its whereabouts. Locomotion is divided into two categories: active and passive movements. Passively, mobile nanodevices are transported by flow processes in a medium and do not require any further external influences. Possible examples are nanorobots, which move through an organism together with the blood flow, diffusion, or Brownian motion.

Similar to communication, actively mobile nanodevices use a special actuator component. The environment is influenced by the intention of locomotion. The following list shows various active and passive ways of locomotion in liquid media. Some are biologically motivated, while others have classical, mechanical models:

- external application of magnetic fields (Wang, 2012),
- Brownian motion,
- diffusion,
- blood flow
- bubble propulsion,
- DNA walker,
- chemotaxis,
- flagella, and
- molecular motors.

A summary of locomotion at the nanoscale can be found in the work of Wang and Gao (2012). Of the technologies listed, passive locomotion is of particular interest to this book. All sufficiently small structures are subject to Brownian motion, diffusion, and blood flow.

It can be difficult to manually and specifically position a nanorobot in the human body without producing unwanted side effects. For example, the target area could be difficult to reach or only partially resilient due to previous illnesses. In the case of passive locomotion, it can at least be assumed that the organism is adapted to the process to a certain extent.

4.4.2.5 Memory M

The memory M describes the ability of a nanodevice to hold information persistently. Memory is of particular interest with regard to information processing I since it is only through memory that programs or configurations can be used, which in turn enables controllable or complex behavior.

Despite the obvious utility for many applications, memory is by no means necessary or self-evident. Circuits, for example, do not require memory and are still able to perform complex calculations. However, memory is usually required for data collection and transmission.

One way of realizing memory is based on molecules such as DNA. Unlike the binary alphabet, DNA provides a base four alphabet. Information can be stored and encoded – in the form of DNA – for long periods of time. The lifetime of some information stored by DNA-based memory variants far exceeds that of magnetic tapes, CDs or hard drives (Mammana et al., 2011).

4.4.2.6 Energy Supply E

The energy supply component E describes how the sensors and actuators of a nanodevice are supplied with energy. The power supply E is one of the few components necessary for nanodevices. A nanodevice cannot be assumed to be reliably and continuously connected to a power source. An internal power source such as a battery or miniature power plant may be necessary.

DNA-based systems use the thermal energy present in the communication channel or medium to form or break stable bonds. DNA tiles thus harvest energy from the environment and thereby supply the self-assembly process with energy. The advantage is that no dedicated power supply component needs to be manufactured. Parts of the environment in living organisms are adapted to provide a synergistic environment for binding reactions. Similar conditions are easily reproduced in an artificial environment such as a Petri dish.

There are numerous other processes that can be used to harness energy from the environment. A common example is a sugar battery, which might use maltodextrin to gain energy (Zhang and Hoshino, 2018).

4.4.2.7 Sensors O

A measurement or observation is a process in which a mostly continuous environmental parameter is recorded and converted into a processable form. A typical example of said parameters is an electrical voltage or an electrical resistance.

Since sensors O often provide continuous values such as voltages or resistance values, an A/D converter may be necessary in order to be able to transform or manipulate the recorded values later using a digital information processing unit I. Although A/D converters can be very complex, they are considered part of a sensor – not part of an information processing unit I.

In the medical field, sensor technology O is divided into laboratory-analytical and apparatus-based processes (Scharringhausen, 2018). Since apparatus-based processes usually require complicated devices, these appear unsuitable for use at the nanoscale.

Laboratory-analytical methods are measuring or detection methods in which the characteristics and significance of selected parameters are determined through the use of sensors. An example is a blood gas analysis.

Laboratory-analytical methods are of particular interest for nanodevices since these deal, among other things, with the measurement of body parameters. They are divided into quantitative – analog measured values – and qualitative methods – truth values.

Immunological methods are a particularly promising special case of laboratory-analytical methods. These can be used at the nanoscale to detect antibodies of the order of approximately 15 nm (Mikhailovich and Lebedenko, 2009). This and other procedures, for example, based on CNT use the concept of complementarity (Staiano et al., 2017). This describes that certain molecules attract or repel each other. Molecules that attract each other often bind to each other, changing their electrical properties. In this way, sensors can be realized at the nano-level. If "markers" are mentioned in the course of the work, a measurable antibody based on complementarity is assumed.

4.4.2.8 Timer T

A timer T enables a nanodevice to make causal statements and implement temporal behavior. Internal clocks are an integral part of macroscopic calculators. However, their availability at the nano-level cannot be assumed. Accordingly, the investigation of the power of devices without an understanding of time also makes sense. There are three different types of understanding of time with different powers in nanodevices:

1. relative understanding of time through the happened-before relation (Lamport, 1978),
2. relative time understanding, which allows a nanodevice to determine a time difference between two events, and
3. absolute understanding of time through, for example, time stamps or clocks, which count in defined time steps and provide this information.

It is generally known that an absolute understanding of time cannot be implemented in distributed systems (Lamport, 1978). Since nanonetworks are distributed systems at the nanoscale, it follows that no absolute understanding of time is possible here either. However, it is possible for nanonetworks to meaningfully interact without being able to retrieve the global system state.

4.4.3 Nanonetworks as Markov Decision Processes

In this section, we analyze the differences between simple single-agent-fully observable and multi-agent-partially observable-joint-reward systems as a basis for modeling nanonetworks (Braun et al., 2021; Lau et al., 2022a). The presented system is based on the principle of *maximum expected utility* (MEU) where a utility function assigns a quality to possible next states (Russell and Norvig, 2020). We use a random variables R which takes discreet values from $\text{ran}(R) = \{r_1, \dots, r_m\}$. When the random variable R assumes a value of its range $r \in \text{ran}(R)$, an *event* occurred, which is called $R = r$ or r in short. If the random variable models actions, it is called a *decision random variable.*

4.4.3.1 Markov Decision Processes

A *Markov Decision Process* (MDP) describes a sequence of decision problems in a fully observable environment. An MDP has a Markov-1 transition model and time-independent, cumulative rewards.

Definition 4.8 An *MDP* is a tuple (S, A, T, R) consisting of:

- a random variable S that describes an agent's environment,
- a decision random variable A_s that represents a set of actions for each state $s \in \text{ran}(S)$ as its range,
- *transition model* $T(S', S, A_S) = P(S' \mid S, A_S)$ (if $a \notin \text{ran}(A_s)$ for a given s, then $P(s' \mid s, a) = 0$), and
- a *reward function* $R(S)$.

To *solve* an MDP, we use a set of sequential actions, called *policy* $\pi : \text{ran}(S) \mapsto \text{ran}(A_S)$.

In addition to the state, the performed actions may also influence the reward, modeling, e.g. costs. Additionally, one can specify a *discount* factor $\gamma \in]0, 1]$, which indicates how relevant future rewards are. The factor corresponds to an interest rate of $1 - \gamma/\gamma$. The utility of a state $s \in \text{ran}(S)$ is given by the Bellman equation,

$$U(s) = R(s) + \gamma \max_{a \in \text{ran}(A_s)} \sum_{s' \in \text{ran}(S)} P(s' \mid s, a) U(s'). \tag{4.1}$$

The sum adds up the *expected utilities* of each state s', given by the probability of reaching s' from s by applying an action a. We call π^* an *optimal policy*,

$$\pi^*(s) = \arg \max_{a \in \text{ran}(A_s)} \sum_{s' \in \text{ran}(S)} P(s' \mid s, a) U(s').$$

In this case, the agent always chooses actions a that maximize the expected utility. We suggest *value iteration* as a technique to find policies, but there are other possible techniques (Oliehoek and Amato, 2016). Value iteration uses a modified Bellman equation 4.1,

$$U_{t+1}(s) \leftarrow R(s) + \gamma \max_{a \in \text{ran}(A_s)} \sum_{s' \in \text{ran}(S)} P(s' \mid s, a) U_t(s').$$

The algorithm loops until a predefined quality is reached, which is represented by an allowed error ϵ.

An intuitive example of a classic MDP can be seen in Figure 4.9. The presented grid represents the entire state space the nanodevices can operate in. As MDPs operate on single agents, only a single nanodevice is present. The gray square represents an obstacle that the nanodevice has to navigate around. For navigation, the nanodevice can move in all cardinal directions, represented by the operations "↑, →, ←, ↓."

In this scenario, the state of a nanodevice would be its position in space. Based on that position and global knowledge about the surroundings, the nanodevice can make a probabilistic decision on the optimal action to perform to maximize the reward it can get. Due to the Markov property, the state history does not matter to the nanodevice.

In Figure 4.10, the scenario is a bit more complicated and an additional charging station (north west lines) and a sand pit (north east lines) have been added. Based on this, the behavior of a nanodevice has to be adapted as it might have to frequently recharge or avoid the sand pit that it cannot leave anymore. Similar constraints might very well occur with autonomous nanoscale systems.

An example of a policy can be seen in Figure 4.11. For each possible state, the nanodevice knows what action to perform to maximize its reward. The nanodevice periodically measures its own state and repeatedly performs actions. In this case, the state is just the position in space and the final policy could be implemented as an If/Else tree.

The quality of a policy could be defined by the number of actions a nanodevice has to perform to reach the final state. To find good policies, a value iteration algorithm or any other kind of program search could be performed.

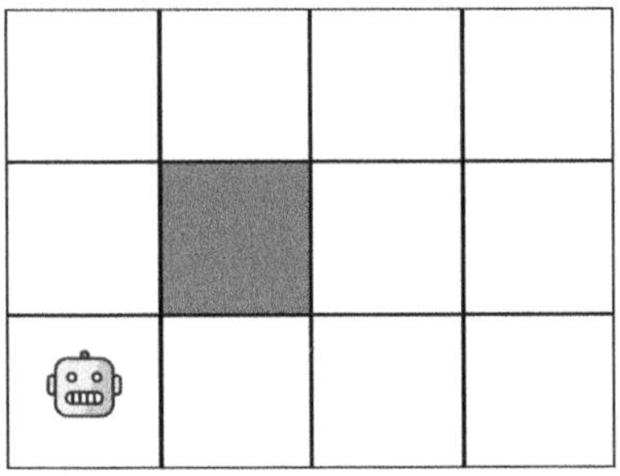

Figure 4.9 The state space for a MDP, including an obstacle (gray) and an agent. *Source*: Florian-Lennert A. Lau.

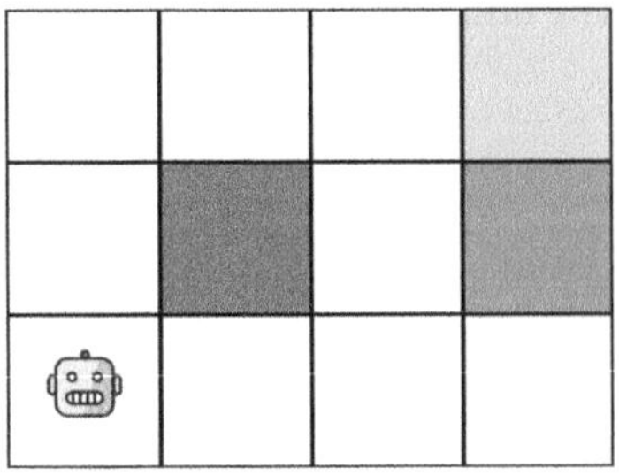

Figure 4.10 MDP with a sand pit (middle right cell) and charging station (top right cell). *Source*: Florian-Lennert A. Lau.

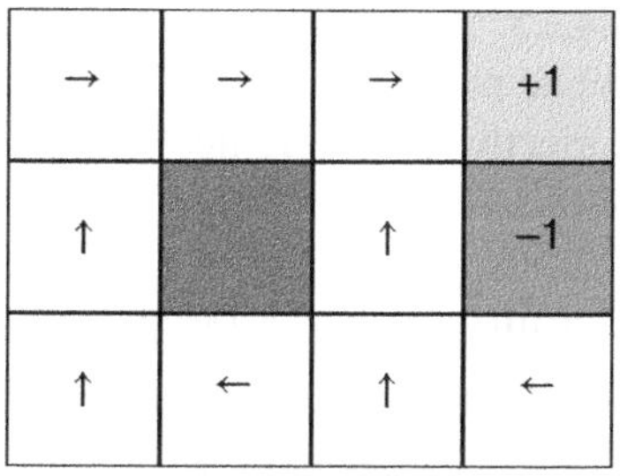

Figure 4.11 Example policy that solves the MDP.
Source: Florian-Lennert A. Lau.

4.4.3.2 Partially Observable MDPs

In nanoscale environments, it is usually not possible to accurately determine the own state. We call such environments *partially observable* and represent this via modifications to the original MDP definition.

Definition 4.9 A *POMDP* is a tuple (S, A, T, R, O, Ω).

- S, A, T, R are the components of an MDP,
- O is a random variable that represents the range of possible observations, and
- $\Omega(O, S) = P(O \mid S)$ is a *sensor model.*

A *belief MDP* is a tuple (B, A, T, R, O, Ω). It is a POMDP where the partially observable S is represented by a random variable B with an infinite set of belief states as range. A *belief state* $b \in \text{ran}(B)$ describes a probability distribution over S. We further replace S, S' in T, R and Ω by belief state variables B, B'.

Just like the reward function, the sensor model may consider actions too.

Solving a POMDP requires us to solve the corresponding belief MDP. However, an optimal policy now maximizes the probability of success as the state may no longer be determined with certainty. A policy may be represented by `if...then ...else` statements, where the if statement contains observations that lead to conditional actions. A conditional policy p is recursively constructed and has depth d,

$$U_p(s) = R(s) + \gamma \sum_{s' \in \text{ran}(S)} P(s' | s, a) \sum_{o \in \text{ran}(O)} P(o|s') U_{p.o}(s'). \tag{4.2}$$

a is the initial action in policy p and $p.o$ refers to the subpolicy of depth $d-1$ for observation o that follows after a. Equation 4.2 represents a value iteration algorithm.

A simple example of a POMDP can be seen in Figure 4.12. The previously available global information about the state has been replaced by local state information that is subject to a potential error.

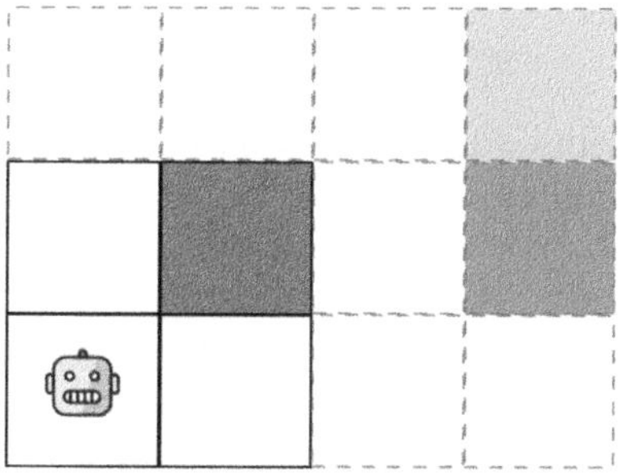

Figure 4.12 Example scenario for a POMDP. A nanodevice has only local information about the environment.
Source: Florian-Lennert A. Lau.

4.4.3.3 DecPOMDP

While partially observable environments are closer to the nature of nanonetworks, they do not correctly reflect the need for collaboration among nanodevices. A DecPOMDP models a set of agents working jointly toward a common goal, a scenario relevant to nanoscale medical systems.

Definition 4.10 A *DecPOMDP* is a tuple $(\boldsymbol{I}, S, \{A_i\}_{i \in \boldsymbol{I}}, T, R, \{O_i\}_{i \in \boldsymbol{I}}, \Omega)$, with

- $\boldsymbol{I}$ being a set of N agents,
- S being a random variable with a set of states as range,
- A_i being a decision random variable with a set of local actions as a range for each agent $i \in \boldsymbol{I}$, with $\boldsymbol{A} = \times_{i \in \boldsymbol{I}} \text{ran}(A_i)$ being the set of *joint actions*,
- $T(S', S, \boldsymbol{A}) = P(S' \mid S, \mathbf{A})$ being a transition model,
- $R(S)$ being a reward function,
- O_i being a random variable with a set of local observations as a range for each agent $i \in \boldsymbol{I}$, with $\boldsymbol{O} = \times_{i \in \boldsymbol{I}} \text{ran}(O_i)$ being the set of *joint observations*, and
- $\Omega(\boldsymbol{O}, S) = P(\boldsymbol{O} \mid S)$ being a sensor model.

Each agent $i \in \boldsymbol{I}$ has a local policy π_i. The solution to a DecPOMDP is a *joint policy* $\boldsymbol{\pi} = (\pi_i)_{i \in \boldsymbol{I}}$.

In a DecPOMDP, each agent has its own set of actions and possible observations,[1] whereas the state and reward function are joint. The joint state is usually assumed to not be fully observable, even if combining all local observations. This is also the case for nanoscale systems. If the joint state and reward function can be split up into mostly independent subspaces per agent, the DecPOMDP decomposes into a set of POMDPs that can be solved individually. The joint reward function encodes that the set of agents receives a reward from the environment as a team in contrast to individual rewards for each agent. The generalization of a DecPOMDP is a partially observable stochastic game in which each agent has its own reward function $R_i(S)$, which possibly conflicts with other agents' rewards.

A straightforward way to find a solution to a DecPOMDP is to build conditional plans for each agent using value iteration and eliminate dominated plans considering all agents. This approach quickly runs into memory problems because of the explosion of agent numbers, possible states, actions, and observations. To formalize the space requirements in a DecPOMDP, we set up a simple lemma, which makes the combinatorial explosion by the agent numbers explicit.

Lemma 4.1 The worst-case memory required for specifying a DecPOMDP of Definition 4.10 is exponential in N.

Proof: The transition model $T(S', S, \boldsymbol{A})$ and sensor model $\Omega(\boldsymbol{O}, S)$ of a DecPOMDP of Definition 4.10 have sizes S_T^{dec} and S_Ω^{dec}, respectively, that lie in

$$S_T^{\text{dec}} \in O(s \cdot s \cdot a^N) \qquad \text{and} \qquad S_\Omega^{\text{dec}} \in O(s \cdot o^N), \tag{4.3}$$

with $s = |\text{ran}(S)|$, $a = \max_{i \in \{1,\ldots,N\}} |\text{ran}(A_i)|$, and $o = \max_{i \in \{1,\ldots,N\}} |\text{ran}(O_i)|$, which is exponential in the number of agents N.

For cases, in which the agent set carries a structure in the form of partitions of indistinguishable agents, we present lifted DecPOMDPs as a step toward making the problem tractable.

1 $A_i = A_j$, $i, j \in \boldsymbol{I}$, is possible but not mandatory. The same holds for the sets of local observations.

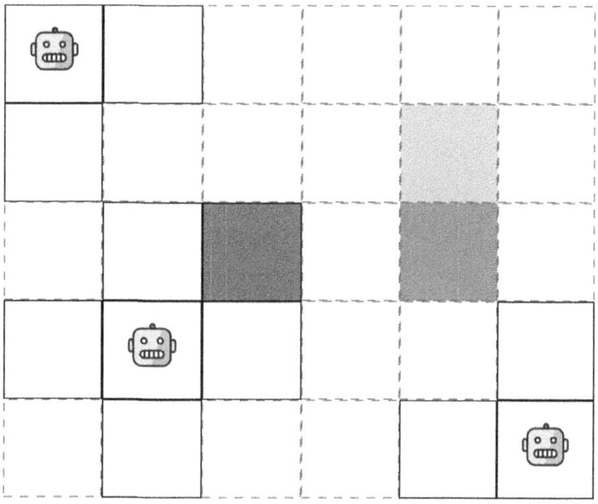

Figure 4.13 Example for a DecPOMDP. Unlike before, we now have a number of nanodevices operating in the same environments. *Source*: Florian-Lennert A. Lau.

Again, an example can be seen in Figure 4.13. Unlike before, several nanodevices operate in the same environment and have to coordinate their actions. They potentially hinder or assist each other. Each of them has only local information about the system, and all the nanodevices have a local as well as a global/joint policy they try to achieve.

4.4.3.4 DecPOMDP with Communication

Since some nanodevices have a means of communication, the last extension to define here is regarding the sequential decision-making problem is that of a DecPOMDPcom, which extends a DecPOMDP with explicit means of communication. The messages in a DecPOMDPcom can be considered in three different dimensions,

1. as another action to perform by a nanodevice,
2. as another observation to perceive by a nanodevice (emitted by fellow nanodevices instead of the environment), and
3. as a basis for communication cost.

Because of this special standing of communication, we deal with it as an own component to be added to a DecPOMDP, even though a DecPOMDPcom (with explicit communication) can always be transformed into a DecPOMDP (with implicit communication).

Definition 4.11 A (*DecPOMDPcom*) model $\mathcal{M}$ is a DecPOMDP tuple extended with

- $\boldsymbol{C} = \{C\}_{i=1}^{N}$ being a set of random variables with identical atomic messages for communication including a null message $\perp_c$, with $\text{ran}(\boldsymbol{C}) = \times_{i=1}^{N}\text{ran}(C)$ being the set of possible joint messages, and
- Γ being a cost function, which specifies the cost of transmitting an atomic message.

Communication is usually considered to be instantaneous and noise free; dealing with other settings is a part of future work. A message $c_i \in C$ refers to an atomic message sent by a nanodevice $i \in \boldsymbol{I}$ as a broadcast to all other nanodevices. A null message indicates that a nanodevice does not want to transmit any information. As defined above, each nanodevice has the same set of messages available to send out, which is in contrast to how actions are handled where each nanodevice may

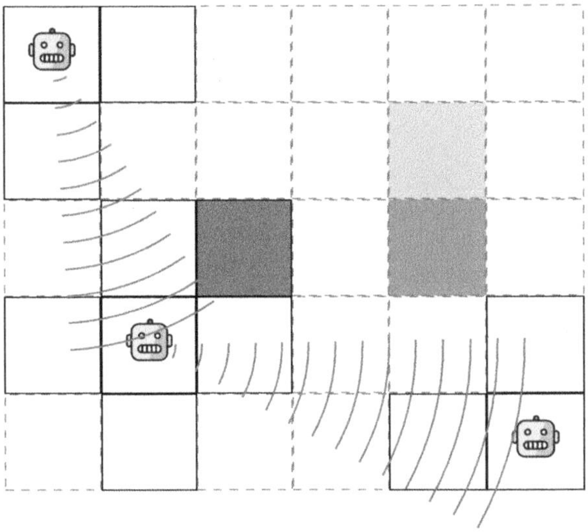

Figure 4.14 Example DecPOMDPcom. Unlike before, the different devices can manipulate and measure special environmental properties that represent communication. *Source*: Florian-Lennert A. Lau.

have its own set of actions available. Of course, one could also apply this idea to communication, with each nanodevice having a random variable C_i with a range of locally specific messages.

Given messages, there is the option to include them in the functions of the DecPOMDP. A reasonable candidate is the function $R(S, \boldsymbol{A}, \boldsymbol{C})$ that rewards or punishes certain communication apart from any cost the messages may. The transition function describes how the environment may change given its own inherent behavior and the agents' actions. The nanodevice's action may be influenced by a joint message, but given the action, the message and environment are independent of each other. The sensor function specifies the probability of perceiving an observation o out of the environment. This percept is also independent of any message. Therefore, incorporating a joint message here is less reasonable, although possible.

One needs to also specify the purpose of the messages, i.e. their semantics. One possibility is for each nanodevice to share its local observations with all other nanodevices to enable a joint state estimation. Another possibility is to communicate intent regarding the actions to be performed. Depending on the semantics, solving a DecPOMDPcom may need to incorporate an update function where, depending on the information shared, actions might be chosen.

An example can be seen in Figure 4.14. The nanodevices in question can now sometimes communicate via terahertz waves and thus exchange messages for the purpose of coordination.

4.4.3.5 Mapping DecPOMDPcom to Components

This part transfers the device and network definitions to the DecPOMDPcom setting. Both the formalism and the nanonetwork have an environment in which they are represented by a random variable $S = \mathrm{S}$. We start with a component that does not have a direct counterpart in the model, namely, memory M. Then, we consider components with a direct connection like sensors O and actuators A. This includes locomotion L and communication C. We end with a discussion of power supply E, timer T, and information processing I, which get a reflection by the functions and solutions of a DecPOMDPcom model.

The first component to consider is memory M. Memory is an important resource in any solving of an inference problem such as decision making. For any MDP-based problem, the model and problem themselves are considered to be specifiable given the available memory, as is each policy (locally for the agents and globally during computation). A theoretical analysis regarding space requirements then may provide an upper bound on memory requirements for given solution approaches. There exists a solution approach to DecPOMDPs called memory-bounded dynamic programming that constrains its representation to available memory (Seuken and Zilberstein, 2007).

The next components to consider are the sensors O and actuators A. The sensors find a direct counterpart in the random variables O_i encoding the observations available to each agent i. These observations come from sensors in O in this nanoscale setting. During solving a DecPOMDPcom, all possible signal sequences are considered that O can emit up to a certain length τ. When acting out a policy, the signals that sensors from O generate (after possibly some preprocessing) become the observations for the agent upon which it decides on its next action. The sensor model Ω allows for modeling noise or accuracy regarding signal as a representation of a certain environment feature. Analogously, the actuators map to the decision random variables A_i and encode the actions available to each agent i. These actions need to be carried out by actuators in A. When acting out a policy and an agent has picked its next action given the observations from the sensors, the agent activates the actuator to which an action belongs. Locomotion is a special kind of action that, within the formalism of DecPOMDPcom, is treated as another possible action.

The last remaining component with a direct connection is communication C, for which we have introduced the communication variant of DecPOMDPs, namely, DecPOMDPcom. Without communication, a DecPOMDP covers all necessary components. The communication part is rather loosely defined in DecPOMDPcom and can therefore encode a multitude of communication to be broadcast between agents via acoustic, electromagnetic, or molecular communication. The communication random variable C that is identical in this setup for all agents would then have a range of all possible messages that nanodevices could exchange. If no costs are associated with messages, the cost function Γ can be set to zero.

Next, we turn to the power supply E as a component without a direct counterpart at first glance. However, its effects can be reflected in the functions and policies of a DecPOMDPcom. Having a power supply implies a need for energy. The supply may get drained during acting and may therefore need to be stocked up. The recharging can be modeled as part of available actions and may actually yield a reward to encourage recharging at appropriate intervals.

A DecPOMDPcom is a formalization of a sequential decision problem, therefore, including a temporal dimension that is represented through discrete time steps t. In the transition function T, time finds its representation through the change from state s (at time point t) to state s' (at time point $t+1$), which is encoded precisely in T. When acting out a policy, time elapses by having actions performed and observations made in each time step. When picking the next action from a policy based on an observation history, the history refers back to the last t time steps of observations. Hence, a timer finds its way into the formalism of DecPOMDPcom.

The last remaining component is that of information processing I. It basically describes a program or acting routine for a device and therefore corresponds to a local policy π_i that is the solution to a DecPOMDPcom. Each agent in $\boldsymbol{I}$ gets one part in the joint policy, which can be represented as a tree and basically qualifies as a look-up table, requiring an agent to have no further processing power during acting.

Solving DecPOMDPcoms (finding policies for nanodevices) is a computational task and will be explained in detail in Chapter 5. Next to the possible solutions, the complexity of the underlying problem of programming nanonetworks and nanodevices is analyzed in detail.

4.5 Nature Adaptation

Now that we have a good understanding of *what* nanostructures exist, how they behave, and how they can be formally defined, we analyze the different concepts for possible instantiations. Mathematical models are usually so general that they allow for many possible physical manifestations. We start with the most intuitive approach of adapting already existing structures to the needs of our desired applications. Overall, there are several naturally occurring structures like cells, bacteriophages, liposomes, and viruses that we discuss successively.

The first and likely most general approach is the modification of cells. Every complex organism consists of extremely large amounts of cells that are supplied with nutrients via neighboring capillaries. No cell is further than three neighbors away from the next vessel. As an example, the human body is estimated to consist of around 30 trillion cells on average.

An example of a cell-based nanodevice can be seen in Figure 4.15. The figure shows many of the important organelles of a human cell. Some of them are annotated with components we have already discussed before. As an example, the core of a cell contains the DNA and acts as a "control station" from which the inner life of a cell is orchestrated. It is not too farfetched to assume that the internal processes might be modified to represent the information processing component. By supplying fitting RNA sequences from the outside, it might even be possible to run primitive programs "on" the cell core.

Similarly, DNA and RNA might be used for storing information in the cytoplasm of a cell, thus realizing the memory M. While the energy supply is generally regarded as a considerable problem, cells are usually supplied with energy E via mitochondria. They already work at the nanoscale and might supply other, artificial parts of a modified cell with *adenosine triphosphate* (ATP) – the energy currency of cells.

Additionally, all cells have a membrane that is permeable for some substances using gap junctions that might be repurposed to act as actuators A, sensors O, or a communication mechanism C. In addition, most cells are sensitive to certain substances like insulin, which trigger an opening mechanism. Such sensitivities might be repurposed to act as a sensing mechanism. Finally, some cells have a flagellum, which might be used as a means of locomotion.

While it remains unclear how and if such modifications are indeed possible, many cells at least supply some of the necessary components that constitute nanodevices. Some scientists

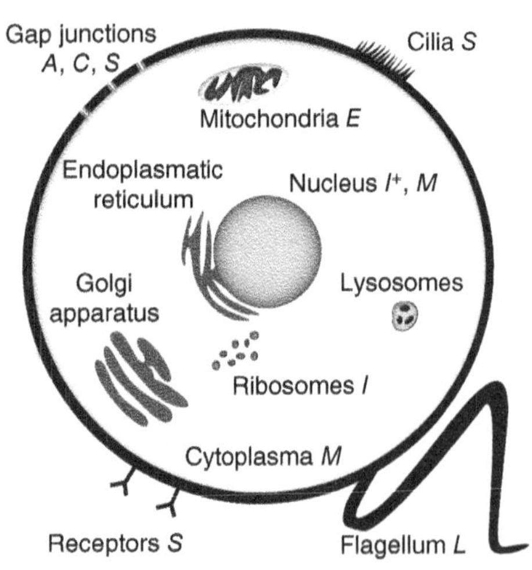

Figure 4.15 Schematic representation of a biological nanorobot. The organelles of a cell are used to implement the functionalities of a nanorobot. *Source*: Büther et al. (2017)/with permission of Springer Nature.

rightly argue that a bacteria-based nanonetwork might be among the first primitive nanonetworks that can be created and tested in wet-lab experiments (Cobo and Akyildiz, 2010). In fact, the modifications to existing bacteria might be so small that showing the feasibility of a multi-hop information exchange might even be a bit underwhelming.

Similar principles may also be applied to bacteria instead of cells. They might be even better candidates, as displayed in Figure 4.16. Unlike cells, prokaryotes are single-cell organisms that already act autonomously in their environment and they were successful enough that at least 30 000 named bacteria exist today (Dykhuizen, 2005).

Unlike complex cells, prokaryotes have a simpler structure. (Eukaryotic) Human cells have a complex cell wall and a cell nucleus that is isolated from the rest of the cell. This structure makes the internal environment of a cell more stable and controlled and allows for more complexity. Prokaryotic cells do not have this advantage and are generally more primitive. The DNA just swims in the cytoplasm.

Yet, prokaryotic cells have the advantage that they function autonomously. They almost always have a flagellum and can move along concentration gradients of specific substances via a process called chemotaxis. In addition to that, we might not need the complexity of eukaryotic cells to fulfill basically all the needs of the majority of suggested medical applications. The DNA might still serve as information storage, and certain bacteria can communicate via plasmids. These are rings of extrachromosomal DNA that bacteria naturally use to communicate with each other. Plasmids are the reason why prokaryotic nanonetworks might be among the first primitive nanonetworks as we can externally influence the behavior of bacteria through them.

Just like cells, bacteria of all kinds have the advantage that they naturally occur in living organisms. This likely allows for much easier waste disposal as well as biocompatibility.

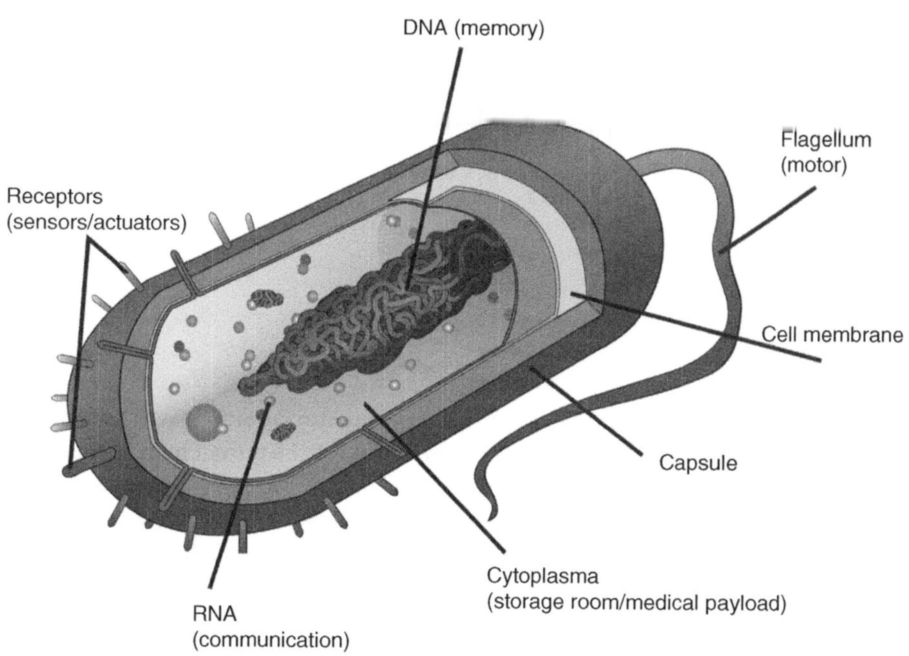

Figure 4.16 Schematic representation of a prokaryotic nanorobot. The individual organelles are converted into functional units of nanorobots. *Source*: Stelzner et al. (2016b)/European Alliance for Innovation (EAI).

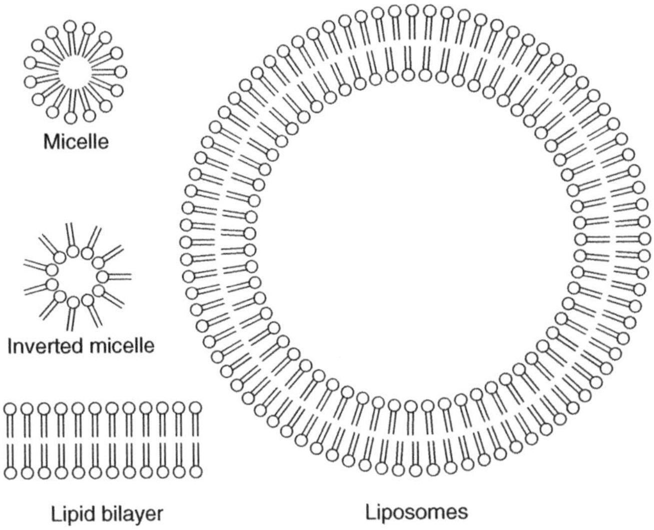

Figure 4.17 Internal structure of liposomes and micelles. Both are based on lipids that are hydrophobe and try to maximize their distances to watery environments. *Source*: Florian-Lennert A. Lau.

Yet, another set of promising, naturally occurring structures are liposomes and micelles. An example can be seen in Figure 4.17. Liposomes are already used in a variety of supplements and medical applications as carrier molecules. Some substances can be stored inside and thus reach parts of the human body that are typically unreachable, as filter organs like the liver usually destroy any unknown substance.

As some cells allow liposomes to travel through their membrane, they might be a suitable candidate to implement a targeted drug delivery system.

All of the mentioned biological components have in common that only small modifications are necessary. While this might be suitable for primitive medical applications, they do not seem to allow for the same level of freedom and controllability that we are used to form macroscopic computers. It remains an open question if and to what degree such building blocks might be utilized in more complex nanonetworks. Yet, due to many *immediate benefits*, research in those areas attracts a lot of funding.

4.6 Miniaturization

An intuitive alternative to nature adaptation is the further miniaturization of CMOS technologies manufactured through photolithography. CMOS technologies are the prevalent technology for computer chips and basically the only known method of mass manufacturing. Oftentimes, CMOS circuits consist of a number of NAND gates. An example of a circuit can be seen in Figure 4.18. In this case, it consists of OR gates, AND gates, and INV gates. In combination, they form a Turing-complete logic. A possible nanodevice based on CMOS technologies has already been presented in Figure 1.4.

For many researchers, further miniaturization based on the same technology is the most popular. This has the simple reason that a lot of the already existing knowledge might be applicable at the nanoscale with only slight modifications. This avoids a lot of problems in learning in an entirely new field of research.

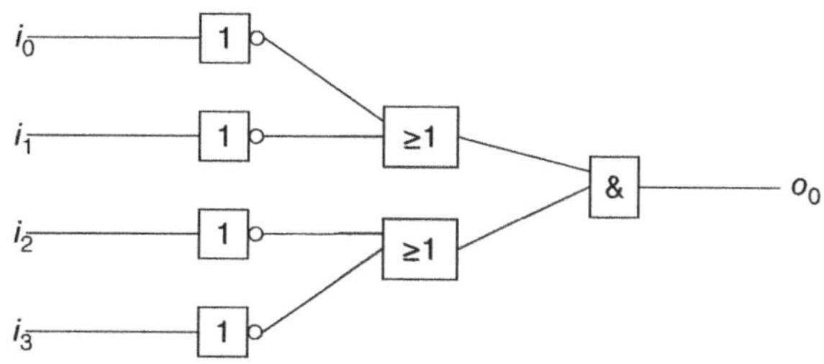

Figure 4.18 Example of a CMOS circuit with four inputs and one output. *Source*: Florian-Lennert A. Lau.

While this sounds like an enticing idea, there are potential problems. Since Moore's law was discovered, the computational capability grew exponentially, while the size of the involved components shrank similarly. If that trend continues, the emergence of nanodevices will only be a matter of time. Yet, in many areas, Moore's law has slowed down dramatically already or has come to a complete halt (Haron and Hamdioui, 2008).

This has several reasons. From a certain scale onward new quantum effects make the further downward-scaling difficult. Additionally, there is a hard limit on how small certain components can be. As an example, the biggest atom is about 0.3 nm in size and it is impossible to manufacture any structure smaller than that. Atoms are the "pixels" of reality. As of right now, transistors of size 7–14 nm can already be produced (Windeck, 2019). As a result, we cannot ever expect more than a further reduction in size than roughly a factor of 20.

CMOS-based technologies are one of the cornerstones of nanonetworks, and many researchers work based on the assumption that there will be sufficiently small chips one day. While this is not entirely clear, it is a sufficiently good assumption that legitimizes further research.

4.7 Self-assembly

The construction paradigm that is likely most foreign to most readers is self-assembly. Thus, this section is a bit longer as the principle is a potential candidate that avoids most of the problems of the other approaches.

To fully understand self-assembly, it is beneficial to grasp the concept of crystal formation. The process takes place on practically all scales of reality. Atoms assemble into molecules and molecules into even more complex structures or even living things.

In living cells, for example, the DNA consists of a double helix, which in turn consists of the bases adenine, cytosine, guanine, and thymine. Following certain (local) rules, these bases form bonds with each other. In fact, all living things are more or less self-assembled and, by a fairly general definition, could be called "crystals."

A clear example of self-assembly is snowflakes, as displayed in Figure 4.19. At low temperatures, snowflakes essentially form from two basic building blocks. They are water molecules and dust particles, which starts the crystal formation process. Over time, adding more water molecules forms a crystalline structure.

With nanodevices, the process can be essentially the same – only the building blocks differ. One possible building block is DNA. DNA building blocks are available almost everywhere in the human body, and DNA tends to be structured in a certain way. This behavior can be exploited to create specific structures. It is not far off to assume that the process can be exploited to construct nanodevices or even nanonetworks (Zhirnov and Herr, 2001).

Figure 4.19 Several examples of self-assembled snowflakes. *Source*: Wilson Bentley/Wikimedia Commons/Public Domain.

Based on the idea of using DNA for construction purposes, theoretical models were designed that can be used to test on the computer what DNA building blocks must be like in order for the desired structure to grow from DNA.

4.7.1 DNA Origami

DNA origami as one of the several techniques for creating nanoscale structures was first introduced by Paul Rothemund in 2006 Rothemund (2006). The technique is based on a long DNA strand that is connected to itself at various points by DNA staples, as illustrated in Figure 4.20 (a). *DNA brackets* are short strands of DNA that bind to two designated locations on the long DNA strand, thereby folding it. The base sequences of the clamps and the long strand can be used to precisely determine the shape of the resulting object. Figure 4.20 (b) shows how the single strand is folded in rows. The possible results are shown in Figure 4.20 (d). As proof of concept, Rothemund produced nanoscale structures in the shape of smileys, stars, and rectangles. Figure 4.20 (c) and (d) show this in detail.

DNA origami has been extended to the third dimension in a number of ways. One of the simplest techniques is to stack multiple layers of DNA structures and staple them together into a three-dimensional structure (Ke et al., 2009a). The robustness and simplicity of the DNA origami process make it one of the most popular tools for creating structures at the nanoscale. Even the controllable DNA box from Figure 2.12 by Andersen et al. could be made using DNA origami (Andersen et al., 2009). A large number of structures and complex devices can be created on the basis of this process.

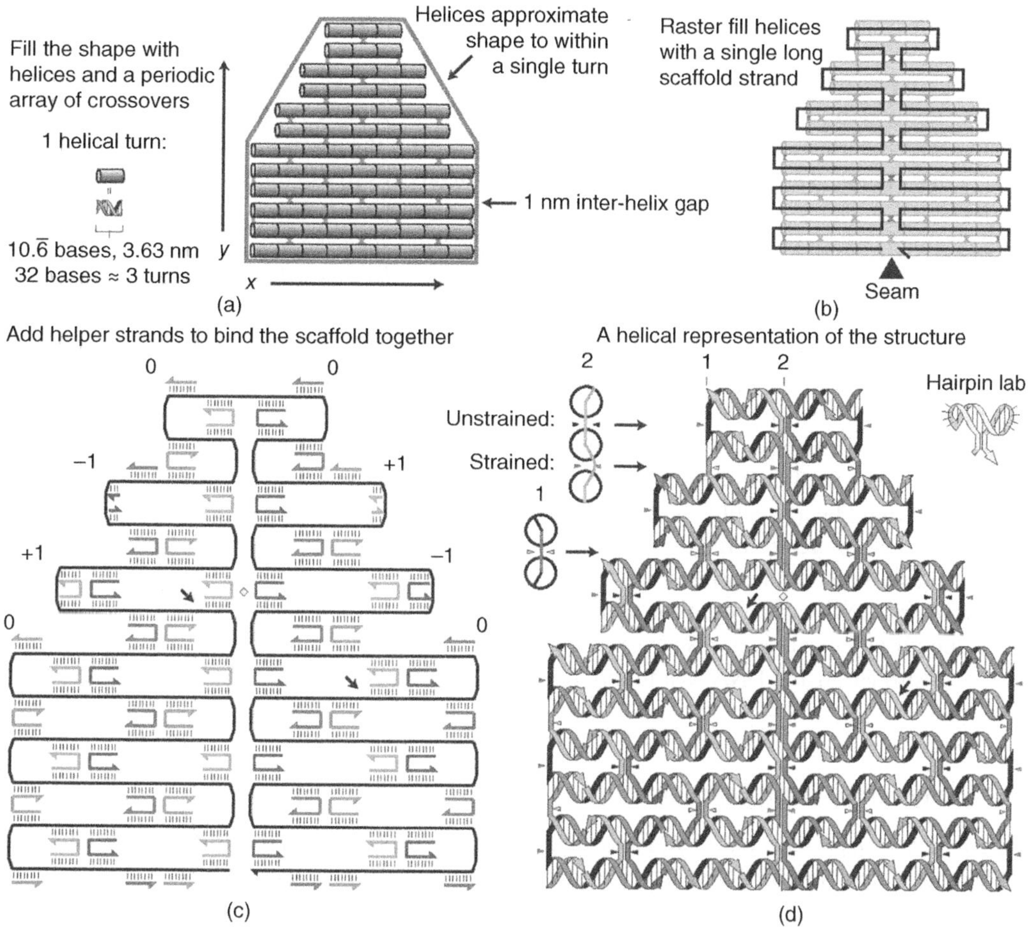

Figure 4.20 Different phases of DNA origami and how the single strand changes over time (Rothemund, 2005). (a) Fill the shape with helices and a periodic array of crossover. (b) Raster fill helices with a single long scaffold strand. (c) Add helper strands to bind the scaffold together. (d) A helical representation of the structure. *Source*: Rothemund (2005)/with permission of Springer Nature.

4.7.2 DNA Templating

The self-assembly properties of DNA are of interest in a variety of different areas. Among those we have already analyzed, DNA can also be used as a tool for engineering novel (nano-)materials (O'Reilly et al., 2017).

In chemical science and chip design, precise and accurate control of nanoscale structures is a fundamental necessity that is hard to achieve. While it is difficult to achieve such levels of precision with man-made top-down tools, evolution led to the emergence of DNA that allows for the necessary precision via self-assembly. DNA is the basis for a vast number of more complex molecules that are known and used today.

Chemical engineers usually search enormous molecular libraries with more than 10^8 entries until they find potential candidates for designing new materials or pharmaceuticals. The entire process may be sped up by exploiting the special properties of DNA. Due to its programmable nature, it might even be possible to create entire "molecular factories" that create the desired output molecules.

The field of research in which all this and more is investigated is called *DNA-templated synthesis* (DTS) (O'Reilly et al., 2017). It has developed to a point where complex and complicated structures like DNA-based nanoscale robots have been created to control the assembly of molecules. It is not unlikely that chemical engineers will create entire artificial cells to use them as factories for novel products.

In addition to using DNA as a catalyst to manufacture molecules, it can also be used for a variety of different purposes. As mentioned in Chapter 1, one of the biggest challenges in nanoscale manufacturing is the precise placement of components in two or more dimensions. This problem exists in both chemical engineering and chip design. Especially, in chip design, it is necessary to place components with nanoscale precision to maintain the logical integrity of memory or logical chip.

As of right now, the only way to mass-manufacture chips is by using photolithography in a top-down approach. Photolithography is limited to manipulating the surface of a material in limited ways as we learned in Chapter 2. DNA templating of logical gates, wires, and other electrical components could be a feasible bottom-up alternative (Pate et al., 2014; Ma et al., 2010). It might even be possible to create three-dimensional chips using this technique, which is currently not feasible on a mass-manufacturing scale.

In summary, DNA can be used for all kinds of purposes where precision is key. This is just as much true for chemical reactions as it is for computer chips. It should even be possible to create novel metasurfaces using DNA as a foundation, and the number of yet undiscovered purposes is likely high.

4.7.3 Polymerase Chain Reaction

The *polymerase chain reaction* (PCR) is one of the most important tools in the artificial reproduction of DNA molecules. For a long time, it was known that DNA was reproduced in living organisms, but only with the discovery of the PCR by Kary Mullis was able to take the process efficiently outside of an organism (Mullis, 1994). The PCR takes place in three steps, which can be repeated indefinitely:

1. denaturation,
2. Primer annealing and
3. elongation.

All components required to reproduce DNA are in a common medium. This includes free nucleotides (the four bases), primers and polymerases. *Primers* is short DNA strands that bind to parts of the DNA strand to be replicated. They enable polymerase to take up their work. *Polymerase* is an enzyme that catalyzes the development of a complementary strand into a single strand.

In the denaturation phase, the DNA segment to be reproduced is heated to 94–96 °C, which are typical values. At high temperatures, the bindings of the base pairs become unstable and the strands separate. The phosphate desoxyribose backbone, on the other hand, binds stronger and is still stable even at 96 °C.

During primer hybridization, the temperature of the medium, in which all components are located, is reduced to the extent that the primers can be stuck to the DNA to be reproduced. Due to their structure from bases, the primers can only adhere to certain places of the DNA to be replicated.

As soon as the primers are firmly bound, the elongation phase begins. This is where the polymerase starts on the primers and begins with the construction of a complementary strand so that the polymerase is also not destroyed during denaturation, and polymers from the bacterium thermal aquatics are used. These bacteria naturally occur in geysers and are particularly heat resistant (Perl et al., 2000). According to the length of the strand to be replicated, roughly 30 seconds are required for 500 base pairs.

These three phases are repeated until the desired amount of DNA strands has been created. After each cycle, the number of DNA segments doubles. Due to exponential growth, an enormous amount of DNA can be synthesized in a short time. PCR is used for the production of almost all artificial DNA modules.

4.7.4 Tile-based Self-assembly

Self-assembly systems can be used for both computing and constructing nanostructures. The idea behind DNA-based manufacturing processes is that you "grow" nanodevices like crystals. In that case, one would only have to provide the building blocks and wait. Since DNA is a molecule found in the human body, the components can be expected to be easily biodegradable than, for example, structures based on CNT or heavy metals. Le Bailly (2016) predicted that DNA-based technologies will play a crucial role in the near future.

Theoretically, numerous different building materials can be considered for self-assembly as long as their binding behavior can be programmed. As a result, DNA is not the only possible material – the theoretical models presented below work for any material. Furthermore, tiles are not limited to squares or rectangles, even though these simple structures make many initial experiments easier.

From Section 4.7.5 onward, we define several mathematical self-assembly models that we use extensively in the following chapters. We first define the Wang tile model, which represents the historical forerunner for many later models. This model is also known as the "domino problem," which is often used in theoretical computer science when trying to prove that a problem does not halt.

This is followed by the definition of the abstract tile-assembly model (aTAM)/kinetic tile-assembly model (kTAM). They were first introduced by Eric Winfree in his dissertation Winfree et al. (1998). The kinetic version is a commonly used model for simulating realistic self-assembly processes. It is based on the aTAM, also conceived by Winfree, which represents a clear simplification and is mainly used to investigate theoretical questions. Then, the two-handed tile-assembly model (2HAM) is presented, which can be used to investigate parallel self-assembly processes. Finally, a combination of the kTAM and the 2HAM is investigated, which can be used for the realistic analysis of a variety of interacting assemblies.

4.7.5 From Wang to Holliday

Wang tiles are the underlying model from which self-assembly systems later emerged (Wang, 1990). Figure 4.21 shows an example. On the left, you can see three Wang tiles, which can be used to completely fill a layer without leaving any gaps.

Wang tiles occupy adjacent positions in the discreet space $\mathbb{Z}^2$ if their colors match. The colors of the Wang tiles are analogous to the TAS glues presented in the following section (see Section 4.7.5.1) – only the strength of the glue is not represented. Based on Wang tiles, the general tiles can be defined.

DNA tiles describe a class of molecules with special properties. They consist of DNA molecules whose double strands have been woven together to create a stable "crystal." In contrast to

Figure 4.21 Wang tiling of a two-dimensional plane. *Source*: Florian-Lennert A. Lau.

conventional DNA, these crystals have several open ends in different directions. The open ends allow binding interaction with other tiles that have corresponding open ends with complementary strands.

The binding behavior of DNA tiles can be compared to pieces of a puzzle. Two puzzle pieces only fit if they have corresponding recesses that fit exactly into one another. The only difference is that DNA tiles assemble themselves when they match.

There are numerous variations of DNA tiles. In the two-dimensional case, these are about 2 nm high. The length and width can be chosen at will (Lin et al., 2006).

The *double-crossover tile* (DX tile) was first introduced by Seeman and Tsu Ju Fu in 1993 Fu and Seeman (1993). Figure 4.22(a) shows a schematic representation. You can see two double strands that have been woven together.

Parts of the double strands merge into the other ones so that a stable union is created, which does not denature even at temperatures of approximately 70 °C.

Because there are several open ends on the left and right side of each tile, a large number of tiles can assemble to form a large crystal, for example, to partially fill the two-dimensional plane (Winfree et al., 1998). For more complex interactions or computations, however, tiles with open ends in more than two directions are required.

The so-called *triple-crossover tile* (TX tile) was discovered by Winfree et al. (1998) and LaBean et al. (2000) (see Figure 4.22(b)). It is an extension of the presented DX tile. The TX tile is created by weaving three double strands together, resulting in a very stable DNA molecule. The open ends of a TX tile are spaced farther apart, allowing for the construction of other shapes.

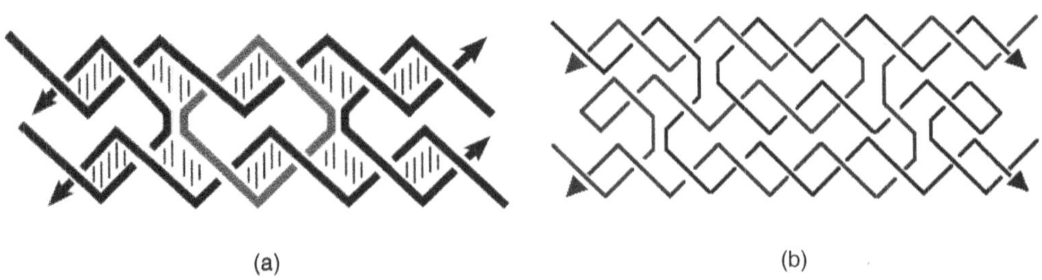

Figure 4.22 DX and TX tiles as a schematic representation. (a) Schematic representation of a DX tile (Commons, 2019a). (b) A schematic representation of a TX tile. *Source*: (a) Commons (2019)/Wikimedia Commons/CC BY 2.5; (b) Wysocki (2019)/with permission of Florian-Lennert A. Lau.

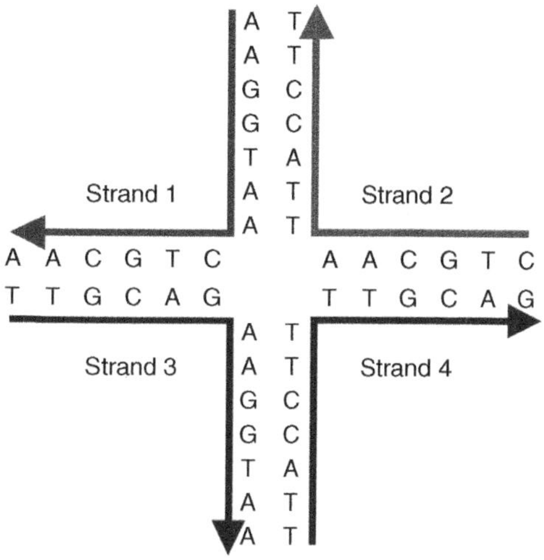

Figure 4.23 Schematic representation of a DNA tile with glues in four directions. *Source*: Adapted from Commons (2019b).

Robin Holliday predicted the existence of the *Four-Arm Junctions* or *Holliday Junctions* named after him as early as 1963 (Holliday, 1974). Although this type of molecule occurs naturally in the human body, it was not until two decades later that Nadrian C. Seeman succeeded in producing it under laboratory conditions (Kallenbach et al., 1983).

Holliday junctions are of particular interest for DNA-based nanodevices and nanonetworks because they have open ends in four different directions. That is enough to be able to implement complex binding behavior. In addition, the length of the open ends and thus the strength of a potential bond can be varied.

Figure 4.23 shows a schematic 3D representation of the molecular structure of a holiday junction. We can see that four different individual strands are involved in the process. Each strand is connected to two other adjacent strands. The open ends of the tiles can contain any sequence of base pairs. In this way, it can be ensured that only certain tiles can enter into bindings with one another.

4.7.5.1 Abstract Tile-assembly Model

This section presents definitions and examples for self-assembly systems based on the work of Eric Winfree, which is an extension of the previously presented Wang tiles (Winfree et al., 1998). Tiles are the atomic components of self-assembly systems. The binding interaction of tiles leads to the formation of assemblies. Assemblies can be used for the modeling of nanosensors and nanorobots as well as for the modeling of nanonetworks, message molecules, receptors, and ligands.

The formalization used in this book is partially based on the work of Lathrop et al. (2007). For a detailed definition of self-assembly systems and an overview of the most important results, refer the work of (Patitz, 2014). Let us start the definitions with the tiles themselves.

Definition 4.12 A *n-dimensional tile* t_n is an object in $\mathbb{Z}^n$ with unit length and 90° angles. A *side* of a tile t_n is defined by a vector $u_t \in U_t \subseteq \mathbb{Z}^n$. U_t is a set of unique direction vectors. The vector $-u_t$ describes the opposite side u_t in the same dimension. The vector u_t has exactly one entry that is not equal to 0 and has the value 1. The sides of a tile are defined by the following relation:

$$\text{side} : t_n \mapsto \underbrace{U_t \times U_t \cdots \times U_t}_{n}.$$

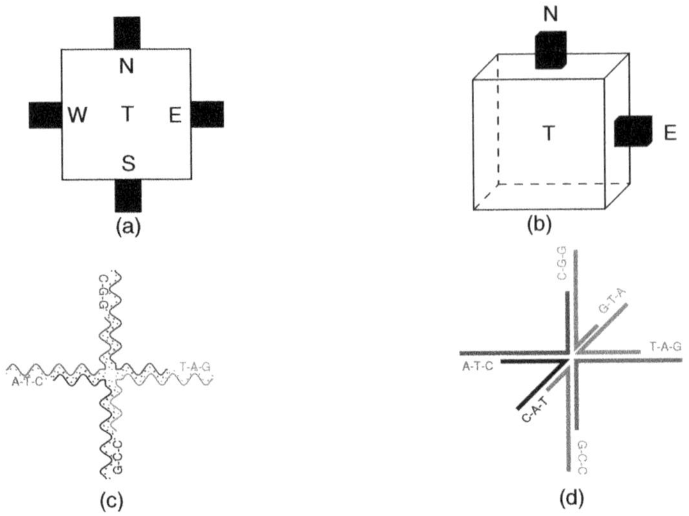

Figure 4.24 Examples of a tile type in 2D and 3D – mathematically and biologically modeled. The mathematically modeled glues are represented by black boxes, and their color is represented by a label. The biologically modeled glues are represented by base sequences. *Source*: Florian-Lennert A. Lau.

A 2D tile t_2 is a square in $\mathbb{Z}^2$ and a 3D tile t_3 is a cube in $\mathbb{Z}^3$. Figure 4.24 (a)–(d) show examples of 2D and 3D tiles. Both graphic models with a focus on the mathematical components and with a focus on the biological components are presented.

In the following, the dimension is largely omitted. Unless stated otherwise, tiles are two dimensional.

Each side of a tile can have any number of glues, which are modeled by a strength.

Definition 4.13 The *temperature* τ of a self-assembly system describes the minimum glue strength s required for a stable bond.

Two tiles *bind* to each other if they are adjacent and have glue of matching color, and the strength with all neighbors is at least τ. The process can be traced in Figure 4.25. A suitable label of a DNA tile can be realized by a suitable complementary strand of sufficient length.

The temperature τ models the physical temperature. When the temperature increases, molecules move faster. The additional energy makes it more likely that bonds will be broken or unable to form stably.

Definition 4.14 Two tiles t and t' of type T and T' *bind correctly* at temperature τ if the following conditions hold:

(1) t and t' are adjacent, and

(2) $\exists\, u_i \in U_t$: $\text{label}(\text{glue}(t, u_i)) = \text{label}(\text{glue}(t', -u_i))$

$\wedge\ \text{strength}(\text{glue}(t, u_i)) \geq \tau$

$\wedge\ \text{strength}(\text{glue}(t', -u_i)) \geq \tau$

Bindings that violate condition (1) or (2) are called *false* or *errors*.

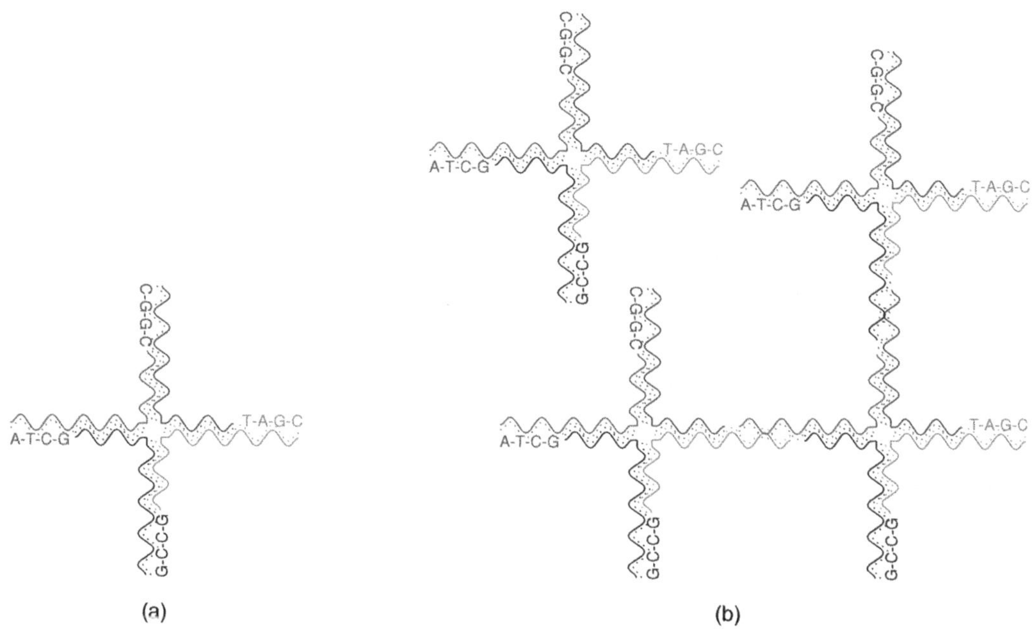

Figure 4.25 (a) Schematic representation of a DNA tile. (b) An assembly of 3 DNA tiles (bottom right) that is in the process of binding with a fourth tile (top left). *Source*: Florian-Lennert A. Lau.

In order for tiles to bind together, they must be rotated to match each other. In biological systems based on DNA, this happens automatically. Mathematical modeling requires tiles not to be rotatable.

Due to the intrinsic properties of DNA, errors are unavoidable (Winfree et al., 1998). However, the number of errors can be reduced if special tile types are used. In the work of Patitz (2014), procedures are described that greatly reduce error probabilities in two dimensions.

The *binding strength* s between the tiles t and t' corresponds to the strength of the correct glue involved. The strength s of a sticker is determined by the number of black boxes on each side. In Figure 4.24, all glues have thickness $s = 1$.

The *absolute strength* of a tile t is the sum of the strengths between t and its correct neighbors. Incorrect glues from neighbors do not contribute to the absolute strength. Binding tiles with insufficient strength or wrong labels are called errors.

Definition 4.15 An n-dimensional *tile assembly*, or *assembly*, is a partial function $\alpha : \mathbb{Z}^n \mapsto T$, where T is a tileset of n-dimensional tile type is T_n. An assembly α is τ-stable if no tile $t_n \in T$ can be removed from the assembly without removing glue of thickness $\tau \in \mathbb{N}$.

Definition 4.16 The *border* of an assembly α is a subset of α. The border includes all tiles with at least one free neighboring site.

Only edge tiles of an assembly can interact with free tiles and thus make the assembly grow.

Definition 4.17 The *growth front* of an n-dimensional assembly is a subset of sites of $\mathbb{Z}^n$. A site is part of the growth front if and only if it is unoccupied and is an adjacent site to a border tile. In addition, a glue of minimum strength 1 to the neighboring site is required.

Growth front sites change as tiles are added or removed from the assembly. We assume that at each discrete time, exactly one tile is added or detached from the assembly.

The initial assembly α_0 at time 0 is called *seed assembly* or *seed tile* σ. Starting at σ, tiles are added non-deterministically to the assembly on the growth front. Because of the inherent non-determinism of tile-based self-assembly systems, it can be difficult to create tilesets that grow into desired shapes.

Definition 4.18 A *tile assembly model* (*TAM*) is a tuple $\mathcal{T}_\tau = (T, \sigma, \tau)$, where T describes a finite set of tile types – also called *tileset*, σ is a seed assembly, and $\tau \in \mathbb{N}^+$ describes the temperature of the tile-assembly model.

Definition 4.19 $\mathcal{A}[\mathcal{T}]$ describes the set of all terminal assemblies that can result from a TAM in finite time. An assembly $\alpha \in \mathcal{A}[\mathcal{T}]$ is *terminal* ($\alpha_\square$) if no τ-stable tiles are added to the assembly can.

Definition 4.20 Let α_i be an assembly created by a TAM. An *assembly sequence* of a TAM is a sequence $S = \langle\alpha_0, \alpha_1, \ldots\rangle$, where α_{i+1} is derived from α_i by adding a correct tile to α_i. If the sequence is finite, the last element is called the *result* $\alpha_\square$ or *terminal*.

An example of an assembly sequence can be seen in Figure 4.26. Starting at σ, at each succinct step in time, another fitting tile is added to the assembly. The assembly sequence thus contains information about all the prior versions of a terminal assembly.

The aTAM is the simplest self-assembly system presented here. It is used to answer theoretical questions about self-assembly and crystal formation. No errors are modeled in the aTAM and parallelism is not intended.

Definition 4.21 Let T be a set of tile types that cannot be rotated. Let σ be a finitely sized seed assembly. An *aTAM* is a tuple $\mathcal{A} = (T, \sigma, \tau)$ with temperature τ. As an additional restriction, only a single tile may be added to the assembly at any one time. Tiles cannot be removed from the assembly.

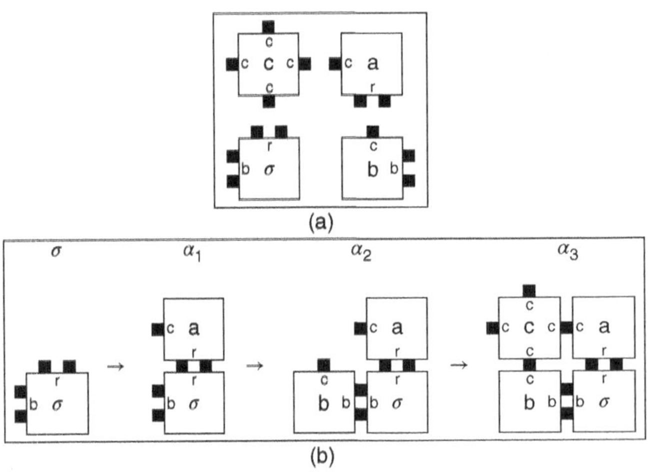

Figure 4.26 (a) Example of a tileset of a 2D-TAS. (b) Assembly sequence for tileset (a). The seed tile is called σ, and the temperature is 2. $S = \langle\alpha_0, \ldots, \alpha_3\rangle$ shows the three necessary steps up to the terminal assembly (Lau et al., 2019). *Source:* Lau et al. (2019)/with permission of Elsevier.

4.7.5.2 Kinetic Tile-assembly Model

The *kTAM* is an important self-assembly system for this book. The kTAM is similar to the basic functionality of the aTAM. Exactly one tile is treated in each discrete simulation step. In contrast to the aTAM, once bound tiles can detach from the assembly again. It is also possible that faulty bindings occur – which is not modeled in aTAM. The incorrect behavior of self-assembly systems is examined in detail in Section 4.8. Due to these two core differences, the simulation result of a kTAM is significantly closer to reality than is the case with aTAM.

The addition and detachment of tiles are determined by a forward rate and a backward rate.

Definition 4.22 Let $\mathcal{T}_\tau$ be a TAM, σ be a non-empty seed assembly, and T be a tileset. The *forward rate* r_f at which tiles t are added to the assembly is defined by $r_f(t) = k_f e^{-G_{\mathrm{mc}}}$. k_f describes the timing of the system. G_{mc} describes the energy required to fix a tile, where $G_{\mathrm{mc}} > 0$.

The forward rate assumes that at any point in time, there is an infinite supply of tiles of each type and the concentration of tiles is the same everywhere at any point in time. G_{mc} depends exclusively on the tile concentration and can be changed at will. This simplification allows for more efficient analysis. In addition to the possibility of binding in the wrong places, it is also possible for tiles to detach from the assembly again.

Definition 4.23 Let $\mathcal{T}_\tau$ be a self-assembly system, σ be a seed tile, and T be a tileset. The *backward rate* $r_{r,b}(t) = k_f e^{-bG_{\mathrm{se}}}$ describes the rate at which bindings in an assembly can break. G_{se} describes the energy costs required for this. b describes the number of bonds that a tile has already formed.

Since the rate falls exponentially with the number of bonds formed, the more bonds a tile has formed, the less likely it is that it will be removed. The G_{se} parameter can be adjusted in wet-lab experiments by manipulating the ambient temperature.

Definition 4.24 Let σ be a seed assembly and T be a tileset. Furthermore, let $r_{r,b}$ be a reverse rate and r_f be a forward rate. A *kTAM* is a tuple $\mathcal{K} = (T, \sigma, r_{r,b}, r_f)$.

The temperature τ of a kTAM correlates with the ratio $\frac{G_{\mathrm{mc}}}{G_{\mathrm{se}}}$ and is henceforth also referred to as temperature in the context of kTAMs. The correct position of a tile is preferred in the kTAM by the factor $e^{G_{\mathrm{se}}}$. As a result, a tile is more likely to bind in a correct location.

4.7.5.3 Two-handed Tile-assembly Model

In contrast to the aTAM and the kTAM, the *2HAM* is defined without a seed tile or seed assembly. At any point in time, all tiles and assemblies can interact with each other as long as the temperature constraint τ and the shape of the assemblies allow it. The absence of a seed assembly reflects reality in DNA experiments much more realistically.

The most important definitions for 2HAM systems are shown below. These are described in detail in the work of Patitz Patitz (2014).

Definition 4.25 Two assemblies α and β are *disjoint* if for all locations of α and β it holds that $\alpha \cap \beta = \emptyset$. An assembly consists of at least one tile.

An assembly in 2HAM is either a tile or an assembly for which the following applies: Two assemblies α and β join *correctly* if the resulting assembly is τ-stable and α and β are disjoint.

Definition 4.26 The state S of a tileset T is a multiset of assemblies for which the following applies:

- All assemblies $\in S$ are τ stable. At least τ glues must be removed for the assembly to break.
- All assemblies $\in S$ can be formed from T by correct union.

The S state of T is described by a multiset since it must be possible to combine assemblies with themselves. A maximum of two copies of each assembly are required.

Definition 4.27 A *2HAM* is a tuple $\mathcal{H} = (T, S_0, \tau)$, where T describes a set of tile types, S_0 is a start state, and τ is a temperature constraint.

Definition 4.28 Let $\mathcal{H} = (T, S, \tau)$ be a 2HAM. The assembly sequence of a 2HAM is a sequence of states $S = \langle S_0 \dots S_k \rangle$ with $k \in \{1 \dots \infty\}$. S_{i+1} is created from S_i by correctly forming all possible unions of assemblies $s, s^{'} \in S$.

Definition 4.29 Let $\mathcal{H} = (T, S_0, \tau)$ be a 2HAM. A state S_i is *terminal* if S_{i+1} is equivalent to S_i after applying all possible unions.

4.7.5.4 Two-handed Kinetic Tile-assembly Model

The following definitions are presented in the work of Kaussow (2022) and were worked out together with Florian Lau. As with the 2HAM, there is no specified seed tile in the *kinetic two-handed tile-assembly model* (kTHAM), as this is more realistic. In each time step, all existing tiles and assemblies can interact with each other. This change is achieved by the union no longer having to be τ-stable. Likewise, the definition of a state must be changed so that it no longer has to be τ-stable, but instead contains the number of the respective tiles.

Definition 4.30 The *correct union* of two assemblies α and β in the Kinetic Two-Handed Assembly model is a union where α and β are disjoint.

Definition 4.31 A *state* $S = \{A_s, C_s\}$ of a tileset T is a tuple of a multiset of assemblies and a vector for their number in the ktHAM. For this tuple:

- All assemblies A_S can be formed from T by correct unions.
- All assemblies A_S are unique, and their frequency of occurrence is defined by the vector C_S.
- The number of items in C_S is equivalent to the number of assemblies in A_S.

The forward rate and the reverse rate correspond to the kTAM. As with the kTAM, the temperature τ of the system is correlated with the ratio $\frac{G_{\mathrm{mc}}}{G_{\mathrm{se}}}$.

Definition 4.32 A *ktHAM* is a tuple $\mathcal{K} = (T, S_0, r_f, r_{r,b})$, where T is the tileset, S_0 is the start state, r_f is the reverse rate, and r_{r_b} is the forward rate.

All existing assemblies and their number in the kTHAM form a sequence. An example can be seen in Table 4.3.

Definition 4.33 An *assembly sequence* in kTHAM is a sequence S of states $S = \langle S_0 \dots S_k \rangle$ with $k \in 1 \dots \infty$. A state $S_i + 1$ arises from the state S_i by separating or correctly joining two assemblies. Assemblies that only consist of one tile cannot be separated.

Table 4.3 The lines reflect the frequencies of the assemblies and the assemblies are in the first line.

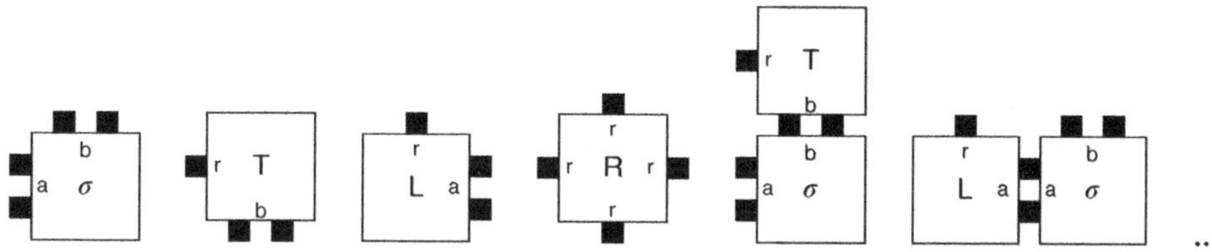

	10	10	10	10	0	0	...
0/1 → 4	9	9	10	10	1	0	...
0/2 → 5	8	9	9	10	1	1	...
0/2 → 5	7	9	8	10	1	2	...
4 → 0/1	8	10	9	10	1	1	...
⋮	⋮	⋮	⋮	⋮	⋮	⋮	⋮

The first column gives the index of the assemblies that have been modified.
Source: Lau et al. (2019)/with permission of Elsevier.

A state in the kTHAM indicates how often an assembly occurs in the medium. Each assembly has an equal chance of being chosen to interact with a random assembly in the next step. This means that the probabilities only change due to the frequency with which each tile/assemblies occur.

The sums of the vectors $C_{S_{i+1}}$ and C_{S_i} differ at most by 1. If two tiles or assemblies join, the total number of assemblies is reduced by 1 since a new assembly is created from the two original assemblies. The opposite applies to the breaking of an assembly.

If an assembly has been created from different original tiles, the tiles into which it breaks are selected at random. This selection is modified by the resulting bindings, so an assembly is more likely to break down at a position where few bindings exist.

4.8 DNA Errors

While mathematical models often assume simple circumstances, it is rarely the case that reality behaves perfectly in line with them. To solve any complex problem, the usual approach is modularization into manageable modules. This reduces the complexity but usually introduces an error. In the end, scientists have to hope that the sum of the partial solutions is also a solution for the original problem.

In the same way, models like the aTAM generalize and abstract and thereby leave out part of the complexity. In this case, errors are not properly modeled and the focus is on fast and efficient approximate assembly generation.

The kTAM is often used when the results of DNA experiments need to be predicted in a way that is close to reality. In the kTAM, it is possible for defective binding to occur, reflecting the natural behavior of DNA.

There are three types of errors that regularly occur in experiments with DNA tiles. The kTAM models two of these error types and is therefore comparatively realistic (Patitz, 2014).

Definition 4.34 A *growth error* is an error where a tile is added to the assembly in such a way that at least one side of the tile has a neighbor with improper glue.

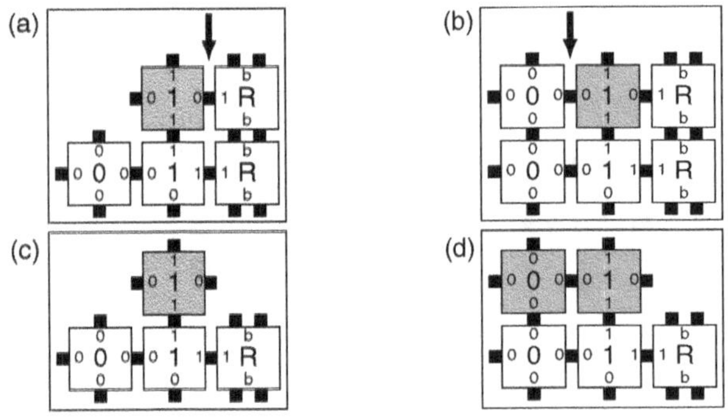

Figure 4.27 (a) Growth error on the gray tile. (b) Growth error imposed by newly added neighbors. (c) Facet error on the gray tile. (d) Facet error imposed by newly added neighbors. *Source*: Florian-Lennert A. Lau.

A growth error is called *fixed* if it is highly unlikely that it can be solved by adding more tiles. This is the case as soon as a correct absolute bond strength in terms of temperature is reached. Figure 4.27 (a) shows the growth error on the gray tile as an example.. The tile highlighted in gray incorrectly binds with the correct glue at the bottom and an incorrect glue on the right. Figure 4.27 (b) shows the same assembly one step later. Another tile was added to the assembly, which is why the temperature requirement of 2 for a stable bond is met for the gray highlighted tile.

Definition 4.35 A *facet error* is an error where a tile binds to the assembly with insufficient strength but has no bad bindings.

A facet error is also referred to as *fixed* if one or more added tiles increase the absolute bond strength of the tile to such an extent that detachment of the tile is unlikely. Figure 4.27 (c) and (d) show an example of this process. The tile highlighted in gray binds to the assembly with just one glue. In the next time step, a now suitable tile binds next to it, which leads to the necessary absolute binding strength being achieved and it is unlikely that the wrong tile detaches again. The behavior described can lead to errors since there can be several tiles with a glue of strength 1 and label 1, of which only one is intended for one position. This should only form a stable bond as soon as a number of neighboring glues with the temperature τ are present. If an incorrect tile is bound to an unintended position due to a facet error, the assembly will no longer grow into the intended structure.

The third type of error is called a nucleation error. This error occurs when an assembly emerges without a seed assembly σ present.

Definition 4.36 A *nucleation error* occurs in self-assembly systems when an assembly is formed without a seed tile being present or involved.

In addition to the three sources of error mentioned in self-assembly systems, there is a fourth source of error in the three-dimensional version. It can happen that parts of the growth front are completely enclosed by tiles. This behavior would prevent enclosed positions from being filled with tiles under realistic conditions.

Since most experiments with DNA tiles were carried out in 2D, this error is little researched, but theoretically to be expected. For this book, this source of error is only of minor interest.

The aforementioned types of errors are mainly special to tile-assembly systems. In addition to that, DNA is subject to several other errors due to its very nature. DNA is subject to *mutation*, and it sometimes happens that one of the originally intended bases is replaced by another. In nature, there are other mechanisms to either avoid or compensate for such errors. One example would be the HSP90 molecule, which is one of the most common proteins in human cells (Wegele et al., 2004).

This molecule assists in the folding process of other proteins, stabilizes them against heat stress, and protects them against protein degradation. HSP90 seems to even "overwrite" small mutation errors by folding desired protein properly nonetheless. As such assisting structures usually do not exist in a controlled environment, regular errors must be handled in other ways.

4.9 Error Correction Mechanisms

While many error correction mechanisms naturally emerged through evolutionary processes, few of them can be directly used in tile-assembly systems. From Section 4.9.1 onwards, various techniques are presented that can be used to reduce the number of potential errors in tile-based self-assembly systems.

The easiest way to prevent errors in the kTAM is to adjust the ratio $\frac{G_{mc}}{G_{se}}$, which can be compared to the temperature τ of an aTAM. If the ambient temperature is set very precisely, errors can become almost infinitely unlikely to occur. For example, tiles could bond stably from a temperature of 37 °C with a strength of two. If this threshold value is set precisely, then practically no errors occur since all incorrect bindings are immediately broken again. However, this leads to a slowdown of the assembly process. For this reason, in wet-lab experiments, the temperature is regulated a little further to achieve an acceptable compromise between the error rate and speed.

Independently of one another, various scientists have proposed other processes that can be used to counter errors in assembly processes. One possibility is to replace tiles with $k \times k$ blocks so that their semantic behavior remains the same (Hopfield, 1974). This is also called *kinetic error reading* or *block replacement*. Block replacement strategies can be used to reduce growth errors and facet errors (Chen and Goel, 2005; Winfree and Bekbolatov, 2003). Two examples are $k \times k$ *Proofreading* and *Snaked Proofreading*.

4.9.1 $k \times k$ Proofreading

$k \times k$ proofreading is the simplest block replacement strategy. Each tile of an original tileset is replaced by a $k \times k$ block of tiles (k block). An example of such a replacement can be seen in Figure 4.28 (a). The top tile is replaced with the block shown below. For the sake of simplicity, the value 2 was chosen for k. The adhesives of the new tiles are selected in such a way that the same logical adhesive structure is provided on the outside. For example, a glue a with an edge

Figure 4.28 (a) The tile type is replaced by a 2 × 2 block of proofreading tile types. (b) The tile type is replaced by a 2 × 2 block of snaked proofreading tile types. *Source*: Florian-Lennert A. Lau.

length of $k = 2$ is replaced by two glues a and a'. Inside, at $k = 2$, four new, unique stickers $x_1 \dots x_4$ are provided. These are unique per k block.

After replacing all tiles with k blocks, several errors must occur in order for the original semantics to be corrupted. In general, the number of growth errors that occur can be reduced by a factor of $1/k$ (Chen and Goel, 2005; Winfree and Bekbolatov, 2003). With an edge length of k, k^2 errors must occur to falsify the original semantics.

4.9.2 Snaked Proofreading

Another improvement of the $k \times k$ proofreading is the snaked proofreading (Chen and Goel, 2004). Snaked proofreading is also a block replacement strategy. In addition to replacing a tile with a k block, the internal glue structure will be adjusted. The process is shown in Figure 4.28 (b). In contrast to $k \times k$ proofreading, only three internal glues are required for $k = 2$. These are arranged in such a way that facet errors are prevented. The bottom left tile of the k block has no internal glue to the right. This prevents facet errors in this direction. The tiles of the k block can only complete one after the other to form a k block.

In this way, growth errors are reduced similarly to $k \times k$ proofreading. In addition, facet errors in the y dimension are prevented.

In two-dimensional self-assembly systems, block replacement strategies can make growth errors arbitrarily improbable. Facet errors are more difficult to reduce because two dimensions only allow a reduction in one dimension.

Since three-dimensional self-assembly systems are of interest for nanonetworks anyway, a block replacement strategy in three-dimensional TASs is shown, which prevents facet errors in all dimensions and makes growth errors arbitrarily improbable.

4.9.3 3D Snaked Proofreading

In the following, we generalize the snaked proofreading method from two dimensions to three dimensions. The procedure was developed by Philipp Bende, Florian Lau, and Stefan Fischer Bende et al. (2019). They presented the proof and generalization of the snaked proofreading method for arbitrary block sizes k.

First, the procedure is described to give the reader an intuition of the procedure. Then the necessary lemmas for the proof are presented. This is followed by a complete induction to prove that the procedure works for any positive natural edge length of the block replacement.

The algorithm creates a $k \times k \times k$ snaked proofreading block for each tile of a given assembly. Three-dimensional $k \times k \times k$ snaked proofreading blocks are also abbreviated as k *block* in the following.

The algorithm creates a snaked proofreading tileset that satisfies Theorem 4.1. The algorithm (see Bende et al., 2019) is called to create a k block for each tile of a target assembly sequence. A *target terminal assembly sequence* (original or target assembly sequence) is a unique kTAM assembly sequence. A tile type is generated for each intended position within a k block. All glues are defined for this tile type. The sides that already have an adjacent glue are treated first, and an identical glue is defined. Then it is checked whether the considered site is external. *External* sites reflect the logic of the replaced tile. If the said glue does not allow a facet error and a unique assembly sequence is granted for the block, then a glue is set according to the origin logic. Glues are then defined along a unique path of the block assembly sequence. After that, stabilizing glues are applied everywhere where no facet errors can be caused.

By replacing each tile of an assembly with a k block, growth errors are reduced analogously to the two-dimensional variant. To reduce facet errors, the glue's structure of each k block is adjusted so that only every other tile has a glue in the same direction in each direction. Glues are thus applied in a three-dimensional checkerboard pattern.

In order to guarantee bounds on errors, it must be ensured that each k block is composed in a unique way with a high probability. For this, an assembly sequence of the original tileset must be assumed. This is necessary because the number of existing neighbors of a tile has an influence on the adhesive required internally in the block. In order to ensure the uniqueness of the block composition, the number of internal adhesives must not be too high. This is ensured by assigning each block a unique start and end position.

Theorem 4.1 The algorithm produces a $k \times k \times k$ snaked proofreading tileset for an original tileset with target assembly sequence $S = \langle \alpha_0, \dots, \alpha_\square \rangle$ using $O(nk^3)$ tile types. Growth errors are reduced by a factor of $1/k^3$, where k describes the side length of the block and n represents the number of entries in the desired assembly sequence. Furthermore, facet errors are prevented.

Proof: The algorithm is proved by a construction scheme for arbitrary but fixed values k. This creates a snaked proofreading k block for each tile in the original assembly sequence S. The k block satisfies the requirements of Lemma 4.3–4.5:

Lemma 4.2 Each k block created by The algorithm has a unique start and end point and a unique k block assembly sequence.

Proof: A *k block assembly sequence* (snaked path) describes the order in which individual tiles are added to the k block (see Figure 4.29 (a) and (b)). To ensure uniqueness, The algorithm adapts the internal glue structure of each k block. For each neighbor with glue already present, the glue strength on the snaked path is reduced by 1. The result is a unique k block assembly sequence.

Lemma 4.3 A k block that satisfies Lemma 4.2 reduces growth errors by a factor of k^3.

Proof: Since each tile of the original assembly sequence is replaced by a k block, k^3 errors must occur for a block to be completely wrong. Since a block replacement strategy is also used, the guarantees are equally good (Winfree and Bekbolatov, 2003). This makes growth errors less likely by a factor of k^3.

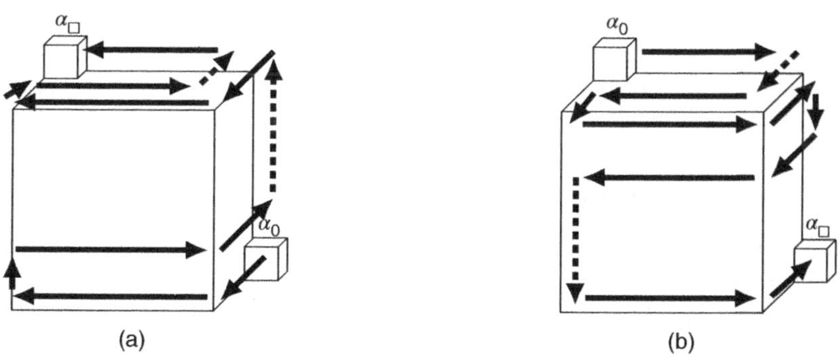

Figure 4.29 Snaked path for two different starting positions. A $k+1$ block can be generated from a k block by adding an additional layer in each dimension. The outer stickers need to be adjusted. (a) Refers to even edge lengths, and (b) refers to odd edge lengths. *Source:* Florian-Lennert A. Lau.

Lemma 4.4 k blocks prevent facet errors and provide $O(k)$ external glue.

Proof: To prove that k blocks effectively prevent facet errors, we refer to the glue structure of every tile that has ever been on the edge. All tiles that are not part of the snaked path do not have a neighbor that has glue in the same direction. This includes tiles in the k block and external tiles.

Since only every second tile has a glue on the outside, facet errors cannot occur. The structure of the snaked path also does not allow for any facet errors.

The number of glues representing the original logic thus amounts to $\frac{1}{2}k^2$ and corresponds to the class $O(k)$. The individual glues are set according to the original strength.

Lemma 4.5 A unique k block assembly sequence exists for each entry of the original assembly sequence S.

To prove this, it is shown that there is a 1:1 correspondence between the original assembly sequence and the assembly sequence generated by the algorithm.

Proof: Let $S = \langle \alpha_0, \dots, \alpha_\square \rangle$ be the original terminal assembly sequence of the tileset T. Let $S' = \langle \alpha'_0, \dots, \alpha'_\square \rangle$ be the assembly sequence generated by the algorithm. Let k be the edge length used for the block replacement strategy and n be the number of entries in S.

The algorithm ensures with high probability that the side of a k block that serves as a growth front for a neighboring k block must first be completed before the new k block can be started.

This allows each assembly in S to be substituted with the block replacement strategy's k^3 assemblies to create S'. Lemmas 4.3 and 4.4 reduce the errors. Lemma 4.2 guarantees uniqueness.

The possibility of substituting the individual assemblies with k block assembly sequences means that the new tileset has a high probability of growing into the desired shape.

The following proof shows an approach that creates snaked proofreading k blocks that fulfill Lemmas 4.2–Lemma 4.5.

Proof: Let $2 \leq k \in \mathbb{N}^+$ be an arbitrary but fixed edge length for the block replacement. In order to show that the strategy is valid for any k, it is shown by induction how a $k+1$ block can be generated from a k block. It is started with a value $k \geq 2$.

Induction hypothesis: The algorithm produces correct k blocks that satisfy Lemma 4.3–4.5.
Induction start: A 2 block can easily be created according to Figure 4.30 (a) and (b):

The 2 block prevents facet errors since no two adjacent tiles t and t' have external glue that is part of the growth front and points in the same direction. Growth errors are reduced by a factor of $k^3 = 8$.

Induction step: Let $j \geq 3$ be a block replacement edge length.

According to the induction hypothesis, The algorithm can create a $j+1$ block by adding another layer. The process is shown in Figure 4.29.

Figure 4.29 also shows possible snaked paths for even and odd k. The previously outward-facing glues are replaced with checkerboard glues, maintaining a unique snaked path. The glues that are now external retain the glues of the original tile.

Because the algorithm is adjacency aware, the assembly sequence of the k block is unique. The checkerboard pattern prevents facet errors. This follows Lemmas 4.4 and 4.2.

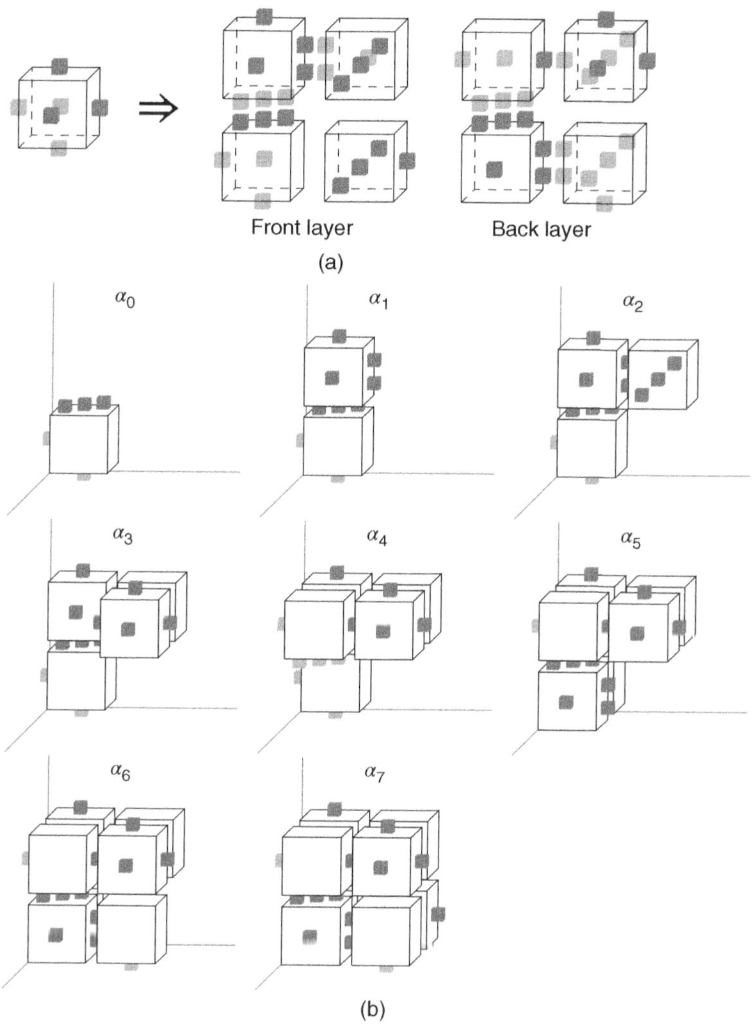

Figure 4.30 (a) $2 \times 2 \times 2$ snaked tileset. (b) Assembly sequence of a snaked block. At the back, left, and bottom, a suitable glue is required from existing neighbors. *Source*: Florian-Lennert A. Lau.

To analyze the complexity and number of calculation steps, proceed as follows: Let n be the number of entries in the original assembly sequence and k be the block length. The algorithm processes each entry of the sequence exactly once and creates a maximum of k^3 new tile types per entry. From this, it follows that a maximum of $O(nk^3)$ tile types are generated, and just as many computational steps are required.

Since the uniqueness of the assembly sequence of each tile has to be ensured, an additional factor of up to 24 can apply. This is due to the different starting and ending positions of the snaked path for each k block. There are eight different starting points and three possible ending points for each starting point.

Proof: From Lemma 4.5, it is known that the new shape is with high probability a scaled version of the original shape. Hence, Theorem 4.1 holds.

4.10 State of the Art of Miniature Structures and Devices

This section gives a brief overview of the functional microdevices and nanodevices that can already be manufactured. We first analyze the advancements in bottom-up manufacturing techniques. Next, we summarize the latest developments in top-down technologies and miniaturization.

4.10.1 DNA Squares and DNA Boxes

Sections 4.10.1.1 through 4.10.1.5 present procedures that create tilesets for squares and cubes. These are partly based on the work of (Rothemund and Winfree, 2000). Squares and cubes are commonly used benchmark structures and can also be used as starting structures for nanosensors and nanorobots. Hollow structures are of particular interest because they can be filled with tiles or medicines in order to later release them in a controlled manner.

First, relatively simple procedures are presented, which are gradually improved. Furthermore, a generalized algorithm is presented that can generate tilesets for any three-dimensional structure.

4.10.1.1 Naive 2D Algorithm

This section specifies the naive generation of filled squares. These can be used, for example, as side walls for nanorobots or nanosensors. Furthermore, it is possible to fold these into a three-dimensional box (Andersen et al., 2009), analogous to the procedure of Andersen et al.

```
Input: edge length n, strength s
Output: tileset T

generateTiles(x,y,s){
  label(t)= l;
  col(t,(1,0)) = x+1,y; str(t,(1,0)) = s;
  col(t,(-1,0)) = x,y; str(t,(-1,0)) = s;
  col(t,(0,1)) = x, y+1; str(t,(0,1)) = s;
  col(t,(0,-1)) = x, y; str(t,(0,-1)) = s;
}

makeTileset(n,s){
  x,y,z = 0;
  T = {};

  while(x < n){
    y = 0;
    while(y < n){
      l++;
      T = T + generateTiles(x,y,s);
      y++;
      }
    x++;
    }
  return T;
}
```

A naive algorithm creates a unique tile type for each position in a square, for which $O(n^2)$ different tile types are required in relation to the edge length n of the square (Becker et al., 2008; Rothemund and Winfree, 2000; Ming-Yang and Ramachandran, 2001). Source Code 4.10.1.1 specifies the procedure. The `generateTiles` function generates a unique tile type for every position of the square with strength s and glue labels `col` corresponding to the neighboring locations in the square. This ensures that all tile types fit together correctly and in only one way.

The `makeTileset` function iterates over all positions of the defined square with dimensions $n \times n$ and calls the previous subroutine.

With this naive approach, tilesets can be created that can be combined into any structure, but these require the largest possible number of different tile types. This can cause problems with more complex nanostructures since the greater the number of different DNA glues, the greater the likelihood of faulty bindings.

4.10.1.2 Naive 3D Algorithm

In this section, an algorithm is presented that, for a given edge length, generates n tilesets for either a hollow cube, a filled cube, or a wireframe cube. A large number of different cubes have already been created from DNA (Andersen et al., 2009; Becker et al., 2008; Ming-Yang and Ramachandran, 2001).

The basic idea of the algorithm is again the creation of a unique tile type for each position of the desired structure. In this way, the number of possible errors can be reduced in the case of small structures since the glues can be chosen differently accordingly. Since the number of realistically distinguishable glue sequences of length k is sufficiently large, it is not a problem to proceed in this way with smaller structures. In most cases, however, there is interest in structures that consist of as few different tile types as possible.

In three nested loops, each position of a cube is considered exactly once. First, we store whether the current position is part of the inner cube, a wall position, an edge position, or a corner position in a variable. In this way, we control whether the respective position should be assigned a tile type and can create hollow, filled, or wireframe cubes.

Figure 4.31 (a)–(c) show intermediate steps and an example result of a simulated assembly process. A tileset for a $5 \times 5 \times 5$ hollow cube is created and displayed in three different states. The different shades of greycorrespond to a unique tile type. The glues themselves have been omitted for clarity.

4.10.1.3 3D Linear Runtime Algorithm

A procedure for more efficient squares is presented in the work of Rothemund and Winfree (2000) and can be extended to three dimensions (Wysocki, 2019). The idea is that only the bottom edge

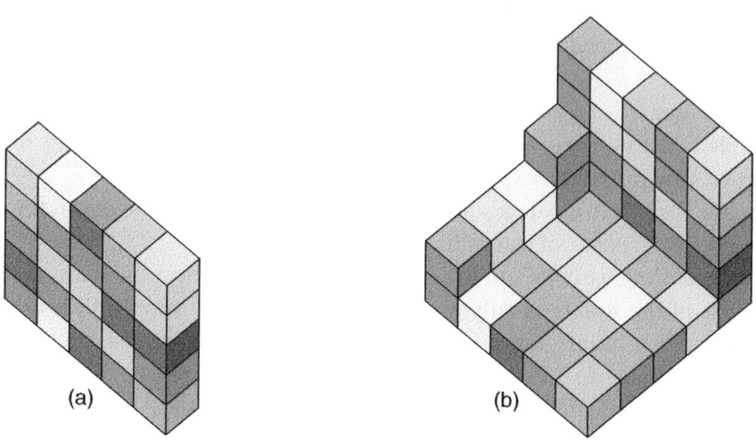

Figure 4.31 Hollow cube of edge length 5 in three stages generated in a 3D aTAM simulator. Each position has a unique tile type. *Source*: Florian-Lennert A. Lau.

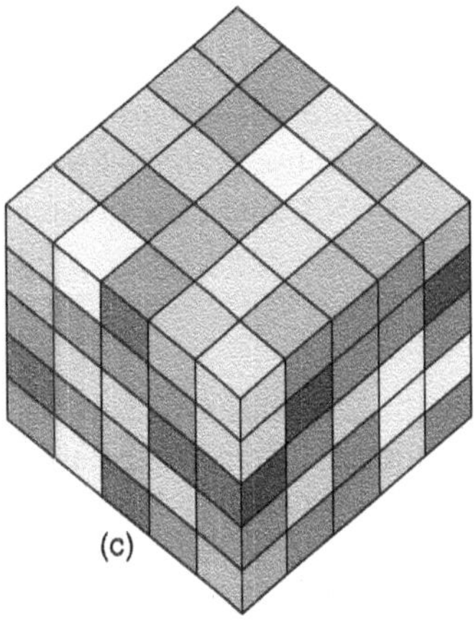

Figure 4.31 (*Continued*)

of a square requires different tile types. The remainder can be filled with identical stacks of tiles to form a square.

The procedure would be even more efficient if a row of individual tile types were created first, then a square from stacks, and then a cube from stacks.

The problem with this approach is that the cube is not hollow and is therefore unsuitable as a transport container. To efficiently create a hollow cube, a core structure of tile stacks is first created, as shown in Figure 4.32 (a) and (c). They have a special glue on the sides so that surfaces can grow from stacks of tiles (Figure 4.32 (b)) to form a hollow cube (Figure 4.32 (d)).

Theorem 4.2 The 3D triple-stack algorithm produces a tileset that produces correct $n \times n \times n$ cubes using $O(n)$ tile types.

Proof: The correctness of the approach follows from Figure 4.32. The size of the tileset is the sum of the core and the faces. The core consists of five stacks with a maximum length of n. The sites consist of a series of six different stacks in total, which are repeated within one area. The overall complexity, therefore, corresponds to the class $O(n)$.

In this way, a linear number of tile types suffice for a three-dimensional cube. It is also possible to use fewer than $O(n)$ but this requires an additional constant number of tiletypes.

4.10.1.4 Constant Runtime Algorithm

Theoretically, it is even possible to create a square or rectangle using only a constant number of different tile types. These can be seen in Figure 4.33 to the left.

Just five different tile types are enough to create a square of arbitrary edge length. However, the simplicity and small size come at the expense of external control.

Figure 4.33 shows a possible resulting assembly to the right. Three tiles σ control the diameter of the square. In theory, the size of the square is only limited by the number of supplied tiles σ. If external conditions allow for it, an arbitrary number of tiles can diagonally attach and thereby grow the square.

The remaining tiles A and B assist in the growth of the middle diagonal. The tiles β_1 and β_2 fill the remaining parts of the square that are limited by the previously grown diagonal. If they are introduced at a later point, the growth of the diagonal is stopped by the tile β_1 that allows for no further stable bindings at temperature 2 and fits the same site as a tile σ. Thus, only five different

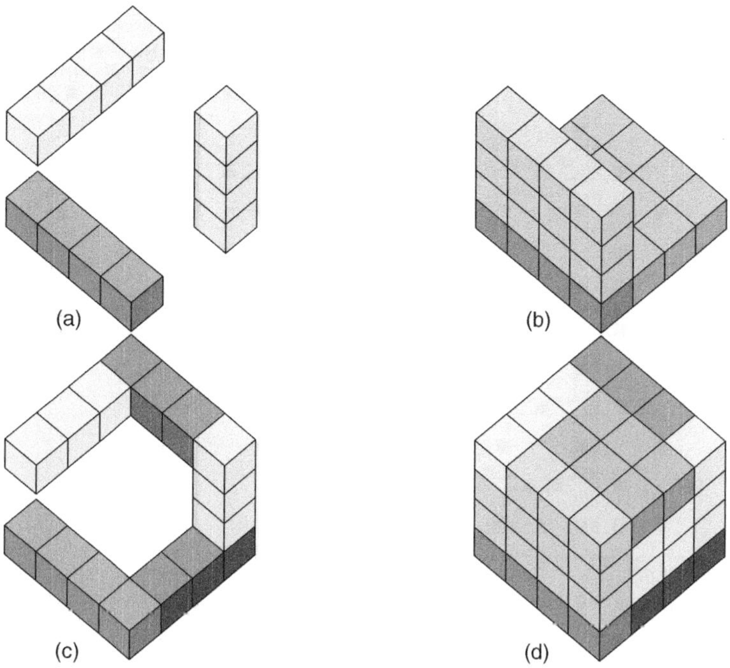

Figure 4.32 Conceptual steps of an algorithm for generating hollow cubes in $O(n)$. *Source*: Florian-Lennert A. Lau.

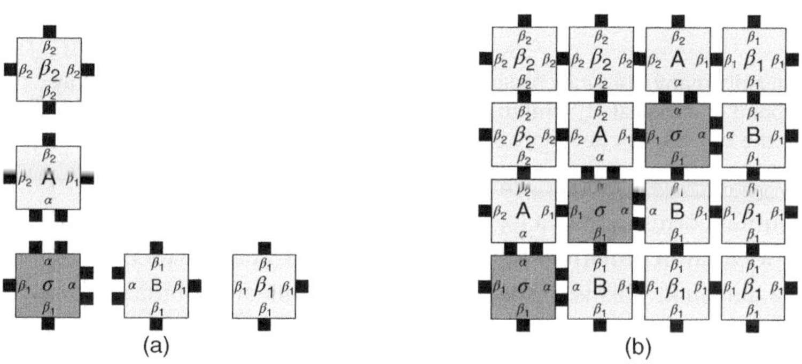

Figure 4.33 (a) Constant tileset for squares. (b) Finished square of size 4. *Source*: Florian-Lennert A. Lau.

tile types and smart external control allow for the generation of squares of arbitrary size. A slightly modified version is capable of creating rectangles.

4.10.1.5 Pragmatic Logarithmic Runtime Algorithm

Through the skillful use of space-limited binary counters, a square of tiles at temperature 2 can be generated by a logarithmic number of tile types (Rothemund and Winfree, 2000). The binary counter is embedded in the square. This forms up to a specified number and then completes the growth in the y-direction. In this way, using 22 additional binary counter tile types, a square can be created from $O(\log n)$ tile types in relation to the edge length n of the square.

Figure 4.34 shows an example. Essentially, the square consists of an integrated, space-limited binary counter, which can generate a rectangle of fixed height using logarithmically many tile types.

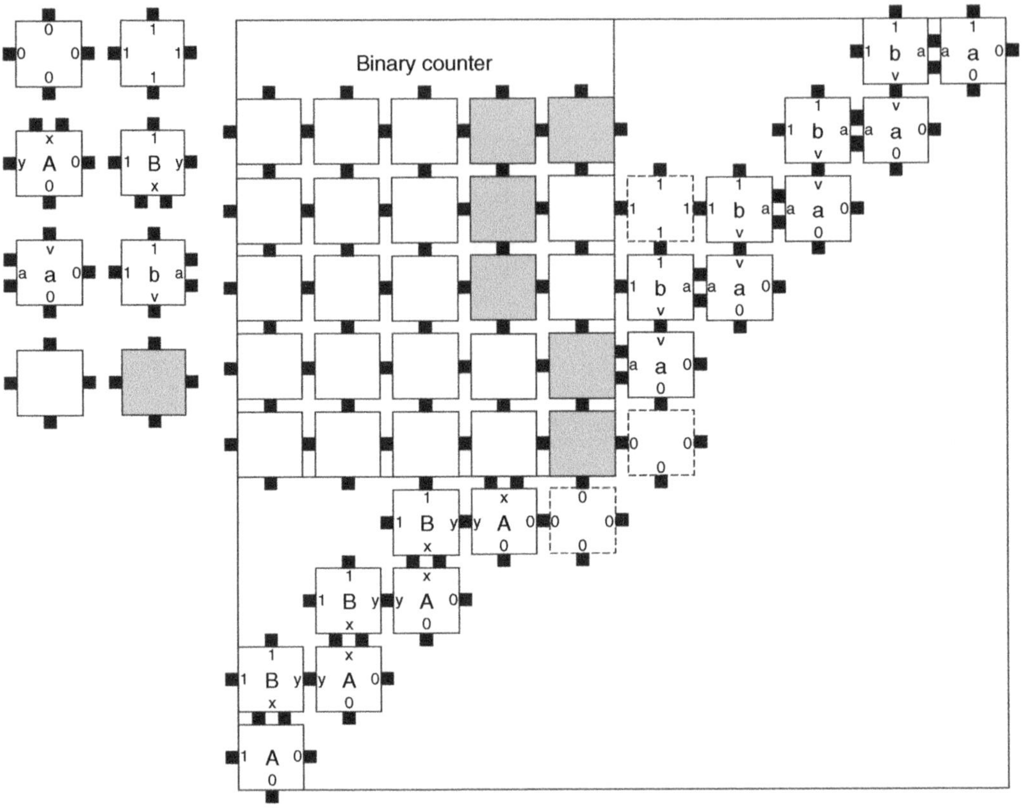

Figure 4.34 Square of logarithmically many tile types based on an embedded binary counter. The gray tiles in the binary counter correspond to a logical "1," and the white tiles to a "0." It can be seen that it counts up in double rows so that the counter can be height limited. *Source*: Florian-Lennert A. Lau.

The gray tiles in the binary counter correspond to a logical "1" and the white tiles to a logical "0." It is incremented in double rows to limit growth in height. Tiles of type A/a and type B/b bind to the binary counter and form a diagonal. This is limited by the dimensions of the binary counter. The remaining space is filled by the type 1 (left half) and type 0 (right half) tiles. The square can only continue to grow in the x or y direction once the previous row has formed.

The method can be generalized to three dimensions by creating a semantically identical tileset for each side of a desired cube, which can form as soon as the last tile of an outer side of the square has been bound. The original square receives glue on the top or at the bottom in the third dimension. The self-assembly process is repeated six times until a hollow cube of edge length n is formed.

The tile complexity is still $O(\log n)$ because a maximum of six times the number of tile types has to be used. For a square with an edge length of 50, 28 tiles are required. A corresponding hollow cube would require at least 168 tile types. The linear algorithm presented previously would require 1152 tile types for the same structure. From an edge length of 3–4, the additional constant number of tile types for the binary counter is compensated and the logarithmic 3D algorithm is more efficient.

Since the logarithmic 3D algorithm is the most efficient of the approaches presented here, it may be used for modeling nanosensors $\mathcal{N}_{Se}$ and nanorobots $\mathcal{N}_R$.

To use the cube as a controllable container, one of the six squares is used as the opening mechanism. This is only attached to one side of the cube. The opposite side receives exactly the amount of glue corresponding to the temperature $\tau - 1$. On the opposite side, there is another glue of thickness $\tau - 1$. In this way, it is possible for another molecule to bind more strongly and thus open the box.

While DNA origami seems like a more efficient approach for generating controllable boxes, we now define tile-based nanosensors and nanobots.

Definition 4.37 A *tile-based nanosensor* $\mathcal{N}_{\text{Se}}^{i,k}$ is a tileset that assembles into a hollow cube of size $2 < i \in \mathbb{N}^+$ composed. $k \in \mathbb{N}^+$ describes the size of the block replacement strategy if one is used. Unless otherwise specified, $k = 1$ and can be omitted. If i is clear from the context or is arbitrary, this can also be omitted. A nanosensor $\mathcal{N}_{\text{Se}}^{i,k}$ has a receptor R that binds correctly with certain markers.

Definition 4.38 A *tile-based nanomachine* $\mathcal{N}_{R}^{i,k}$ is a tileset that assembles into a hollow cube of size $2 < i \in \mathbb{N}^+$ composed. $k \in \mathbb{N}^+$ describes the size of the block replacement strategy if one is used. Unless otherwise specified, $k = 1$ and can be omitted. If i is clear from the context or is arbitrary, this can also be omitted. A nanobot $\mathcal{N}_{R}^{i,k}$ has a receptor R that correctly binds to at least one message molecule.

A tile-based nanosensor or nanorobot must have an edge length of at least three in order for a cavity to form. A tile-based nanosensor $\mathcal{N}_{\text{Se}}^{12,2}$ thus describes a size 24 nanosensor since each original tile was replaced by a size 2 block to avoid errors.

According to the presented definition, nanosensors and nanorobots differ in the sensor component. A nanorobot can recognize message molecule through a simple binding reaction. With nanosensors, the process is implemented differently and differs depending on the application. In both nanosensors and nanorobots, proper binding to the receptor can trigger a mechanism. Depending on the application, this can involve opening nanosensors or nanorobots to release tiles or medicines.

4.10.2 DNA Origami Boxes

While it is theoretically possible to create DNA boxes from tiles, it is likely not feasible. DNA tiles are likely more suitable for computational purposes as we explore in Chapter 5. DNA origami boxes, however, seem more suitable for creating boxes or other hollow objects that might contain medication or other payloads.

Many proposed applications in the nanonetworking community depend on the conditional presence of certain molecules. This includes construction, computation, and communication mechanisms. As this problem is well known, scientists created boxes from DNA using the previously presented DNA origami technique. This or a different vessel based on DNA or other materials is a cornerstone for many envisioned technologies related to nanonetworks.

4.10.3 Microbots

Apart from DNA-based nanodevices, there are some advancements regarding artificial, microscale robots. One of the first notable applications has been presented as early as 2000 (Iddan et al., 2000). The latest advancements in WSN and digital technologies made it possible to manufacture the first pill camera. This type of endoscopy allows for a painless imagining of the entire human gastric system. The patient simply swallows the camera that records and transmits the recorded images to an external device.

While the device itself is not even close to nanoscale in size, it is still one of the first autonomous systems deployed in the human body. The device itself is only a few centimeters in size.

Devices that are much closer to the nanoscale are the microrobot introduced in the work of Miskin et al. (2020) and Reynolds et al. (2022). We have already seen an example in Figure 2.11. The biggest challenges in-body microelectronics currently face are the combination of integrated circuits and actuators. Yet, in the work of Reynolds et al. (2022), they managed to create microrobots of size 100–250 μm which can move around and are powered by light. As of right now, their main function is to move around, which is not yet sufficient to fulfill a meaningful medical task.

While the presented robot is a very interesting proof of concept, there are many remaining challenges in the human body. As soon as a device is introduced into the body, it will be coated by a number of different substances. A possible way to circumvent this is an ultrasonic actuator that shakes the device to remove the coating. However, the combination of both is no longer nanoscale.

While we have no functional microdevices yet, the trend of miniaturization does continue and it will be interesting to see what future developments might have in store for us. In addition to that, there are many other concepts for nanodevices in existence. Most of them are in a very early stage and without many practical applications. Yet, they serve as a suitable proof of concept for what might be possible.

4.11 Simulation

There are several tools available for the simulation of nanoscale construction processes. They are necessary as biological processes can be both complex and complicated. The most well-known examples would be from medical, biological, or pharmaceutical fields where scientists try to predict the folding behavior of molecules or predict the properties of new medications. One of the most famous examples would be the "Folding at Home" project, where everyone can offer computational resources to help simulate protein folding behavior (Bowman, 2000). Pharmaceutical corporations often use artificial neural networks to try and predict how possible new medications might perform.

As both areas are well covered, this section mainly focuses on DNA-based approaches, where either DNA origami or self-assembly is simulated. DNA origami is well suited for creating arbitrary shapes, and the process is frequently used. Tile-based self-assembly is well suited for computational purposes.

4.11.1 ISU TAS

The simulator ISU TAS by Matthew J. Patitz supports the evaluation of three different models (Patitz, 2009). The aTAM, the kTAM, and the 2HAM can be efficiently simulated. Since these three models can be used to evaluate different aspects of nanonetworks, the ISU TAS program is a suitable candidate. The simulation of numerous assemblies in a network is not offered. Furthermore, three-dimensional assemblies are not supported.

The aTAM is the simplest form of a TAS. In each time step, a random matching tile is added to the assembly at a random location on the growth front in ISU TAS. In each time step, several tiles can be suitable. The aTAM can be used in the course of the evaluation to check whether the desired structures are growing. Multiple seed tiles can be examined at once. This applies to both the aTAM module and the kTAM module presented in the following paragraph.

The aTAM module is used to check particularly large assemblies as this takes too much time in the kTAM module.

The kTAM module of the ISU TAS simulator can be used to investigate error behavior under realistic conditions. The kTAM uses the same tileset as the aTAM module. The temperature τ is set by rates that describe how likely it is to detach or add tiles. Suitable values correspond to temperature 1, 2, or 3. The software does not support larger values. However, since only two- and three-dimensional systems are of interest in the course of this work, the selection is sufficient.

The 2HAM module also uses the same tileset, but it is based on a different set of axioms. In the 2HAM, we not only observe a single assembly but try to understand what new assemblies may be formed given all the available tiles and assemblies at each moment in time. As soon as no more new assemblies are possible, the simulation stops. The 2HAM in the ISU TAS can only be simulated at temperature 1 or 2.

The 2HAM module is used to explore scenarios with no dedicated seed tile. Each tile can take on the role of a seed tile and potentially start an assembly process. However, the assembly process is only examined like in the aTAM. Advanced tools are required for a more thorough analysis. However, the module is sufficient to examine unwanted interactions of intermediate products of an assembly process.

In Section 12.4.7.3, the kTAM implementation as well as the 2HAM implementation is partly used to evaluate the designed tilesets. In rare cases, the aTAM module is used. The syntax of the software is also explained in Section 12.4.7.3.

However, it should be noted that the ISU TAS software is classified as outdated. It does not do justice to modern scientific requirements regarding expandability and the use of software patterns. In addition, the program code is classified as poorly commented and difficult to understand, which is why we use an alternative software called NetTAS.

4.11.2 Xgrow

Eric Winfree's Xgrow program was developed for simulating DNA-based computations only (Xgr, 2003, DNA). The simulator was written in the C programming language and optimized for Windows systems. It is primarily used to study failure behavior at the nanoscale in tile-based self-assembly systems.

The simulator has some extensions and tools based on them:

- *Xtile*: A tool that transforms tilesets. An example is proofreading tilesets (XTi, 2009).
- *R-TAM*: A tool for simulating the thermodynamic effects of DNA (Fujibayashi and Murata, 2007).

Xgrow implements the aTAM and the kTAM. It can simulate numerous assemblies at once as long as they don't interact with each other. The simulator is also many times faster than other simulators such as ISU TAS since numerous computations are pre-calculated in a quad tree. However, this ensures that only assemblies of a predefined size can be simulated.

Tilesets from the Xgrow software can also be used in the ISU TAS. Since the ISU TAS software is more modern and therefore more accessible for many scientists, the Xgrow software is often considered outdated.

4.11.3 NetTAS

The NetTAS simulator is among the most recent additions to a series of simulation tools for tile-assembly systems (Kaussow, 2022). Like all tile-assembly simulators, the NetTAS can analyze both construction and computational processes. However, tile-assembly systems are likely better suited for computational and communication purposes, while DNA origami is better suited for construction. Nonetheless, both methods are based on DNA and thus compatible.

Most of the other tools require the installation of some dedicated software. NetTAS has been programmed in TypeScript and can be accessed as a web application in the browser. Thus, every system that runs a browser also runs the NetTAS simulator.

Another advantage of the NetTAS simulator is its algorithmic efficiency and readability/documentation of code. Most academic software has severe weaknesses in this domain and is difficult to extend and maintain. NetTAS further avoids certain errors at temperature 3 that might occur in the ISU TAS.

Furthermore, NetTAS introduces another model in addition to the aTAM, kTAM, and 2HAM. This new kTHAM model combines the 2HAM and the kTAM and thereby more realistically models self-assembly processes. This allows for a much better simulation of *small* nanonetwork-related applications. In the general case, both the 2HAM and the kTHAM are computationally infeasible and only work for comparably small tilesets that are nonetheless sufficient for many resource-constrained applications.

Additionally, at any given time, the external parameters can be changed and it is thus possible to accurately reflect external changes in the environment. Finally, the simulator offers much better support for publications by allowing for CVS exports for histograms, error rates, and tikzpictures of tilesets/assemblies. It is also possible to automate the simulations in many different ways by specifying various termination conditions.

4.11.4 caDNAno – DNA Origami Simulation

As early as 1982, Seeman et al. proposed DNA as a possible building material at the nanoscale (Seeman, 1982). Since then, numerous advancements have taken place in that area. In 2006, another milestone was reached by the DNA origami procedure from Rothemund Rothemund (2006). The method is suitable for creating a variety of different 2D and even 3D structures of around 100 nm in size (Douglas et al., 2009a).

The process is based on folding a long strand of DNA using DNA staples. However, designing such strands and fitting staples to create desired nanoscale shapes can be tedious and prone to errors. Hence, the previous simulation of such processes can lead to a significant speed-up. For simple 2D structures, (Birac et al., 2006) introduced a program for testing and rapid prototyping. One of the most well-known software frameworks for designing and prototyping 3D DNA origami is caDNAno (Cadnano, 2017; Yoo et al., 2018; van Dijk and Bonvin, 2009).

The program uses a honeycomb-like lattice structure to create 3D structures. Each helix has three neighbors that are stapled together as explained before. Thus, it is possible to create almost arbitrary 3D structures at the nanoscale. The method has been successfully tested by creating seven different rectangular blocks (Douglas et al., 2009b).

In addition to the base module, caDNAno has several extensions that allow for the prototyping of different methods. Partner projects are CanDo, MrDNA, oxDNA, mMaya, and scadnano. All of them allow for the more efficient creation of 2D and 3D shapes that could be useful in a variety of applications. As of right now, self-assembly on the basis of DNA might be the only feasible method for the mass manufacturing of nanoscale structures that require a high degree of precision.

4.12 Summary

As we have learned, there are several possible paradigms for construction at the nanoscale. Modifying cells or bacteria might be the first method that works in practice, but it might be limited in several ways. There are strict limitations on how much we can modify as evolution relies on gradual change. Furthermore, the actual changes might be so small that the "artificiality" of such a nanonetwork might be debatable.

Electromagnetic technologies seem still quite far away from realization as of right now. The closest technologies we have are microbots without any significant functionality and pill cameras of several centimeters in scale.

Bottom-up approaches based on DNA are likely the most realistic construction method right now. In addition to the construction properties, DNA technologies can also be used for communication and computation. DNA origami seems especially suitable for creating all kinds of structures and templates. Tile-based self-assembly seems most suitable for computation and communication.

While there are many possible DNA errors, it seems like they can be mitigated in many ways. The current state of the art seems to indicate that DNA is one of the most popular candidates for nanoscale construction. It is one of the only known materials that allow for the required precision due to self-assembly properties.

5

Computation

This chapter introduces the fundamental knowledge of how to perform and understand computation at the nanoscale. We first analyze how state can be represented at the nanoscale. Based on that, we formally define rules of state transformation, which are also called *programs*. Finding suitable programs for networks of possibly millions of devices can be extremely challenging. Then we introduce the very basics of complexity theory that are necessary to represent the prevalent resource constraint at the nanoscale. Complexity classes can be used to assess the difficulty of given operations/programs by putting them in relation to similar methods. Next, we analyze a number of nanoscale applications with regard to the necessary basic operations. Then we analyze a number of computational models regarding their potential use and feasibility at the nanoscale. Finally, we analyze computations based on self-assembling particle systems in considerable depth and summarize the chapter.

5.1 State at the Nanoscale

In addition to the required computing power, nanoscale memory is essential for many proposed applications. Numerous scenarios require nanodevices to be able to store their position, state, or simply a timestamp of a measurement.

A memory problem of particular interest is the addressing of individual nanodevices in a network. It takes at least a logarithmic amount of memory (relative to the size of the network) to store an address, and it is not clear that nanodevices can do that – but more on that later in Chapter 6.

Yet, memory is also required for simple scenarios. For example, in order to compute an average value, numbers must be compared in pairs and an intermediate result must be stored and updated. The same applies to operations like MAX, MIN, THRES, and MAJOR.

If we talk about state in computer science, we usually talk about numbers. Numbers are a well-researched kind of state that behaves predictably. Furthermore, many mathematical tools are available for manipulation.

As there are no "bare numbers" available in reality, they have to be represented by some physical component. Possible examples are transistors that can assume one of the two states. The two states could also be interpreted as on/off, and in that case, a transistor would be equivalent to a switch. Transistors are controllable and either allow a high or a low current to pass through. "High voltage" usually represents a logical "1"/"True," and "low voltage" represents a logical "0"/"False." By having several transistors interact with each other, several digits can be combined into a number and all combinations of bits of a fixed length are now possible.

Nanonetworks: The Future of Communication and Computation, First Edition. Florian-Lennert A. Lau.

Such a "1/0" decision is also called a *bit*, and it is the fundamental unit of information. In reality, it is not possible to transfer less than one bit of information at a time. However, through the aforementioned chaining of bits, it is possible to transmit way more than a single bit of information. In such a case, the information contained in a series of bits is usually equivalent to the length of the number. Mathematically speaking, it is thus defined as $\log_2(n)$ if n is the length of the number.

In classical computers, laptops, tablets, or smartphones, the state is always represented in the same way. The state of a device is usually the entire content of the main memory, but all the small caches, graphic card memories, etc., can also be seen as part of the state. In the simplest sense, the entire content of those memories can be interpreted as one long binary number. A binary number of length n can assume any of 2^n possible states and can thus be used to distinguish/designate just as many different objects. Such a number can be changed and manipulated by the CPU or GPU according to the rules specified in programs. As it is the easiest to physically implement a number system based on the digits "1" and "0," most computers work with numbers in binary representation. However, it is possible to create computers that can work with any kind of number system, given that there is a suitable physical component that can represent those numbers.

If we now want to save or store a piece of information, we have to take it in whatever representation it was and manipulate physical matter in a way that the number information can be retrieved at a later time. In the early days of computer science, simple punch cards were a popular medium for storing binary information – either there was a hole in the paper or there was not. Technically, many potential candidates for physically realizing state exist, but not all of them are equally well suited. Some require too much energy, some are too big, some are too costly or inefficient to read, and some are too volatile. A good physical representation of state fulfills all the criteria.

Now we can transfer that knowledge to the nanoscale. The first thing to notice is that most transistors nowadays are already only a few nanometers in size. The latest research suggests transistors as small as 7.7 nm. While transistors are surely small enough for nanoscale applications, they usually require enormous amounts of energy relative to their size to function. Thus, it is questionable if it is possible to continue using the already established technologies.

A possible alternative form of state that provably works is DNA. DNA in its natural form already encodes the genetic information of all living beings, and a single chromosome contains as much as 50–300 MB of information while still only having a diameter of 2 nm. Unlike transistors, DNA works based on a different mechanism. While each transistor can assume one of the two states, each position in a DNA helix can contain one of the four possible bases (see Chapter 4). DNA-based state can thus be interpreted as a quaternary number.

Unlike most modern storage media that only last for several decades, DNA can last for thousands of years, given suitable external conditions. However, storing digital information in the form of DNA takes a lot of time. Depending on the method used, appending a single base to a DNA strand can take between a single second to many hours. Storing a movie in DNA could take decades of time as of right now, and the 2019 record for information stored as DNA was more than 1 GB (Pluta, 2019). That said, reading the information is usually much faster and the entire topic of DNA-based storage draws a lot of attention from researchers. It is to be expected that this technology will be much more efficient in the future.

That said, most technologies like that still rely on heavy machinery and can only be used for nanonetworks in a limited sense. More research is necessary to see if DNA in its simplest form can be a suitable information carrier in nanonetworks.

Luckily, other DNA molecules like DNA tiles might circumvent some of the aforementioned problems. Each tile can either encode information in its glues or marker. Information stored in the glues can be used for direct interaction with other DNA structures at the nanoscale without much external effort. Thus, it is possible to autonomously manipulate information in many different ways. Information stored as a marker could be a fluorescence or radioactive marker that makes it easier to transmit information to the macroscale.

While the information stored in DNA tiles is much less space efficient, other advantages make up for it. They do not require much energy, come with a free addressing mechanism, and require little space. As we will see later in this chapter and in Chapter 12, it should be possible to create entire autonomous DNA-based nanonetworks. That said, there are challenges when it comes to the tile-based information storage. While "1" and "0" are typically mutually exclusive in transistor technologies, that is not the case with DNA tiles. A and $\neg A$ can theoretically co-exist. The designer of such a system must carefully engineer a system that avoids such conflicts as information is usually encoded by specific strands of DNA. Furthermore, as with all distributed systems, it is not easy to tell if some tile is "not yet" available or if it never will be available. Nonetheless, DNA tiles likely have many more advantages than disadvantages.

5.2 Computation

In the most general form, a computation is simply a manipulation of state according to a fixed set of rules. Those rules are often captured in a so-called *program*. As it is a bit tedious to work with transistors or DNA strands, computer scientists usually model the physical information storage as numbers.

Most of the time, the state manipulation described in a program happens intending to solve a problem. In theoretical computer science, two different types of formal problems for nanodevices are of particular interest. These are decision problems and functional problems. Underlying these types of problems is the computational problem:

Definition 5.1 A *computational problem* is a tuple of inputs, also called *instances*, and outputs, also called *answers*. Furthermore, there is a mapping of instances to answers.

Instances can be any data structure. Binary numbers are popular instances in classic computers. Many decision problems, on the other hand, are modeled as language. A decision problem is part of a language if it evaluates to "1." A decision problem can thus be understood as a set of "yes" instances. Importantly, problems must be encoded in a data structure. The following problem definitions are limited to aspects that are important for nanonetworks.

Definition 5.2 Let $\mathcal{D} = \Sigma^*$ be an *instance set*, also called a language. A *decision problem* is a subset of Σ^* which evaluates to "1."

Decision problems are of particular interest for nanonetworks since information about the solution to such a problem can be directly communicated through, for example, binding reactions. As a result, hardly any additional interpretations are necessary. An example of a decision problem is whether there are a significant number of markers for a disease or, in the simple, mathematical case, whether a number is even or odd.

Definition 5.3 Let $\mathcal{D}$ be a set of problem instances. A *function problem* is a mapping from $\mathcal{D}$ to $\mathcal{D}$.

The result of a functional problem cannot be communicated by a single bit. The answer to a problem can contain any number of bits. An example of a function problem is the addition of two numbers, where $\mathcal{D} = \mathbb{N}$.

5.3 Complexity Theory

Now that we know about state and computations, it is time to put things into perspective. Complexity theory is a scientific discipline that concerns itself with sorting and classifying computational problems according to the required resources. A complexity theorist analyzes certain algorithmic problems and tries to determine the minimum necessary computational power to solve them. As you might imagine, this is especially useful at the nanoscale, where resource constraints are even more imminent than with regular computers.

It is important to notice that complexity theory concerns itself with *complexity* and not just with *complicated* problems. A complicated problem is often hard to understand because it is difficult to describe and many rules interact with each other. However, a computer might still be able to "sort things out" algorithmically with relative ease. A complex problem on the other hand is intrinsically hard to solve due to the sheer number of possible solutions.

A well-known example of a complex system is Langton's ant thought experiment (Gajardo et al., 2002). Imagine there is an ant on an infinite chessboard plane that is completely colored in white. The ant acts according to only two rules:

1. At a white square, turn 90° clockwise, flip the color of the square, and move forward one unit.
2. At a black square, turn 90° counter-clockwise, flip the color of the square, and move forward one unit.

As there are only two simple rules and a simple state space, the problem is not very complicated. However, it is indeed a simple example of a complex system. If you repeatedly apply those two rules, the result is completely unpredictable. One might assume that the chessboard shows a relatively simple pattern, but for the first 10 000 steps, there seems to be no symmetry whatsoever. After roughly 10 000 steps of random seeming behavior, the ant starts creating an infinite "highway" structure of 104 repeating steps. That said, the only known way to find out about this behavior is through simulation. Complex systems can be simple seeming but still computationally infeasible – there are no shortcuts to predict the future of such a system.

Similarly, regular problems can be deceiving in their apparent simplicity. Complexity theory mathematically analyzes such problems to find out their true difficulty and resource requirements. Based on that, it is much easier to design nanodevices or nanonetworks that are suitable for a task.

5.3.1 Complexity at the Nanoscale

Although nanonetworks were proposed as a computing paradigm as early as 2008, the functional scope of individual nanonodes has not been specified very much (Akyildiz et al., 2008). However, it can be assumed that there must be at least one possibility for communication. Otherwise, no network can be set up. This shows that coordinated actuators and sensors and thus rudimentary possibilities for information processing must be available in order to enable meaningful applications at all. However, it remains unclear how the mentioned technologies can be realized. The small size of the nanonodes implies severe resource constraints. This refers to both computing power and memory sizes.

Unfortunately, for numerous scenarios, it has hardly been formalized how complex individual nanonodes have to be in order to be able to solve given problems (Staples et al., 2006; Amato et al., 2010). The required capabilities of nanorobots vary widely. There is some talk of nanoparticles with infinitesimal complexity. Other publications write about nanorobots, which have a similar power as today's CPUs (Cobo and Akyildiz, 2010).

Many scenarios only explain in a very abstract way what needs to be done to solve a problem. Sometimes formulas, abstract operations, or algorithms are given, but the power of a nanonode is often not specified. Sometimes it is even overrated. Likewise, the consequences of the required power on the feasibility of the machines are not part of the elaborations.

Less complex nanonodes are probably easier to realize than complex ones. Simple, potential components of nanodevices have already been realized (Amlani et al., 1997; Benenson et al., 2004). However, the combination of those building blocks into complex structures, as in modern CPUs, is still largely unexplored, and various sources of error, such as the precise placement of components, make research progress difficult. Because although CPUs are getting smaller and smaller, the combination of all elements of a CPU fills several square centimeters.

This leads to the question of what are the lowest possible requirements for individual nanonodes to be useful for a given scenario. In other words: What is the lower bound of the required complexity of a nanonode so that all requirements made in a scenario can be fulfilled? A simple nanonode might be easy to realize but could miss crucial computational capabilities.

This problem was dealt with in Lau et al. (2017) in order to allow for an estimation of the feasibility of proposed scenarios. In addition, a precise definition of the machine model enables simpler and more precise simulation.

5.3.2 Reductions

In order to identify requirements for nanodevices, problems from proposed scenarios must be sorted according to their complexity. *Complexity classes* are used for this. They are a model from theoretical computer science – more precisely from complexity theory.

The short introduction given here refers to Lau et al. (2017).

As a deep analysis of the topic is beyond the scope of this book, the interested reader is advised to have a look at the work of Clote and Kranakis (2002).

The complexity of systems is determined depending on a system parameter such as the available memory or the runtime required for an operation. For this work, complexity classes that refer to the available working memory, also called *space classes*, are primarily of interest. In the classic computational model, the simple Turing machine, there is an unlimited amount of this. This is not a realistic assumption for nanodevices, which is why only those complexity classes that are extremely space constrained are presented here.

Put simply, complexity classes contain problems that can be solved by a particular machine model. For example, the complexity class L contains all those problems that can be solved by a Turing machine with a logarithmically space-limited working tape. Inputs and outputs are not considered. The logarithmic ratio refers to the length of the encoded input, taking into account factors that are omitted in the Landau analysis. A nanodevice that is at least as powerful as a certain machine model can therefore probably also solve all problems of the respective class. In the case of circuits, it must be possible, at least with a similar effort, to create a nanodevice that solves the related problems.

In addition to the L class, two other circuit classes are also examined, which may be of interest for nanodevices. Circuits are of particular interest because they can perform computations without memory. Some publications already deal with circuit-like systems (Nielsen et al., 2016). These circuits essentially consist of Inv, Or, and And gates. Together with the constants, these three gate types form the atomic components of the classes AC^0 and NC^1 (Clote and Kranakis, 2002). The class AC^0 is additionally characterized by the fact that Boolean circuits, which are described by it, have polynomial size and constant depth. The metrics are relative to the number of inputs. NC^1 circuits

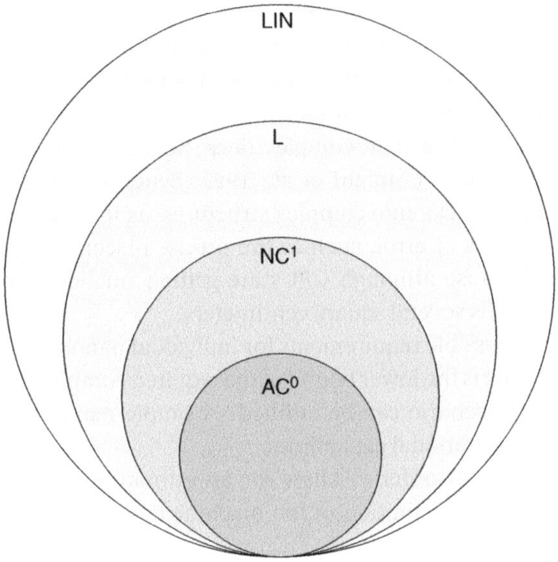

Figure 5.1 Relationship between different complexity classes that focus on the required memory size: LIN, L, NC^1, und AC^0 (Lau et al., 2017). *Source*: Lau et al. (2017)/with permission of Association for Computing Machinery.

have the additional restriction that each gate may only have a maximum of two inputs. However, they may be logarithmically deep and support parallelism, which is a valid assumption for nanonetworks of potentially millions of participants.

Complexity classes can contain each other. For example, class NC^1 contains the entire class AC^0, both of which are contained in L. The inclusion ratio is illustrated in Figure 5.1. The inclusion is easy to see since AC^0 gates with any number of inputs can be replaced by a logarithmic deep tree structure, which gives the same result.

This makes it clear that problems can lie in many classes. The problem Log_2 is, e.g., solvable by AC^0 circuits and L Turing machines.

Consequently, if a nanodevice is at least as powerful as a L Turing machine, it can also solve the Log_2 problem.

Reductions are used to sort a problem into complexity classes or to assign them to them. Figure 5.2 clarifies the procedure using a schematic figure.

A reduction proves that a problem A is at most as difficult to solve as another problem B.

A reduces to B if the following applies.

An instance i_A of a problem A can be converted into an instance i_B of a problem B with little effort. The instance i_B is then solved by a "known" algorithm B, resulting in the solution s_B. This solution can then be converted into a solution s_A for the problem A with little effort.

For example, one can show that the subtraction problem SUB is at most as difficult as the addition problem ADD – i.e. it is in the class AC^0 – by reducing SUB to ADD. To do that, we change the input for SUB to become an input for ADD. This can be done with a negation gate for the second

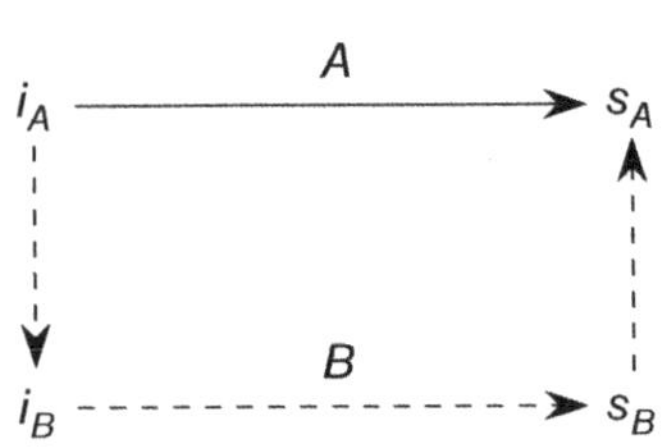

Figure 5.2 Representation of a reduction scheme. A problem instance i_A is transformed into another i_B, which can be solved by a machine model B that we already know. The result of the calculation s_B is then transformed into a result s_A that can be used by machine A (Lau et al., 2017). *Source*: Lau et al. (2017)/with permission of Association for Computing Machinery.

summand. The modified circuit is still of constant depth and uses the allowed class AC^0 gates. In this case, the inverse transformation would be the identity function. It is important that the combined out-and-back transformation effort is done only using allowed means (or less) for the class of problem A. Otherwise, one could solve problems just by the transformation alone.

We now define, various computational models based on the complexity-theoretical foundations are presented. However, first, we analyze the proposed computational problems nanonetworks might face.

5.4 Computational Analysis of Nanoscale Applications

This section describes the approach to finding reasonable scenarios for nanonetworks, which serve as an input for the complexity analysis. The investigation of possible applications makes it possible to make more precise statements about the requirements and performance of individual nanodevices (Lau et al., 2017). Based on the analysis, useful computational models and complexity classes for individual nanodevices are derived. The focus of the scenarios examined is on nanomedicine and communication.

One of the most prominent scenarios is the detection of specific disease markers and subsequent control or reporting of the disease using nanodevices or nanonetworks (Staples et al., 2006; Amato et al., 2010).

Various nanotechnologies are already used in oncology, and new ideas are constantly being integrated from research into industry (Singhal et al., 2010). Nanoparticles with medical functions are among the most popular candidates for future treatments. Some application ideas suggest, e.g., liposomes or modified bacteria to deliver drugs to an application site (Wang et al., 2015; Rudnitzki et al., 2018; Yao et al., 2017).

The procedure can be generalized to the targeted administration of drugs by nanodevices at an application site. Research already done includes treating diabetes (Gu et al., 2013), detecting and fighting inflammation (Staples et al., 2006; Amato et al., 2010; Stelzner et al., 2016b), and using molecular factories to create drugs (Benenson et al., 2004; Anderson et al., 2006).

Even intracellular nanosurgery has been proposed by some researchers (Singhal et al., 2010).

Furthermore, it is often suggested to use nanonetworks to support the body's own immune system, e.g. to support self-healing or to provide temporary relief (Freitas, 2005; Akyildiz et al., 2008; Gu et al., 2013).

Another popular scenario is also the monitoring of health parameters (Akyildiz et al., 2011). The macroscopic capacities are often not sufficient to generate plausible data (Freitas, 2005) over a longer period of time. For effective long-term monitoring of health parameters, patients often have to be monitored in a hospital, which is rarely an option.

Many proposed technologies share the property that a high degree of precision and location awareness is required for actuators to do their job at the right time and place (Freitas, 2005).

Most of the scenarios presented require a minimum amount of computing power so that sensible, conditional decisions can be made (Benenson et al., 2004). Due to the limited computing power per nanodevice and the merging into nanonetworks, it can be assumed that nanocommunication and thus routing and addressing will also be important. For example, point-to-point protocols or hop-count approaches have been proposed in order to be able to work collaboratively despite the low computing power per device (Tsioliaridou et al., 2015; Büther et al., 2018).

More complex procedures require significantly higher computing power of the individual nanodevices (Yu et al., 2015; Atakan et al., 2012). Checksum procedures such as *cyclic redundancy check* (CRC) or encryption procedures such as the *Advanced Encryption Standard* (AES) only require simple operations such as XOR, but there are so many of them that it is not clear whether the computing power of individual nanodevices is sufficient (Peterson and Brown, 1961; Rijmen and Daemen, 2001).

Unfortunately, most publications provide insufficient details to determine the complexity or computational requirements of individual nanodevices or nanonetworks. Although formulas or abstract problems are often mentioned, the complexity or mathematical aspects are left out. Furthermore, the resulting consequences for the design of nanodevices are almost never discussed. This leads to the question of how complex nanodevices must be in order to be able to process the scenarios presented in a meaningful way.

5.4.1 Extraction of Mathematical Problems

Mathematical problems and operations or functions are extracted from the applications presented. These mathematical problems correspond to the requirements placed on nanodevices.

Table 5.1 shows and defines all extracted mathematical operations. In the course of this section, the individual problems are explained and placed in the context of nanonetworks.

5.4.1.1 Arithmetic and Logical Operators

Most of the scenarios presented require at least the ability to arithmetically compute Booleans and integers. This includes the operations addition ADD, subtraction SUB, and multiplication MULT as well as the operation IT-ADD on which multiplication is based. Furthermore, basic Boolean functions like AND, OR, and XOR are needed to make simple decisions that have a Boolean output. Together, arithmetic and Boolean operations form the basis for value aggregation, most routing methods, and complex computations that are composed of basic operations.

In addition, comparators like EQ are required. An example is the signum function SIGN, which determines the sign of a number. Similarly, EVEN/ODD determines whether a number is even or odd. Likewise, a comparison is often made as to whether a threshold value THRES has been exceeded or which value is the minimum MIN, the maximum MAX, or the most frequent element MAJOR of a list. These operations enable complex, conditional behavior and provide the basis for robust, fault-tolerant systems.

More complex scenarios require more complicated operations such as division DIV or determining the modulus MOD of a number. It may also be necessary to find the average AVG of a list or to perform exponentiation EXP or iterated multiplication IT-MULT. The latter two methods are reduced to each other and are equivalent since exponentiation is merely an iterated multiplication.

Many operations become easier in binary representation. For example, dividing by 2 is equivalent to removing the last bit of a number. In Table 5.1, problems that receive a power of two as input are denoted by a 2 in the subscript.

5.4.1.2 Communication

Many of the scenarios shown relate to nanonetworks, in which communication is a necessary prerequisite. An easy example is the routing protocol from the work of Tsioliaridou et al. (2015). Only two-dimensional position information is computed there, which only requires ADD, GEQ, LEQ, and EQ. The hop-count routing from the work of Büther et al. (2018) requires even fewer requirements to get a message to a destination. The operations INC and a comparison GEQ, LEQ, or EQ are sufficient here.

Table 5.1 Formal definitions for problems of interest to nanonetworks.

Problem	Signature	Description
ADD	$\mathbb{Z} \times \mathbb{Z} \rightarrow \mathbb{Z}$	Integer addition
EQ	$\mathbb{Z} \times \mathbb{Z} \rightarrow \{0, 1\}$	Integer comparison
LEQ	$\mathbb{Z} \times \mathbb{Z} \rightarrow \{0, 1\}$	Integer $\leq$
GEQ	$\mathbb{Z} \times \mathbb{Z} \rightarrow \{0, 1\}$	Integer $\geq$
SUB	$\mathbb{Z} \times \mathbb{Z} \rightarrow \mathbb{Z}$	Integer subtraction
MULT	$\mathbb{Z} \times \mathbb{Z} \rightarrow \mathbb{Z}$	Integer multiplication
DIV	$\mathbb{Z} \times \mathbb{Z} \rightarrow \mathbb{Z}$	Integer division
SIGN	$\mathbb{Z} \rightarrow \{-1, 0, 1\}$	Signum division
INC	$\mathbb{Z} \rightarrow \mathbb{Z}$	Integer increment
AND	$\{0, 1\} \rightarrow \{0, 1\}$	Logical AND
OR	$\{0, 1\} \rightarrow \{0, 1\}$	Logical OR
XOR	$\{0, 1\} \rightarrow \{0, 1\}$	Logical XOR
ODD, EVEN	$\mathbb{Z} \rightarrow \{0, 1\}$	Odd/even test
DIV_2	$\mathbb{Z} \times \mathbb{B} \rightarrow \mathbb{Z}$	Division by power of 2
MOD_2	$\mathbb{Z} \times \mathbb{B} \rightarrow \mathbb{Z}$	Modulo with power of 2
INV_2	$\mathbb{Z} \rightarrow \mathbb{Z}$	Binary inverse
IT-MULT	$\mathbb{Z} \times \cdots \times \mathbb{Z} \rightarrow \mathbb{Z}$	Iterated multiplication
MIN, MAX	$\mathbb{Z} \times \cdots \times \mathbb{Z} \rightarrow \mathbb{Z}$	Min/max of inputs
MAJOR_2	$\mathbb{Z} \rightarrow \{0, 1\}$	Binary majority
EXP	$\mathbb{Z} \times \mathbb{Z} \rightarrow \mathbb{Z}$	Exponentiation
THRES_2	$\mathbb{Z} \times \mathbb{N} \rightarrow \{0, 1\}$	Checks last n Bits
IT-ADD	$\mathbb{Z} \times \cdots \times \mathbb{Z} \rightarrow \mathbb{Z}$	Iterated addition
REG_a	$\Sigma^* \rightarrow \{0, 1\}$	Pattern matching REG
PARITY_2	$\mathbb{Z} \rightarrow \{0, 1\}$	Binary parity check
MOD	$\mathbb{Z} \times \mathbb{Z} \rightarrow \mathbb{Z}$	Modulo
AVG	$\mathbb{Z} \times \cdots \times \mathbb{Z} \rightarrow \mathbb{Z}$	Average on n inputs
DFS, BFS	$(V, E) \times V \rightarrow \{0, 1\}$	Depth/breadth search
REACH	$(V, E) \times V \rightarrow \{0, 1\}$	Graph reachability
LOG_2	$\mathbb{Z} \times \mathbb{B} \rightarrow \mathbb{Z}$	Binary logarithm of integers
MEDIAN	$\mathbb{Z} \times \cdots \times \mathbb{Z} \rightarrow \mathbb{Z}$	Integer median

It is assumed that all natural numbers $\mathbb{N}$ and all integers $\mathbb{Z}$ are represented in binary – as $\{0, 1\}^k, k \in \mathbb{N}^+$. Binary powers of two are written as $\mathbb{B}$. A 2 in the subscript corresponds to a power of two as input and an a to a regular expression (Lau et al., 2017; Lau, 2020).
Source: Florian-Lennert A. Lau.

More complex forwarding mechanisms like in the work of Yu et al. (2015) require division, square roots, and logarithms and assume a large amount of environmental information.

In general, it can be assumed that nanodevices must be able to distinguish between different types of messages in order for a network to be formed or maintained. This may require pattern matching, conditional computation, and addressing mechanisms to be used.

5.4.1.3 Complex Operations

In addition to the comparatively simple scenarios already mentioned, *artificial neural networks* (ANNs) have also been proposed as a way of implementing functionalities (Staples et al., 2006). These represent an alternative computational model based on a large number of neurons.

Neural networks are typically represented as directed or undirected graphs, which is why graph algorithms are of interest. In particular, search algorithms such as depth-first search DFS or breadth-first search BFS can be used to find outliers in sensor data. The function REACH for determining reachability can be used for monitoring purposes, for example, to check whether a network is still connected or whether partitioning has occurred.

The use of neural networks at the nanoscale is still largely unexplored and is likely to pose unknown challenges. Nevertheless, the mathematical operations required for the use of neural networks are analyzed later in this chapter in order to enable a general statement about the feasibility.

5.4.1.4 Pattern Matching and Parity

Pattern matching using regular expressions REG_a checks whether a string input matches a pattern. The underlying operation is the comparison of numbers. In the simplest case, an input must be compared with a constant pattern. For example, this can be a simple command set for controlling a nanodevice.

A special case is the parity check PARITY. Here, it is checked whether a binary number consists exclusively of the binary digit "1." This can be used, for example, to check the integrity of a message.

5.4.1.5 Security

Since nanonetworks are often proposed for use in safety-critical environments, the issue of safety becomes more important. Three properties are of particular interest:

- *Integrity*: Ensuring the authenticity of messages on a network. It must be recognizable whether a message has been changed – e.g. by attackers or errors. This can happen, for example, using signatures, parities, or checksum procedures (Peterson and Brown, 1961).
- *Confidentiality*: Ensuring that the content of a message can only be read by the intended recipient. This is usually done using encryption methods such as AES (Rijmen and Daemen, 2001).
- *Authenticity*: Unambiguous determination of the identity of sender and recipient. This can also be ensured by encryption methods in the form of signatures.

The security procedures have in common that above all the operation XOR is used. The same applies to the proven secure one-time-pad method, in which a message is encrypted with a secret binary string using a bit-wise XOR operation (Bellovin, 2011). In addition, PARITY can be used, for example, to determine a deviation from a CRC control string of ones.

5.4.2 Classification in Complexity Classes

With the help of reductions, the mathematical operations can now be sorted into complexity classes. These put the various problems in relation to each other according to their difficulty.

Table 5.2 List of problems sorted by the complexity class.

	Problems from papers	Additional problems
AC^0 Device:	Add (Clote and Kranakis, 2002; Chiu et al., 2001)	Odd, Even
	Sub	Div_2
	Sign	Mod_2
	Inc	Inv
	And	Log_2
	Or	
	Xor	
NC^1 Device:	Mult (Clote and Kranakis, 2002; Chiu et al., 2001)	Min, Max
	Div (Chiu et al., 2001)	Parity (Clote and Kranakis, 2002)
	Exp (Immerman and Landau, 1995)	Reg (Vollmer, 1999)
	Major (Clote and Kranakis, 2002)	Mod
	Thres (Clote and Kranakis, 2002)	It-Mult (Immerman and Landau, 1995)
	It-Add (Clote and Kranakis, 2002; Chiu et al., 2001)	
	Avg	
	Geq	
	Leq	
	Eq	
L Device:	Label	Dfs (Cook and McKenzie, 1987)
	Log mem	Bfs (Cook and McKenzie, 1987)
		Reach (Wigderson, 1992; Savitch, 1970)
		Median
Miscellaneous:	*Addressing*	
	Routing	
	Broadcasting	
	Forwarding	

A device presented below is able to solve all the problems above it. Operations in italics are of unknown complexity (Lau et al., 2017; Lau, 2020).
Source: Florian-Lennert A. Lau.

Table 5.2 shows which problems can be solved by computational models that correspond to the complexity classes AC^0, NC^1, and L. As shown in Figure 5.1, there is an inclusion relationship between the individual classes. For example, class NC^1 also contains all problems of class AC^0 and can solve them as well.

The classification presented here does not have to be final. Only the currently known conditions are shown. Only a few lower bounds for complexity are known. Problems could consequently be assigned to a more specific class in the future.

Furthermore, the classes listed here refer to specific circuits and space-constrained Turing machines and cannot be transferred directly to self-assembly systems. Nevertheless, it has been shown that there are often simple and compact solutions in self-assembly systems for the problems of the lower classes, such as AC^0.

Based on the presented classification, short reductions are now shown, which classify the remaining problems without bibliographical references:

- Inv: The inverter gate is an elementary component of the AC^0 circuits and is therefore obviously computable.
- And, Or: And and Or are also elementary components of circuit classes.

- Xor: The exclusive Or can be determined by a logical Or via an AND link of all possible inputs that contain exactly one "1."
- Sign: Sign can be computed by AC^0 circuits by negating a predefined sign bit by an inverter gate.
- Sub: The subtraction problem is reduced to inverting the sign of a number, followed by addition.
- Inc: Inc is a special case of Add.
- Odd, Even: The problem is just checking whether the last bit of the input is a "1" or "0."
- Div_2: A binary number is divided by a power of two by removing the last k bits, where k describes the power.
- Mod_2: The modulo of a binary number and a power of two is computed by outputting the last k bits.
- Log_2: The logarithm of a binary number is the index of the first "1."
- Avg: The average is computed by adding a list cumulatively, followed by division by the number of items.
- Mod: n mod m can be computed by Sub(n,Mult(Div (n, m), m)).
- MEDIAN: The MEDIAN of a list can be determined by comparing all elements in pairs and counting the larger and smaller ones.
- Geq, Leq, Eq: Comparisons can be made using a Thres computation.

The complexity classes presented include further problems, which can also be computed by similarly powerful devices. Some of them are presented in Table 5.2 in the third column.

Furthermore, the following additional problems can be solved by L devices:

- REACHABILITY/PATH/RELATIONSHIP (Wigderson, 1992; Savitch, 1970).
- TREE ISOMORPHISM (Jenner et al., 2003).
- Bipatit test(Alvarez and Greenlaw, 2000).
- PARITY test (Allender et al., 2009).

5.4.2.1 Uncategorizable Problems

The problem of being able to store a unique identifier (label) and the use of a logarithmic amount of memory (log mem) are not problems in the classic sense. They represent statements about the storage capacity of nanodevices. Both entries are covered by logarithmically space-limited Turing machines.

The italic entries from Table 5.2 are of unknown complexity as their implementations vary widely. However, since they are essential for networks, they are listed in the table. For many of these algorithms, at least a linear amount of memory is required, which is unthinkable in a nanonetwork with millions of nanonodes.

- *Addressing*: In classical computing systems, at least a logarithmic amount of memory is required for addressing messages. If self-assembly systems are used, addressing can be implemented via DNA/glue sequences of fixed length. However, this only solves a simplified version of the problem since the length of the sequences is arbitrary but fixed.
- *Routing*: Routing often requires a priori knowledge about the underlying network structure, which has to be made accessible for nanodevices. The topic is being intensively researched in the nanonetwork community. Particularly good distributed methods for determining distances and paths (distributed Bellman–Ford algorithm) require $O(\sqrt{n})$ large routing tables (Gavoille et al., 2013).
- *Broadcasting and message forwarding*: Both operations can be part of routing mechanisms. This behavior is not represented by circuits or Turing machines.

5.4.3 Landau Notation

In 1976, D.E. Knuth coined the term "Landau notation"/"O notation" in theoretical computer science (Knuth, 1976). Among other things, it is used to describe the Limes behavior of functions (Landau, 1909). In the context of this book, the Landau notation is mainly used to analyze and describe the asymptotic behavior of self-assembly processes and other algorithms or problems.

Definition 5.4 Let f be a function over the real numbers, $x \in \mathbb{R}$, and $a \in \mathbb{R} \cup \{-\infty, +\infty\}$. For $f \in O(g)$, the following applies:

$$\limsup_{x \to a} \left| \frac{f(x)}{g(x)} \right| < \infty.$$

The Limes notation indicates that the growth of the function f is limited by the growth of the function g.

The Landau notation makes it possible to assign any function to a set of similarly rapidly growing functions. For example, a function $f \in O(1)$ does not exceed a constant value. $f \in O(\log n)$ increases by a constant value c when n doubles. $f \in O(n)$ grows about as fast as n. $f \in O(n^2)$ quadruples when n doubles.

In addition to the most well-known O notation, other variations exist. The O symbol is mainly used to describe the worst-case behavior of functions and algorithms. However, due to the expected number of participants in nanonetworks, it might be more useful to analyze the average performance of chosen algorithms. Due to the law of large numbers, the worst-case behavior becomes arbitrarily unlikely, and average cases are more realistic. For those purposes, the Θ notation is used. Similarly to the O notation, a function $f \in \Theta(g)$ means that the function f grows exactly as fast as a function g.

5.5 Computational Models for the Nanoscale

In theoretical computer science, we use computational models to find out the peculiarities, weaknesses, and possible strengths of certain models of computation. This approach is of special interest when physical experiments are not yet possible due to a lack of suitable tools, for example, at the nanoscale. As we have only limited information about the realistic circumstances at the nanoscale, there are many potential candidates for the job.

We now present the computational models: Turing machine, ANNs, circuits, chemical reaction networks, quantum dot cellular automata, and self-assembly systems. These are explained in sufficient depth to justify why they are explored in more detail or omitted from this book. The key criterion for the usability of a computational model in the context of the work is physical feasibility. The search for new computational models at the nanoscale is of interest as further miniaturization based on CMOS technology is believed to be limited. Conventional transistor technologies are already about 7 nm small (Windeck, 2019). Furthermore, a variety of other factors will likely make the use of well-established technologies unlikely, as shown before.

There are many possible models of computation that mathematicians and theoretical computer scientists have come up with over the years. As a computation is technically just a structured manipulation of state, countless approaches can work, but not all of them are equally powerful. Some models like regular expressions or different kinds of automata are not capable of solving all computational problems (Turing, 1937).

In 1937, famous computer scientists Alan Turing and Alonzo Church introduced the *Turing machine* and the *Lambda calculus*. Both believed that their respective models were computationally universal and thus capable of doing anything that a skilled mathematician could do with a pencil on a sheet of paper. Later, Alan Turing proved that both models were indeed equally powerful. It is assumed that from a certain range of functions, all computational models are of identical power. They only differ from each other in subtleties. This idea is now known as the "Church–Turing hypothesis," which is yet unproven but strongly believed to be true.

That said, with harsh resource constraints, such "subtleties" might very well be relevant. As an example, in the "real world," it *does* make a difference if there is a big constant factor in the Landau notation. Practically, $O(30n)$ and $O(n)$ grow roughly equally fast given infinite time, but practically the difference between waiting for 1–30 minutes for a download is very relevant. Thus, we analyze the most proposed models in terms of their advantages and weaknesses.

5.5.1 Nature-Inspired vs. Artificial Models

There are two overarching paradigms when it comes to computational models. There are nature-inspired models and artificial models. Examples of nature-inspired models are ANNs or evolutionary algorithms. Examples of artificial computational models are Turing machines or classical electric circuits.

Nature-inspired models have the advantage that they are often able to find solutions to problems that artificial models cannot find. Many of the nature-inspired approaches are able to traverse an enormous search space and still find a fitting solution to a problem. That solution might not be optimal, but at least it is something. Such methods work surprisingly well for image or sound data. However, the results and the way leading to the results are rarely explainable. In addition to that, some nature-inspired models have the advantage that they work data-driven. Instead of specifying the necessary steps a machine has to perform, the "programmer" instead supplies example data that, e.g., a neural network generalizes an algorithm from. This has the advantage that the often difficult task of finding algorithms can be reduced to the task of finding fitting data sets.

Artificial computational models on the other hand are usually very structured. An algorithm designer specifies each step a computer must perform, and the possible results are well defined and can be understood. They usually work well for complex problems where models like ANN usually fail. Both paradigms have advantages and disadvantages, and we consider candidates from both areas in the following analysis of computational models.

5.5.2 The Turing Machine

We start with the most well-known model of computation in theoretical computer science: the Turing machine. The Turing machine is often used to formally define what a computation even is. Compared to other models, it relatively closely resembles the behavior of a real computer and is easy to understand by most people.

While there are several advantages concerning this model, it is of merely theoretical interest. There are no known physical components at the nanoscale that could allow for a physical implementation of a Turing machine. Nonetheless, the model is still very useful as a point of reference that can be used to estimate how well other models fare in comparison. In many cases, new models of computation are categorized concerning their strengths and weaknesses by showing that the new model is at least as powerful as a Turing machine. Thus, an often tedious "bootstrapping process" can be omitted using reductions.

Formally, a Turing machine can be defined as follows (note that there are many possible equivalent definitions):

Definition 5.5 A *k-tape deterministic Turing machine* (k-DTM) M consists of:

- a *finite* set of states Q,
- an *initial state* $q_0 \in Q$,
- a set $Q_a \subseteq Q$ of *terminal* states,
- a *tape alphabet* Γ with $\square \in \Gamma$,
- an *input tape alphabet* $\Sigma \subseteq \Gamma$ with $\square \notin \Gamma$,
- k *work tapes* (of which one can be designated as an "output tape"), and
- a *program* δ for controlling k read/write heads.

At any given time, a Turing machine is in a state that is also called *configuration*. Given the content of all tapes and the positions of the read/write heads, the machine performs the steps that are determined by the program δ in the current state q.

Definition 5.6 The configuration of a Turing machine M is a tuple of:

- the current state $q \in Q$,
- the content/state of all k tapes and the input tape, and
- the current read/write head positions.

Formally, the configuration of a Turing machine is an element of $Q \times [\mathbb{Z} \to \Gamma]^k \times \mathbb{Z}^k$. $[\mathbb{Z} \to \Gamma]$ is the set of all functions $t : \mathbb{Z} \to \Gamma$.

A configuration $C = (q, t_1, \dots, t_k, h_1, \dots, h_k)$ has a next configuration $C' = (q', t'_1, \dots, t'_k, h'_1, \dots, h'_k)$ denoted as $C \Vdash_M C'$, based on the following definition.

Definition 5.7 Let σ_i be the symbol at position h_i of tape t_i. Let $\delta(q, \sigma_1, \dots, \sigma_k) = (q', \sigma'_1, \dots, \sigma'_k, m_1, \dots, m_k)$. It follows that

$$t'_i(j) = \begin{cases} t_i(j), & \text{if } j \neq h_i \\ \sigma'_i, & \text{otherwise.} \end{cases}$$

$$h'_i = h_i + \begin{cases} -1, & \text{if } m_i = l \\ 1, & \text{if } m_i = r \\ 0, & \text{if } m_i = n \end{cases}$$

In simple terms, the machine can read a symbol from the input tape or of the working tape and manipulate the content of the working tape as well as write onto the output tape. The possible manipulations and movements of the read/write heads are determined by the program δ, which defines the behavior of the machine for any given state. Simple programs can be expressed as deterministic, finite automate, while more complex programs are usually expressed as tables that fully specify the machine behavior for any possible input in each state. Once the machine reaches a terminal or accepting state, the computation is finished and the current state of the output tape can be defined as the result of the computation.

5.5.3 Circuit-Based Computers

Unlike Turning machines, circuits can be physically implemented but are usually hard to define mathematically. Circuits always have a fixed size which necessitates the definition of a new circuit for each possible input size. While the problem can be avoided by defining entire classes of circuit families, the model is nonetheless "unwieldy."

Circuits are the de facto standard in physically implementable computational models. These are usually manufactured on the basis of *complementary metal oxide semiconductors* (CMOSs). Many computational models in research share properties with circuits or fully emulate them.

To understand how nanodevices perform computations, it can be helpful to be familiar with the circuit model. An example is shown in Figure 4.18. The circuit receives certain information in the form of bits (high or low voltage), transforms them in a suitable way, and finally returns a result that can represent a simple yes/no decision. Based on that, a nanodevice can decide on a (joint) course of action.

In the context of this book, we will not examine classical circuits in much detail as they already seem to be approaching the end of their miniaturization possibilities (Haron and Hamdioui, 2008). Furthermore, they are likely the most well-known model of computation and have a variety of possible problems we already analyzed before.

5.5.4 Artificial Neural Networks

ANNs received considerable amounts of public interest in recent years and are suggested for a variety of applications. While the model has already been introduced in the 1980s, it never took off as the computing power of available devices was never sufficient to train those networks properly. The nanonetworking area is no exception, and the use of ANNs is frequently proposed for solving computational tasks at the nanoscale. In this context, ANNs serve two distinct purposes. First, researchers propose using pre-trained ANNs directly on nanodevices. Second, researchers suggest using ANNs for reinforcement learning to identify programs for nanodevices that can be installed on arbitrary hardware. In this case, the result of an ANN might be an IF/ELSE tree that a nanodevice might use for decision making.

We evaluate both ideas in this section and explain the most important definition of ANNs to give the reader a good intuition of the problems and potentials. While most modern ANNs use the same system architecture of, for example, deep learning, we nonetheless present the more general definitions as it is unclear if hardware at the nanoscale is powerful enough to simulate a multi-layer ANN. In the following, we use the definitions from the work of Kriesel (2007).

Definition 5.8 An *artificial neural network* (ANN) is a tuple (N, V, w) with two sets N, V and a function w, where N denotes the set of neurons and V is a set $\{(i, j)|i, j \in \mathbb{N}\}$ whose elements indicate connections from neuron i to neuron j. The function $w : V \rightarrow \mathbb{R}$ defines the weights, where $w((i, j))$ describes the weight of the connection from neuron i to neuron j written as $w_{i,j}$.

Depending on the definition, it is either undefined or 0 for connections that do not exist in the network.

We continue with the definition of the propagation function, which is computed for each neuron. The propagation function takes the outputs o_i of all neurons i that are connected to neuron j. All the associated weights $w_{i,j}$ are processed by the function.

Definition 5.9 Let $I = \{i_1, i_2, \dots, i_n\}$ be a set of neurons for which $\forall z \in \{1, \dots, n\} : \exists w_{i_z,j}$ holds. The way in which the input net_j of j is processed is defined by the *propagation function* f_{prop},

$$\text{net}_j = f_{\text{prop}}\left(o_{i_1}, \dots, o_{i_n}, w_{i_1,j}, \dots, w_{i_n,j}\right).$$

The weighted sum is the most popular: You simply multiply the output of each i by $w_{i,j}$, and sum up the results,

$$\text{net}_j = \sum_{i \in I} (o_i \cdot w_{i,j}).$$

Furthermore, each neuron has an *activation function*. Per default, a signal is not necessarily transmitted through a neuron. A certain threshold/condition has to be met for signal propagation. Oftentimes, a so-called *rectifier/reLU function* is used that only allows a non-negative signal to pass through.

Definition 5.10 Let j be a neuron. The *activation function* is defined as

$$a_j(t) = f_{\text{act}}\left(\text{net}_j(t), a_j(t-1), \Theta_j\right),$$

and combines the input net_j and the old activation state $a_j(t-1)$ into a new activation state $a_j(t)$, based on the threshold Θ.

In most cases, no additional computation is necessary for each neuron. Theoretically, you could define an additional transformation step on the output, which is usually omitted.

Based on these functions, an ANN can be *trained*. In that process, an often large number of examples with known truth values are supplied as inputs to the network. The network computes how "different" it would classify the input compared to the predetermined truth value and uses a *backward propagation function* to modify the network in a way that leads to a successful classification. Repeating the process often enough usually leads to reasonably good predictions/accuracy levels on unknown but similar input data.

In a sense, the network "finds" an algorithm for you, and all you have to provide is a suitable data set. This method is inviting for many people as it oftentimes appears much simpler to collect data than to write a complex algorithm yourself.

ANNs are extremely popular nowadays and are proposed for a variety of possible applications. The appeal of the computational model is quite clear as the "programmer" no longer has to invent algorithms and prove their correctness. Instead, the objective is to collect a suitable data set, and the algorithm results from training the ANN. Thus, a considerable part of the mental load is removed.

Furthermore, trained ANNs require little memory and are extremely fast as the computation is akin to propagating a signal through a circuit. In most cases and for most popular ANN architectures, no memory is used. While this might be a limitation for more complex problems, ANNs are extremely well suited for "natural inputs" like images or sounds.

However, ANNs are currently simulated on circuit-based hardware and require considerable overhead for the emulation. While the human brain uses real neurons, those are not typically available as building blocks for ANNs. Furthermore, ANNs are typically not Turning complete and thus not well suited for solving more complex problems that rely on storing intermediate results (Blum and Rivest, 1992).

Specifically for medical nanonetworks, another set of problems is relevant. In general, people's life and health depend on the proper functioning of a nanonetwork that might be either programmed by an ANN or directly uses pre-trained ANNs on the nanodevices. Thus, it must be 100% clear why and when a network decides on a course of action. Sadly, ANNs are neither explainable nor predictable in their current state. There is always a degree of uncertainty on how the ANN might decide given unknown inputs.

Furthermore, ANNs usually only find approximate solutions for a problem, while exact solutions might be necessary for many use cases, especially given severe resource constraints.

Finally, there are political problems concerning the medical use of ANNs. Most of the time, nanonetworks are assumed to be autonomous systems with little possible interaction after deployment. As a result, a nanonetwork must decide on a course of action by itself. However, medical regulations require that approximate systems like ANN only offer suggestions to medical professionals who then make the final call.

As of right now, ANN-based solutions cannot meet the named requirements and it is unclear if they can be met in the near future. As a result, it seems unlikely that ANNs are a suitable candidate for computation at the nanoscale. This is partially due to shared problems with classical CMOS-based hardware that ANNs are currently designed for.

5.5.5 Quantum-Dot Cellular Automata

This section introduces the fundamentals of quantum-dot cellular automata. Since quantum-dot cellular automata are a form of cellular automaton, they are defined first and quantum-dot cellular automata are derived from them.

Quantum-dot cellular automata (QCAs) use quantum dot cells as elementary components. A quantum dot cell can assume exactly one of the two possible states. The states are determined via the positions of particles in a defined area. The first quantum dot cell was physically realized in 1997 (Amlani et al., 1997).

The model appears advantageous for application at the nanoscale since the size of a quantum dot cell is only a few nanometers. In addition, quantum dot cells use precisely those effects that could prevent CMOS from further miniaturization. For example, if wires are too close together, electrons could tunnel between them.

Consequently, QCAs have a greater potential for future developments (Hänninen and Takala, 2008). QCAs are considered in more detail because at first glance they represent a sensible alternative to proven technologies.

This section presents mathematical definitions for cellular automata. These are based on the terminology (Wolfram, 1984) introduced by Wolfram. The notation used goes back to the work of Buss Buss (2017).

Definition 5.11 The state of a cellular automaton at time $t \in \mathbb{N}$ is an infinitely large, d-dimensional *lattice* L^t with $d \in \mathbb{N}^+$. A cell L_v in the lattice is indexed by a vector $v \in \mathbb{Z}^d$. Each cell L_v can assume any state L_v^t of a finite set of states at time t. The state alphabet is called S. A lattice is an element of $L^t \in S^{(\mathbb{Z}^d)}$.

Wolfram has been instrumental in studying one- and two-dimensional cellular automata (Wolfram, 1984; Packard and Wolfram, 1985). However, in order to be able to place wires on top of each other in all situations, a third dimension is required.

Definition 5.12 Let $v \in \mathbb{Z}^d$. Let N be a function $N : \mathbb{Z}^d \to (\mathbb{Z}^d)^n$ such that $N(v)$ returns a sequence of cell indices of length n. $N(v)$ is called the *neighborhood* of a cell L_v.

This is a very abstract definition. For one-dimensional lattices with $v \in \mathbb{Z}$, Wolfram has defined the neighborhood of a cell L_v by $N(v) = [v - r, v + r]$, where $r \in \mathbb{N}$ is an arbitrary radius.

For two-dimensional grids, the most common neighborhood functions are those that include cardinal neighbors and those that include all diagonal neighbors as well. Figure 5.3 (a) and (b) show the respective examples.

It is easy to see that the definition can also be extended to three dimensions. In the first case, the extensions also take into account the cells in the third dimension above and below. In the second case, the other 18 neighbors are considered.

Definition 5.13 Let $N(v)$ be the indices of the neighbors of L_v. At each discrete point in time t, the state of a cell L_v is updated. The new state depends solely on the neighbors of L_v at time t,

$$L_v^{t+1} = \psi\left(L^t_{N(v)_1}, L^t_{N(v)_2}, \dots, L^t_{N(v)_n}\right),$$

for a function $\psi : S^n \to S$.

Since $N(v)$ describes a sequence of indices, the cell's next state depends not only on the values of the neighboring cells but also on their indices. Each neighboring cell can therefore be weighted differently.

Definition 5.14 A *local transition function* ψ enables a global transition function Ψ. Let L^t be the state of the automaton at time t. The lattice state is defined as follows:

$$L^{t+1} = \Psi(L^t),$$

where the state L_v^{t+1} of each cell L_v in L^{t+1} is determined by applying the function ψ.

The state L^t of a cellular automaton L at time $t \in \mathbb{N}^+$ is completely determined by the initial state L^0.

Definition 5.15 Let $d, n \in \mathbb{N}^+$. Let S be a finite set of states. Let $L^0 \in S^{(\mathbb{Z}^d)}$ be an initial state. Let N be a function $N : \mathbb{Z}^d \to (\mathbb{Z}^d)^n$ and ψ be a function $\psi : S^n \to S$. A *cellular automaton* is a tuple $L = (d, n, S, L^0, N, \psi)$.

We can now modify the stated definitions to express QCAs as a special case of cellular automata. Intuitively, a quantum dot cell can be understood as a cell of a cellular automaton. However, the

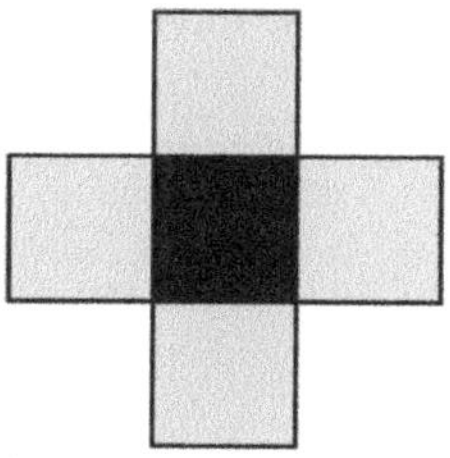

(a)

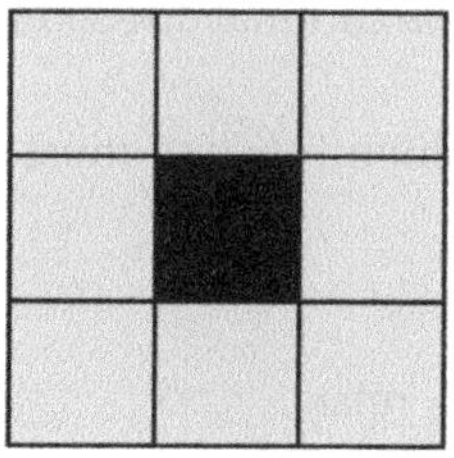

(b)

Figure 5.3 a) A cell that is influenced by 4 neighbors (b) A cell that is influenced by 8 neighbors. The state of the cell in the middle is computed considering all adjacent cells. *Source*: Florian-Lennert A. Lau.

model must be expanded to include a clock generator; otherwise, the signal flow can hardly be controlled. At any point in time, only a subset of all cells should be considered. Figure 5.4 explains how a quantum dot cell works.

Definition 5.16 A *clock* coordinates the state changes of a quantum dot cell and is described by the following phases, which are repeated indefinitely in the specified order:

- In the relax phase, the signal is strong and the tunnel junctions are open. Electrons can move completely freely.
- In the switch phase, the signal is lowered and the tunnel junctions are closed. In this phase, the actual computations take place.
- In the hold phase, the signal is weak and the tunnel junctions are closed – electrons are therefore locked to their positions. In this phase, a cell assumes one of the possible states.
- In the release phase, the signal is amplified and the tunnel junctions are opened. Electrons can start to move freely.

In the hold phase, a cell is firmly anchored in one state. One of the states "0" or "1" is assumed. In addition, a neutral state is required for technical reasons. This is called ϵ. We now use the general definitions of cellular automata to derive the necessary modifications for QCAs.

- *Lattice*: The dimension of the lattice is $d = 3$ because the third level is needed for wire transitions. Furthermore, the number of quantum dot cells is finite. A maximum of three shifts are permitted. All L_v^t with $v \notin \mathbb{Z} \times \mathbb{Z} \times \{0, 1, 2\}$ are empty. A lattice with quantum dot cells that computes a Boolean function is also called *chip*.
- *Alphabet*: The alphabet of QCA consists of $(p, t) \in S = \{-1, 0, 1\} \times C$ with $C = \{\text{relax, switch, hold, release}, \epsilon\}$. p describes the polarization of a cell. "0" corresponds to an unpolarized or no cell. The second part of the Cartesian product describes the clock C. A clock zone ϵ means that changes will never occur. The clock phase of a cell also describes the zone.
- *Neighbors*: The neighborhood $N(v)$ of a cell with index v can be chosen freely. The larger it is, the more realistic the QCA is. Neighbors that are further away have a much smaller influence. $N(v)$ should consist of at least the following elements: $\mathcal{N}(v)$ is the set of all elements $N(v)$,

$$\begin{aligned} &\mathcal{N}((v_1, v_2, v_3)) \supseteq \\ &\{(v_1+1, v_2, v_3), (v_1-1, v_2, v_3), (v_1, v_2+1, v_3), (v_1, v_2-1, v_3), \\ &(v_1+1, v_2+1, v_3), (v_1+1, v_2-1, v_3), (v_1-1, v_2+1, v_3), \\ &(v_1-1, v_2-1, v_3), (v_1, v_2, v_3+1), (v_1, v_2, v_3-1)\}. \end{aligned}$$

  The first four are the neighbors in the cardinal direction. Then comes the diagonals, and finally, the two are cells above and below v.
- *State transition function*: The transition function must take the clock zones into account. The following function is used for this:

$$\text{next}: C \to C$$

$$\text{next}(t) = \begin{cases} \text{switch} & \text{if } t = \text{relax} \\ \text{hold} & \text{if } t = \text{switch} \\ \text{release} & \text{if } t = \text{hold} \\ \text{relax} & \text{if } t = \text{release} \\ \epsilon & \text{else.} \end{cases}$$

A cell in the switch phase is just fully polarized. Since the neighborhood is unclear, ψ can only be specified depending on the choice of neighborhood. L_v^{t+1} is defined as follows, where $L_v^t = (p, c)$:

$$L_v^{t+1} = \begin{cases} (p, c) & \text{if } c = \epsilon \\ (\psi(N(v), t), \text{hold}) & \text{if } c = \text{switch} \\ (0, \text{next}(c)) & \text{else.} \end{cases}$$

The behavior of this clocking is identical to that of Braun (2017).

A L_v cell can be imagined, as shown in Figure 5.4. Two electrons are trapped in a lattice. There is a quantum dot in each of the four corners. These are connected via tunnel crossings. Due to electrical repulsion reactions, electrons are always in opposite corners, allowing for two possible logic states. These are shown in Figure 5.5. Electrons repel each other because of their negative charge. This effect also occurs across cells, which is why one can compute with quantum dot cells. It should be noted that in reality, the cells in the figures can have significantly different distances.

Cleverly placed, logical gates can be formed from the cells. The elementary components here are majority gates and inverters. Together with constant cells, a Turing-complete logic can be formed. Figure 5.6 shows a majority gate. The electrons try to switch to the state that requires the least energy. This leads to the behavior of a majority function on three bits.

Figure 5.7 shows an inverter. The same principle applies to the majority gate. The energetically most favorable state is the inverse of the input.

A transition function coordinates the way in which cells change state. A possible transition function ψ based on the previous definitions can look like this: Two quantum dots of a cell are 18 nm

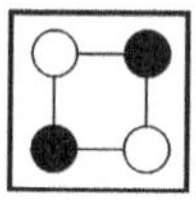

Figure 5.4 Quantum dot cell with tunnel junctions shown as lines (Lau, 2020). *Source*: Florian-Lennert A. Lau.

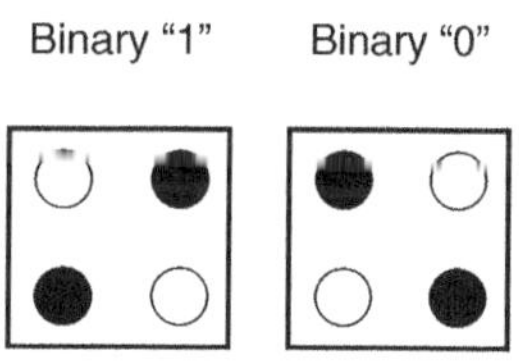

Figure 5.5 Two quantum dot cells in different states. These are interpreted as logical "1" and "0" (Lau, 2020). *Source*: Florian-Lennert A. Lau.

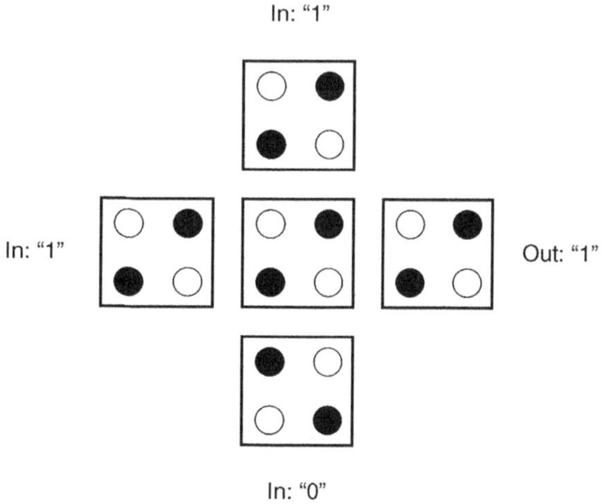

Figure 5.6 Simple quantum cell majority gate. Due to the repulsion reaction between the cells, the cell in the middle tries to accept the maximum of the adjacent cells (Lau, 2020). *Source*: Florian-Lennert A. Lau.

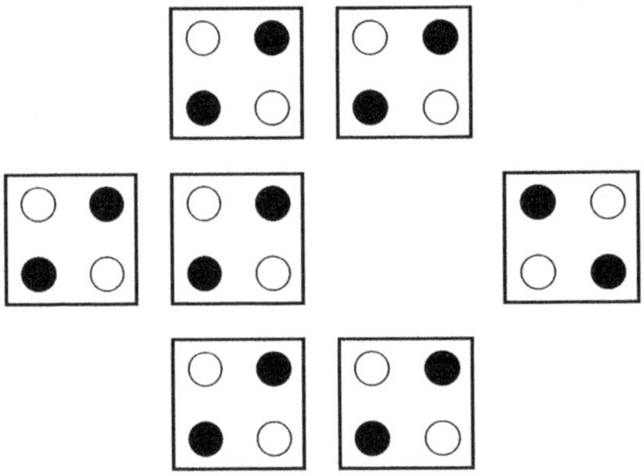

Figure 5.7 Inverter gate is shown. Due to repulsion reactions between the electrons, the output cell takes on the inverse value of the input (Lau, 2020). *Source*: Florian-Lennert A. Lau.

apart, the distance between two cells is 2 nm, and cells in the third dimension are 11.5 nm apart. The QCA Designer software uses the same values (Walus Group, 2009).

The influence of diagonal cells is less than the influence of a cell in the cardinal directions. A cell in the third dimension has a higher influence than a cell in the cardinal directions. The different influences are based on the distances and the number of electrons from each other. The repelling forces decrease with the square of the distance.

Definition 5.17 Let $\mathcal{N}(v)_d$ be a set of indices of diagonal neighbors, $\mathcal{N}(v)_c$ be the set of indices of neighbors in cardinal directions, and $\mathcal{N}(v)_z$ be the set of indices of the vertical neighbors of a cell L_v. A possible transition function ψ can look like the following:

$$\Psi(N(v), t) = 1 \cdot \left| \sum_{i \in \mathcal{N}(v)_c} L_i^t \right| - 0.2 \cdot \left(\sum_{i \in \mathcal{N}(v)_d} L_i^t \right) - 1.35 \cdot \left| \sum_{i \in \mathcal{N}(v)_z} L_i^t \right|.$$

The function guarantees that only "0" and "1" are allowed states in the switch phase,

$$\psi(N(v), t) = \begin{cases} 0 & \text{if } \Psi(N(v), t) < 0 \\ 1 & \text{else.} \end{cases}$$

Although QCAs represent a promising computational model at the nanoscale, there are still many unsolved problems. Although individual quantum dot cells were produced in the laboratory, e.g., their precise placement is a hurdle. Large quantities of quantum dot cells cannot currently be manufactured on an industrial scale.

Furthermore, it is necessary to use a suitable synchronization mechanism such as clocking, which is an unsolved problem in practice. Otherwise, the results of computations cannot be used meaningfully. A grid of quantum dot cells can appear chaotic without clocking.

Additionally, there is a need to amplify signals after a period of time, which requires *repeater* or similar technology. With increasing distance, signals become too weak to cause polarization of neighboring cells.

Another problem is the implementation of memory. Although there are solutions, they are not fully developed (Vankamamidi et al., 2005).

Finally, the placement of quantum dot cells in the third dimension is challenging. It is necessary to place cells above each other in order to be able to realize wire transitions. Without said transitions, for example, gates with numerous inputs cannot be implemented or can only be implemented to a limited extent.

For these reasons, QCAs are not analyzed in more detail in the remainder of this book. The focus of this book is explicitly on currently available, "feasible" technologies.

5.5.6 Chemical Reaction Networks

Chemical Reaction Networks (CRNs) model the behavior of well-mixed solutions (Brijder, 2019). A CRN consists of a finite set of chemical reactions and can be used to perform computations. In the work of Liekens and Fernando (2007), it was shown that some CRNs are Turing-complete, but many abstract components have been used. Those chemicals have no real-world equivalent. The work of Angeli (2009) is recommended for a detailed introduction to the topic. The following terminology is based on the content presented there.

Mathematically, a CRN consists of a finite set of *reactions* $\mathcal{R}_i$ with $i \in \mathbb{N}^+$ and a set of chemical building blocks S_j with $j \in \mathbb{N}^+$, also called *species*.

Definition 5.18 A *reaction* is specified as follows:

$$\mathcal{R}_i : \sum_{j \in S} \alpha_{ij} S_j \mapsto \sum_{j \in S} \beta_{ij} S_j.$$

The components on the left of the reaction are called *reactants*, and $\alpha_{ij}, \beta_{ij} \in \mathbb{N}$. The components on the right side of the reaction are called *products*. A reaction can only take place in the direction of the arrow. Both reactants and products may be $\emptyset$.

The following example shows a possible CRN. Given the six species E, F, P, S, ES, EP and the reactions,

$$E + S \leftrightarrow ES \rightarrow E + P,$$
$$F + P \leftrightarrow FP \rightarrow F + S.$$

The two equations encode a total of six reactions. "$\leftrightarrow$" means that the inverse reaction also exists for a reaction. When there are multiple arrows in an equation, multiple reactions are represented at once by a transitive relationship.

A corresponding Petri net can now be created from the components and reactions mentioned. A node is created for each species. According to the reactions, edges can now be created and provided with transactions, which correspond to the coefficients α_{ij}, β_{ij}.

Due to the transaction rules of the Petri net, conditional behavior can be implemented, which in turn allows arbitrary computations.

The actual computation can then be carried out in a test tube. However, the process is lengthy and reading out results involves significant macroscopic effort. It is also difficult to imagine that CRNs can be used in an environment like the human body since a large number of reacting substances are likely to either interfere with natural processes in the body or be disturbed by them.

In addition, the introduction of abstract species in real implementations poses a problem. These have no real counterparts, and it is not clear that there is always a real replacement. Since the focus of the work is on the timely feasibility of the models presented, CRNs are not examined in more detail.

5.5.7 Genetic Circuits

Genetic circuits are yet another way of performing computations at a very small scale (Kobayashi et al., 2004). For detailed information about a variety of different genetic/biological circuit models, the interested reader is encouraged to have a look at the work of Pasotti et al. (2014).

As the name suggests, the methods emulate the behavior of electronic circuits using biological components. Parts of human cells are modified to allow for the conditional production of specific proteins. Both modifying existing biological pathways and introducing new, artificial biological pathways are possible options.

The entire mechanism is based on the observation that certain regulatory mechanisms in cells are encoded in the DNA. Thus, certain DNA sequences encode logical building blocks like promoters, regressors, oscillators, and many more. These can, for example, incentivize or disincentive certain production pathways in a cell.

Genetic circuits are mainly proposed for the production of biochemicals and biomaterials. While the procedure is capable of computing mathematical functions, the capability of a single cell is limited. The often-used *E. coli* bacteria has a diameter of roughly 1.5 μm and a length of 2–6 μm. Thus, it exceeds the size limitation of the nanoscale by far. As a result, a single cell is often limited to a single-digit number of logical gates. If more gates are used, the natural function of the cell cannot be maintained and the organism dies. Some of these effects might be mitigated by "chaining" multiple cells and introducing something like a distributed computation among many cells at the micrometer level.

While all of this sounds promising and has many potential use cases, the differences and limitations concerning nanonetworks seem rather big. The components are rather big, the process is slow, and the intended use case is a different one. Based on these observations, genetic circuits are not considered potential candidates for realizing the computational components of nanodevices.

5.5.8 Quantum Computing

Much of the following information is based on the work of Homeister (2008). Next to ANNs, quantum computers are another popular area of research attracting considerable public interest. Several big companies like D-Wave are currently refining quantum technologies, and the company released the first 128-bit quantum computer in 2011. Universities all around the globe, especially in the United States and China, are also actively researching new quantum computers.

While the history of quantum computers already begins as early as 1858 with the birth of Max Planck, we will not analyze all of it. During his time, one of the biggest open questions in Physics was finding a formula that describes the relationship between the temperature of a heated black body and the frequency of the emitted light. Max Planck discovered such a formula in 1900, but it contained a constant h that limited the energy of photons. According to his formula, the energy could only assume certain discreet values called *quantums*. All values in between seemed impossible according to the newly discovered formula. Today, this constant h is known as the *Planck constant*.

While Planck initially saw his constant as a "trick," it stayed and revolutionized Physics. In 1918, Planck received a Nobel prize for the initial formula. In 1905, Albert Einstein used that very formula to explain the so-called photoelectric effect, for which he later received his Nobel Prize in 1923.

Based on these discoveries, we can now define the basic building blocks of quantum computers. In general, quantum computers work with *Q-bits* instead of classical bits. A classical bit can only assume the values "1" and "0." A Q-bit can be in "both states" at once, which is called a *superposition*. A classical computer can represent one of the 2^n possible numbers with n bits. A quantum computer can represent all those numbers *at once* with the same amount of Q-bits.

Based on these very different building blocks, new algorithms can be designed that exceed the capabilities of classical computers. As an example, Lov Grover discovered in 1996 that quantum computers can execute faster database search algorithms (Grover, 1996). If the database is not structured in any way, a classical computer would require $O(n)$ steps to find the desired element, while a quantum computer only requires $O(\sqrt{n})$ steps.

While quantum computers can perform many computations in a more efficient way than regular computers, there are also problems. A quantum computer cannot copy the content of a Q-bit, and every performed operation must be reversible as a reversible operation technically does not produce environmental heat as a byproduct. Furthermore, very specific environmental conditions are typically necessary to realize quantum computers. A Q-bit must be isolated from its environment in an often complicated process for it to maintain its quantum properties. Even the simple *possibility to know* the state of a Q-bit already "destroys" a superposition, let alone a direct measurement. Such isolation is likely very hard to implement at the nanoscale, where the precise placement of components is an unsolved problem.

Quantum computers are an extremely interesting topic in themselves, but even at the macroscale, they are still in a very early stage. As a result, we will not analyze quantum computers in more detail as the expected near-term feasibility in (especially medical) nanonetworks is not given.

5.6 Self-assembly Systems

We now go through all the necessary aspects that are needed to create DNA-based nanonetworks are defined.

The idea to use DNA tiles for construction, computation, and communication dates back to the work of Lau et al. (2019).

First, truth values and a communication mechanism based on tiles are explained. This is followed by an explanation of how receptors and ligands can be constructed using DNA. The individual components are then combined into a DNA-based nanonetwork and formally defined. A reference architecture for DNA-based nanonetworks is also presented, and possible flaws in the process are pointed out.

5.6.1 Truth Values in Self-assembly Systems

Numbers are represented in computational models as a concatenation of stateful basic components. Two distinguishable states are sufficient to represent numbers in a space-efficient manner. Most computational models use logical values to represent the numbers "0" and "1." In the case of classic computers, a "0" is often represented by "no voltage" and a "1" is often represented by "high voltage."

There are two ways to express truth values in tile-assembly systems. On the one hand, the presence of a tile can be interpreted as "0" or "1." This has the advantage that no further interpretation is necessary in order to be able to further process results at the nano-level. However, only one of the two truth values can be displayed at the same time. It is not possible to interpret the absence of a tile as the inverse truth value. It is not possible to meaningfully distinguish whether a tile is not present or "not yet" present.

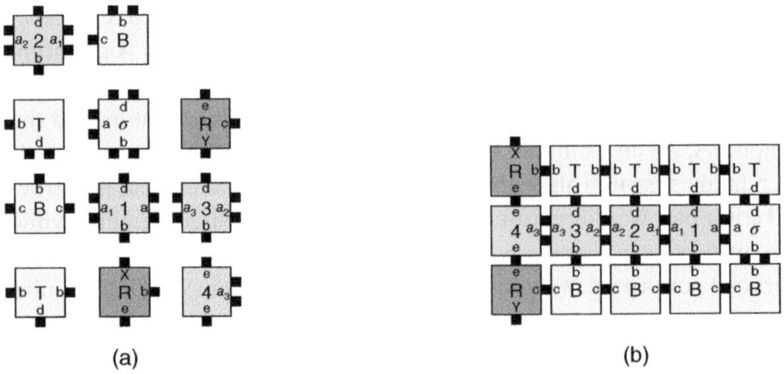

Figure 5.8 (a) Tileset that assembles into a 4-bit AND at temperature 2. (b) Resulting message molecule. *Source*: Lau et al. (2019)/with permission of Elsevier.

The second possibility consists of the representation of truth values by means of glues of tiles. Semantics is assigned to an unambiguously determinable tile, and computations can be carried out according to the glues. Both "0" and "1" can be displayed and processed at the same time.

5.6.2 Message Molecules

In addition to terahertz communication and acoustic communication, molecular communication is a frequently proposed mechanism for communication at the nanoscale (Nakano, 2017). Molecular communication is a hormone-analog method in which particle concentrations are measured in order to transmit information.

This section presents several approaches to implementing molecular communication, e.g., through tile-based self-assembly systems. In this case, special structures created by self-assembly systems that can transmit information are called *message molecules*. A message molecule can be either a single tile or an assembly.

Since self-assembly systems can also be used as a computational model, this approach has the advantage that computations can be integrated into the assembly process of a message molecule. This modifies the prevailing paradigm of computation "inside" nanodevices. Computations are outsourced to the communication channel, which is typically subject to fewer restrictions since space can be available almost indefinitely.

A message molecule can be designed in such a way that specific tiles are required for the complete assembly to occur. If we ensure that these tiles are only available under certain conditions, they can be interpreted as input for a computation. Other tiles that are also required for the computation can be kept in the channel as desired (Lau et al., 2019). In this way, it can be ensured that, for example, a ligand required for a binding reaction is only formed on a message molecule as soon as a computation has been finalized. A ligand is the part of a molecule that can bind to a receptor.

Tiles and assemblies are subject to Brownian motion. This can be used as a distribution mechanism to move tiles to where they are needed. The process is largely random, requiring a large number of message molecules.

Every decision problem (Turing, 1937) can be solved by self-assembly systems from a temperature of 2 (Winfree et al., 1998). Yet, it remains unclear whether nanodevices or nanonetworks will perform computations like macroscopic computers and whether there is enough space for more complex computations (Lau et al., 2017).

Figure 5.8 illustrates the idea of using a tileset for a message molecule that computes a 4-bit AND. Input tiles 1–4 are used as input for the computation, which is distributed by nanosensors under certain conditions. *Input tiles* describe tile-based instantiations of algorithms. In the following, a tileset that describes an algorithmic behavior is also called *algorithm* if there is no risk of confusion.

The colors of the tiles encode different semantic properties. Receptor tiles R are ligands and can bind to nanorobots. A ligand consists of two or more unbound glues attached to a terminal assembly. The remaining tiles are *framework tiles*, which are unconditionally present and prevent further growth of the assembly by not providing any glue to the outside. They also increase the stability of the message molecule.

The tile σ is the seed tile for the assembly and is necessary as a simulation entry point in the kTAM or the aTAM. In the 2HAM, each tile is to be considered a seed tile. The labels in the middle of the tiles can be used to visualize the result of functional problems. In this example, they only serve to distinguish the different tiles. The truth value of a tile is encoded by its mere presence.

Figure 5.8 (b) shows the resulting message molecule. The message molecule requires the presence of all tiles from the tileset shown; otherwise, neither the computation nor the message molecule can be completed. The presented molecule can be extended to k bits. If input tiles are only used, an event is detected. This could be the detection of disease markers. Once detected by sufficiently many nanosensors, a distributed consensus about an event can be computed in a network and false positives are arbitrarily unlikely to occur.

Definition 5.19 A message molecule $\mathcal{M}_\Phi$ is a tileset T that computes a formula Φ and forms a ligand upon completion of the assembly/computation.

5.6.3 Ligands

Ligands are binding sites on message molecules that enable correct interaction with other nanorobots or entities. These are modeled by the tile's glues.

The design process of ligands can be difficult. For example, the variable size and shape of message molecules require additional tile types in order to always create a uniform starting point for the assembly of a ligand. Furthermore, it may only form if a computation has been successfully completed. Figure 5.9 shows an example of a ligand. As this is a temperature 2 system, the two tiles with the marker "R" can only bind once the tile with the marker "C" is present. This in turn depends on the successful assembly of an adjacent, matching assembly – indicated here by white tiles.

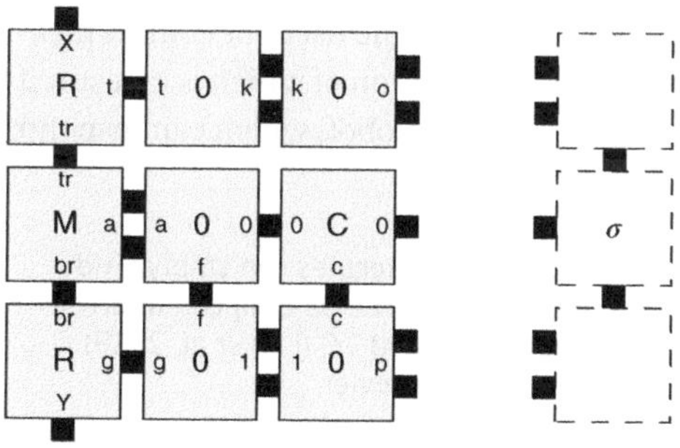

Figure 5.9 Ligand of a message molecule at temperature 2. The tiles to the left with marker "R" can only attach to the assembly if tile C in the middle has a right neighbor. The white tiles with a dashed border to the right represent an adjacent message molecule that has to be finalized before the ligand can form (Lau, 2020). *Source*: Florian-Lennert A. Lau.

5.6.4 Message Molecule Receptors

Receptors are the parts of nanodevices that can bind message molecules. They are modeled by tiles with appropriate glues. Unlike ligands, they are not tied to the successful computation of a function and are therefore static.

Receptors can be of any shape as long as they can bind to their corresponding ligand without interference. The temperature restriction τ must be observed. Furthermore, the glues of a receptor that are available to the outside must be at least one tile apart so that parts of the ligand cannot bind prematurely to the receptor. A receptor can easily be formed for all message molecules.

Figure 5.10 shows two examples of possible receptors of message molecules. The gray squares represent any part of a message with ligand. The black squares represent individual strength 1 glues. The labels of the receptor's glue encode the binding condition. If designed correctly, the receptor can "recognize" the outermost two/three tiles (grey) of the message molecule at once.

Once a message molecule binds to the receptor, it can elicit any response. For example, a container could be opened to pour out medication or a chemical reaction could be triggered that changes the color of the medium to indicate a successful computation. This can be achieved by the message molecule binding to the receptor more strongly than the closure mechanism of the container, thereby opening it.

5.6.5 Medical Example Scenario

One vision of DNA tile-based nanonetworks is their use in the human bloodstream to complete tasks that medicine currently cannot solve. This could be the recognition of disease markers or the direct combating of pathogens by nanorobots. Since such scenarios are complicated by a large number of unknown parameters, the procedure is illustrated by a simplified vision.

A DNA tile-based nanonetwork can be used in a Petri dish in vitro to quickly and reliably identify disease markers, akin to a *liquid biopsy* (Alix-Panabieres, 2020). The components of the network are tested with, e.g., a blood or tissue sample. By mixing both together and given the necessary environmental conditions, the nanonetwork may assemble. Nanosensors and nanorobots have to be assembled and filled with tiles and medicines in an earlier step.

Once the disease marker is recognized, input tiles are poured out and distributed by Brownian motion. The concentration of all components can be arbitrarily high in a Petri dish. However, too high a concentration of the components could cause damage to the human body. It is unclear whether message molecules can assemble at harmless concentrations in the human body.

Once a message molecule is assembled, it behaves like any other molecule used for communication. Calcium ions are often suggested for communication, the concentration of which is measured (Nakano, 2017). After full assembly, the message can be detected by a nanobot, which can respond to it.

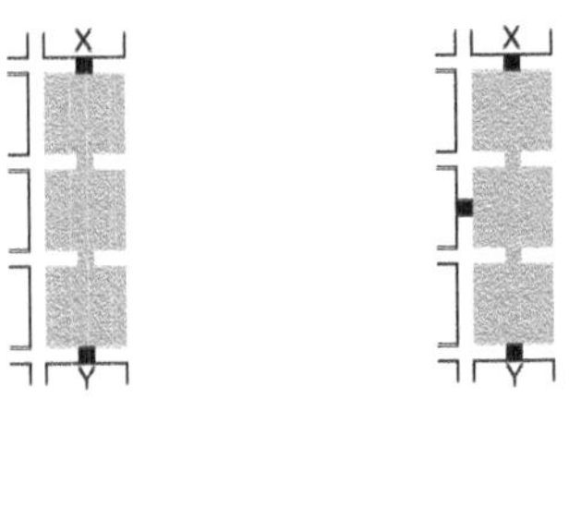

Figure 5.10 Receptors, to which message molecules can stably bind (a) at temperature 2 and (b) at temperature 3. The binding condition is encoded in the open glues with markers "X" and "Y" (Lau et al. 2019). *Source*: Lau et al. (2019)/with permission of Elsevier.

The scenario is shown in Figure 5.11. The scenario features a large number of nanosensors and a nanorobot. There are four different types of nanosensors that emit four different input tile types. The tileset from Figure 5.8 (a) is used.

In the first phase, nanosensors measure whether there are markers for a defined event. If this is the case, two tiles are distributed in phase. These come together in phase 3 to form a message molecule. This can only happen when at least four different types of nanosensors have measured an event. In the process, a 4-bit AND is computed. Once a message molecule has been fully assembled, it can bind to a nanobot in stage 4. Once this binding has taken place, a drug can be dispensed in phase 5.

All presented components can be verified in tile-assembly simulators like the NetTAS or the ISU TAS (Kaussow, 2022; Patitz, 2009), as shown in Section 12.4.7.3.

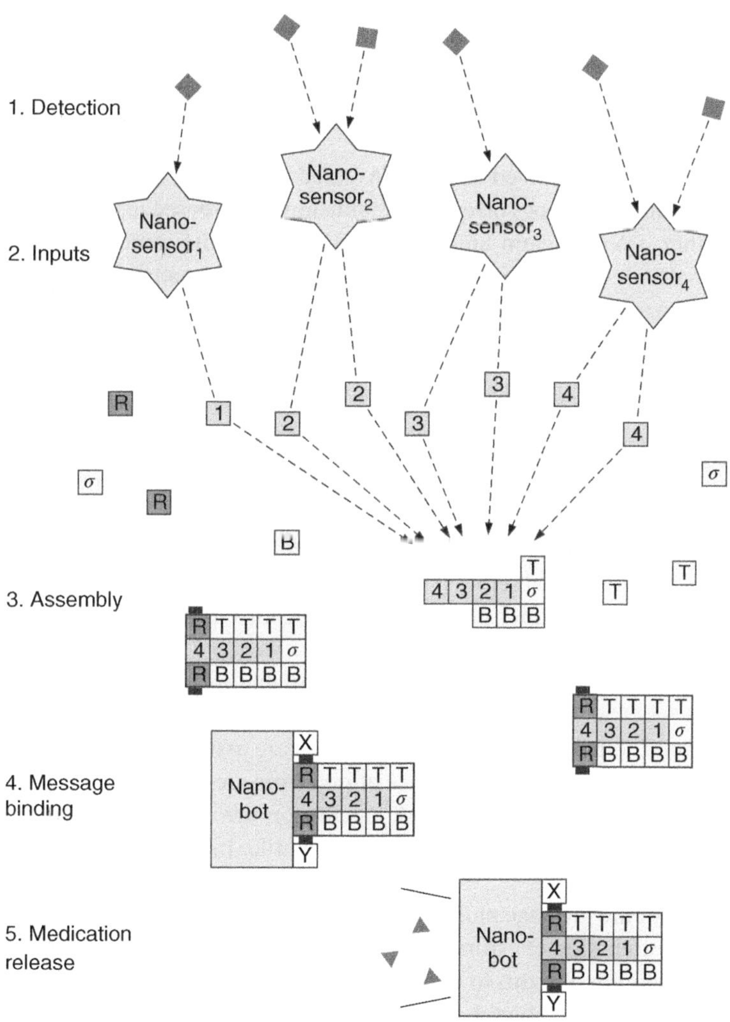

Figure 5.11 Nanonetwork that recognizes markers and then computes a logical 4-bit AND, creating a distributed consensus about detection (Lau, 2020). *Source*: Florian-Lennert A. Lau.

5.6.6 Modularizing the Scenario

The scenario presented in Figure 5.11 illustrates a vision of medical nanonetworks. Under realistic conditions, simpler experiments can likely already be implemented in the laboratory. To facilitate wet-lab experiments, the scenario can be broken down into five modules, which could later be assembled into a whole.

The scenario can be broken down into the following generic phases:

1. The detection of any marker by a nanosensor. Li et al. have already shown that drugs can be released after the detection of markers (Li et al., 2018; Benenson et al., 2004).
2. Storing tiles and medicines in nanosensors and nanobots. A variation of the task has already been solved in the work of Li et al. (2018) and Andersen et al. (2009).
3. The assembly of a message molecule under laboratory conditions. Many experiments with DNA tiles have already been able to generate complex structures (Lee et al., 2004).
4. The binding of a message molecule to a nanobot (Cheng et al., 2005). This is generally described by the 2HAM and kTHAM models but must still be verified in wet-lab experiments.
5. The release of tiles or drugs, see 1 (Li et al., 2018).

Variations of all modules have already been solved using DNA-based methods. It is thus reasonable to assume that the entire scenario can also be solved. In general, any DNA-based technologies can be combined with one another. DNA origami and tiles can be used in the same use case. It is conceivable that message molecules will be implemented by tiles, while nanosensors and nanorobots will be created by DNA origami.

5.6.7 Errors in Message Molecules

As described in Chapter 4, message molecules are also susceptible to growth, facet, and nucleation errors. Facet and growth errors can be significantly reduced using block replacement strategies such as snaked or $k \times k$ proofreading (Chen and Goel, 2005; Winfree and Bekbolatov, 2003). Growth errors are reduced by both methods. For a k block to grow incorrectly, k^2 errors must occur compared to the original case in two-dimensional self-assembly systems. Snaked proofreading also prevents facet errors in one dimension in two dimensions – in all dimensions in three dimensions, as was shown in Chapter 4.

The original logic is retained, which is why both methods are suitable for making message molecules more robust.

Nucleation errors are more difficult to handle. Since numerous tiles are assumed to be present at all times and the self-assembly process is probabilistic, nucleation errors are common and must be considered part of the mechanics. In aTAM and kTAM, nucleation errors are excluded by definition since the self-assembly process is always started from a seed tile. In reality, all tiles are to be considered as potential seed tiles.

For example, the two left columns of the message molecule from Figure 5.8 (b) can form without the seed tile σ if an unfavorable concatenation of errors occurs. This would result in the assembly of a ligand without successful computation. For this to happen, input tiles 3 and 4 must be present and generate a facet error above and below. Although the probability of this happening is small, the behavior cannot be ruled out.

This can be prevented by using the block replacement strategies described in Chapter 4. However, the glue structure of the ligands has to be adjusted in such a way that no part of an incomplete ligand can correctly bind to the receptor.

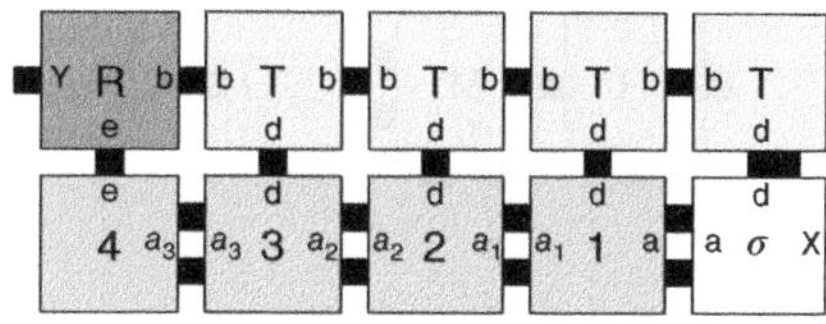

Figure 5.12 Adapted message molecule at temperature 2 that bypasses nucleation errors (Lau et al., 2019). *Source*: Lau et al. (2019)/with permission of Elsevier.

Another possibility is a modified conception of a receptor and message molecule, which requires the presence of the seed tile σ. Figure 5.12 shows a possible adjustment. A suitable receptor can easily be designed. In this way, the presence of the seed tile σ is enforced, and, thus, nucleation errors are prevented. It is thus more difficult to generate erroneous behavior.

5.6.8 Logical Combination of Message Molecules

In order to express any problem as a computation within a message molecule, it is sometimes necessary to logically combine message molecules. This can be made possible by additional, cleverly designed assemblies. Figure 5.13 shows an example of this. The tiles with a dashed border background represent two message molecules N_1 and N_2 with special ligands. The long white strand of tileset F contains different receptors that can bind the two message molecules. The ligand of the combination of both message molecules can only form if both message molecules bind stably. The ligand is identified by the marker "R."

A special feature of the construction is that the first message molecule N_1 provides part of the receptor x_2 for the stable binding of the second message molecule N_2. In this way, the entire assembly fulfills the function of a two-input AND gate. The result can be communicated to other nanobots through the attached receptor, and the ligand forms only when the required number of message molecules are bound to the structure.

The procedure can be extended to any number of input messages. Natural stability restrictions must be taken into account.

5.6.9 Modeling Message Molecules

A message molecule for the problem ADD with four input bits has already been presented in Section 5.6.2. We now, we design message molecules for other sometimes more complicated problems. In addition, we discuss a method that allows us to solve any Boolean formula using message molecules. In the following figures, only the algorithmically relevant parts of tilesets are shown. Framework tiles are omitted for clarity.

We start with the most general procedure and present more specific and smaller molecules for other problems later.

5.6.9.1 Solving The Decision Problem

Although it is known that self-assembly systems are Turing-complete from temperature 2 onward, this result cannot be directly extrapolated to message molecules. In this section, we present a method with which most mathematical problems can be expressed as a message molecule. This demonstrates the potential usefulness of DNA networks.

It is known from theoretical computer science that different types of problems can be transformed into one another. Because decision problems can be easily processed at the nanoscale, these are of particular interest.

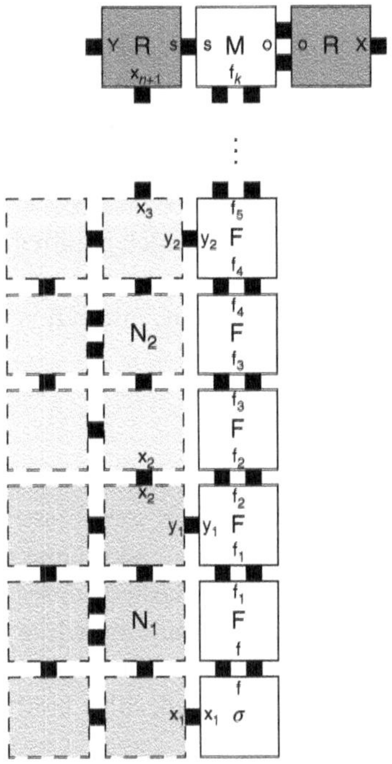

Figure 5.13 Schematic representation of an assembly that combines n message molecules N_n at temperature 2, and it provides a ligand "R" (Lau, 2020). *Source*: Florian-Lennert A. Lau.

Theorem 5.1 For each decision problem modeled as a Boolean formula Φ, a message molecule can be created that performs the same computation and forms a ligand upon successful completion.

Proof: Let Φ be a Boolean formula. According to Quine, it is known that every Boolean formula Φ can be converted into a disjunctive normal form (Quine, 1952). This is done by setting up the truth table for the said formula. Table 5.3 shows an example. Truth tables are generally exponentially large, which limits the procedure to small problems. However, this is likely not a problem as more complex problems are prevented by resource constraints anyway.

For each row that contains a "1" in the last column, we create a subformula. All literals A_n in a line are combined with a logical $\wedge$ and bracketed – also called *clause*. *literals* are Boolean variables and their inverses. All clauses created in this way are linked with the logical $\vee$. The result is a formula Φ' which has a canonical form and returns the same result as the original formula. Such a formula

Table 5.3 Example of a truth table for the formula Φ (Lau, 2020).

ϕ_1	ϕ_2	...	Φ
1	1		1
1	0		1
		⋮	
0	1		0
0	0		1

Source: Florian-Lennert A. Lau.

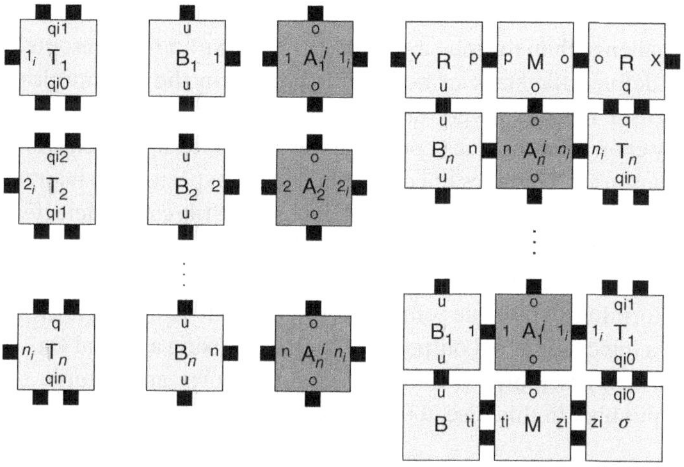

Figure 5.14 General procedure for creating tilesets from formulas in the DNF using the example of the truth table (Table 5.3) (Lau et al., 2020). *Source*: Lau et al. (2020)/with permission of Springer Nature.

is also called *distributive normal form* (DNF) of a formula Φ. This can be further minimized using procedures such as McCluskey McCluskey Jr. (1956).

For a formula Φ' in the DNF, a tileset with identical semantics can be created as follows.

A message molecule is created for each clause. The length of the message molecule corresponds to the number of different literals. Figure 5.14 shows the general procedure. The tiles A_n^i correspond to the literals of the formula Φ'.

The truth value of each literal is reflected either in the marker or by the glues. Only once all literals are part of the message molecule, a receptor can assemble.

Since a message molecule was created for each clause and these are linked using ∨, it is sufficient if any message molecule is completely assembled so that a successful evaluation of "1" can be communicated.

5.6.9.2 *k*-Bit Or

In terms of specific operations, the logical Or is most easily implemented as a nanonetwork. In principle, no computation is required within a message. A single tile is enough to transport information to a nanobot.

The nanosensors contain tiles that have different markers but identical glues. These tiles fulfill both the role of an information carrier and the role of a ligand. The glue strength corresponds to the temperature τ. A receptor consists of only one tile with the strength τ.

In the presented way, it is sufficient if any nanosensor detects an event and releases tiles. Those tiles can bind to the receptor unhindered and without any further conditions. The behavior corresponds to a logical Or in a circuit. As soon as any tile binds to the receptor, a "1" is considered transmitted. In this case, the truth value of a computation corresponds to the presence of a tile.

Definition 5.20 A message molecule $\mathcal{M}_{Or}$ is a tileset that computes the decision problem Or and forms a ligand if the evaluation is successful.

Due to the simplicity of the message molecule, the computation is already completed by the distribution of the input tiles.

5.6.9.3 *k*-Bit Thres

A *k*-bit Thres poses a much greater challenge than the message molecules for And and Or presented so far. The intuitive thought that a defined tile stack of height *k* is present in the communication channel fails because a nanosensor releases a very large, hard-to-control number of tiles. Self-assembly processes are driven by chance, and a stack could thus be assembled upon detection of a marker by the tiles of a single nanosensor. This does not correspond to the expected behavior of a Thres computation over different inputs. Using a single stack, either the same tile can be detected *k* times, which does not correspond to the behavior of a Thres gate in the classical sense, or one of the many possible constellations of input bits can be detected.

In circuits, a Thres is computed by forming all possible combinations of *k* input values, regardless of the order, and linking them using a logical And. All outputs are connected using a logical Or.

This behavior can be adapted for self-assembly systems. Figure 5.15 shows an example of computing a 3-bit Thres over four input bits. In this case, too, the presence of a tile corresponds to

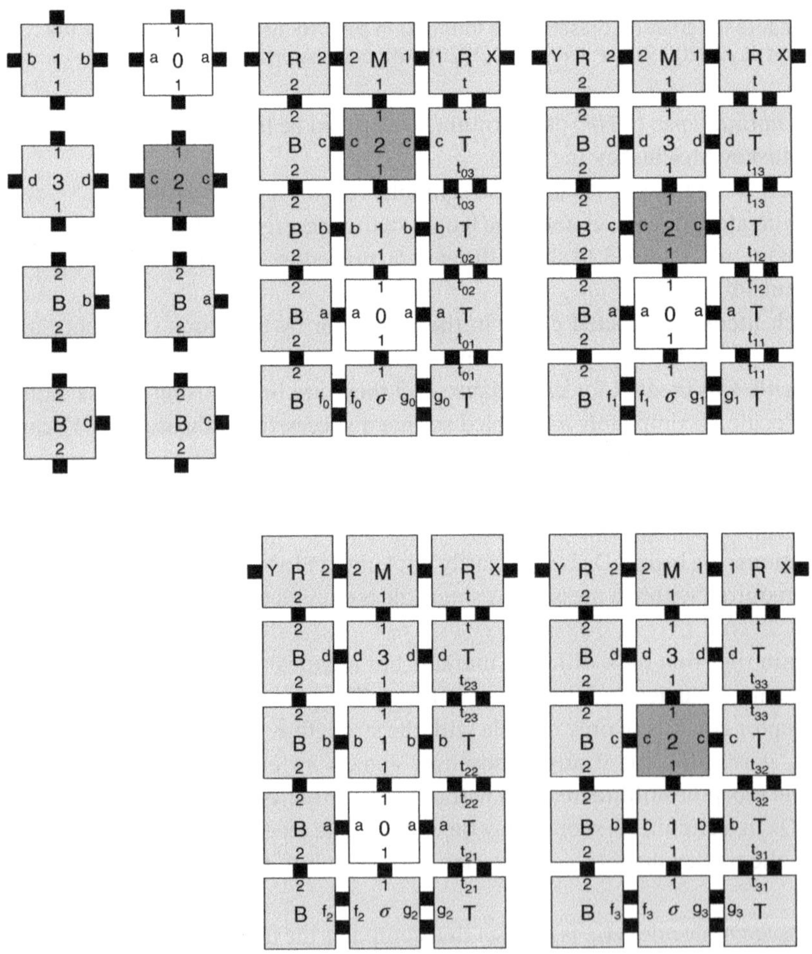

Figure 5.15 Tileset (left) and fully assembled message molecules (right) for the decision problem Thres. The Tileset on the left only contains tiles that can be used more than once. On the right, all four possibilities are shown to reach a threshold value of 3 with four possible inputs. The order is ignored (Lau et al., 2021b). *Source*: Lau et al. (2021b)/with permission of Elsevier.

a truth value of "1." The different shaded tiles correspond to the different binary inputs. All but the numbered tiles form the framework of the various message molecules. The tileset on the left only shows tiles that are used more than once. A total of $\binom{n}{k}$ different AND message molecules are formed, all of which form the same ligand. This covers all possibilities of reaching a threshold value of 3 with four different inputs. It is sufficient for a single message molecule to bind to a nanorobot to implement a logical OR. Combined, this results in a behavior that corresponds to a logical THRES.

One problem is that the number of different message molecules can grow exponentially. With a constant number of glues of length k, the input size for THRES is limited.

Definition 5.21 A message molecule $\mathcal{M}_{Thres}$ is a tileset that computes the decision problem THRES and forms a ligand if the computation is successful.

5.6.9.4 *k*-Bit ADD

In the case of the function problem ADD, nanosensors can emit whole binary numbers as assemblies that are to be added by the nanonetwork. The temperature of the system must be at least 3 for the presented architecture. A tile-based adder can be designed that works at temperature 2 and computes bits in rows (Brun, 2007). However, temperature 2 adders are significantly more complicated, and, for simplicity, the message molecule is designed for temperature 3.

The truth values are adjusted to match the labels/markers on the tiles. This is necessary to display bits with the values "0" and "1" correctly at the same time.

Figure 5.16 shows an example of a composite message molecule (Brun, 2007). White input tiles describe the addition algorithm. These are permanently present in the communication channel. Ocean green tiles form ligands and the framework. The top and bottom binary numbers represent

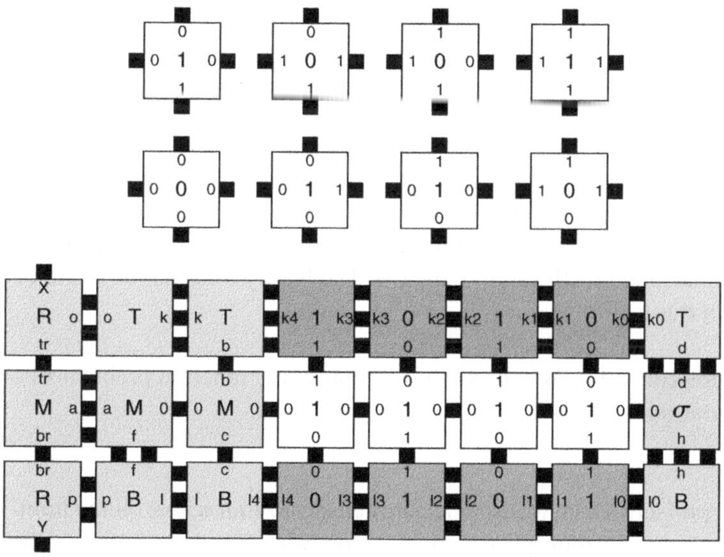

Figure 5.16 Tileset and fully assembled message molecule for the function problem ADD. The white tiles of the message molecule (above) encode an addition algorithm. A message molecule is shown below. The top row with markers 1010_2 and the bottom row with markers 0101_2 encode the input numbers. The remaining tiles are part of the seed assembly or the ligand (Lau et al., 2021b). *Source*: Lau et al. (2021b)/with permission of Elsevier.

the input bits. It may be necessary to append an additional "0" because the result of a binary addition is potentially one bit longer than the input numbers in binary representation. The right glue of already existing input tiles is understood as a *carry bit*.

The algorithm corresponds to the procedure for a written addition of binary numbers. The algorithm's tiles are designed to organize themselves as follows when compared bit-wise:

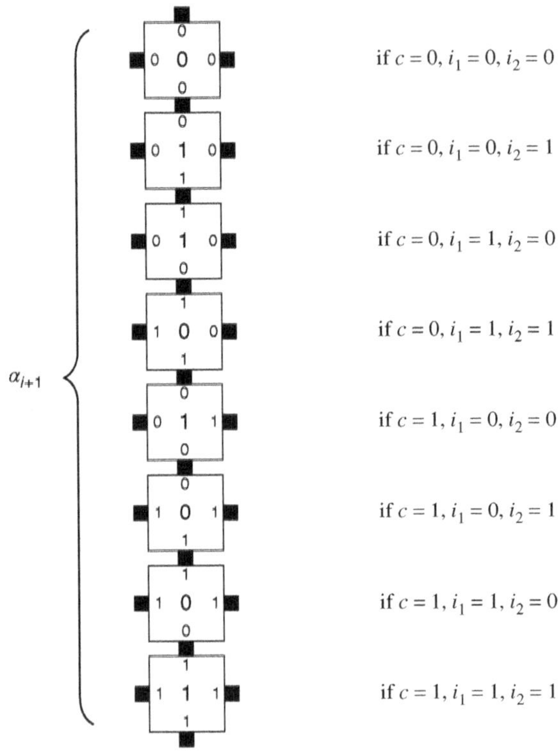

Only once the result has fully formed in the center, a ligand can form. Since the truth values in the event of functional problems correspond to the labels/glues on the tiles, the result cannot be used directly by a nanobot. Additional steps are necessary for further processing.

One possibility is to use three-dimensional tiles. In this case, another glue can be attached to the back of the tile, which corresponds to the truth value of the tile. The binary information could be propagated to other structures in this way.

Definition 5.22 A message molecule $\mathcal{M}_{Add}$ is a tileset that computes the function problem ADD and forms a ligand if the computation is successful.

5.6.9.5 *k*-Bit MULT

In the case of the multiplication problem MULT, we assume nanosensors to emit whole binary numbers in the form of assemblies that are to be multiplied by the nanonetwork. The temperature of the system must be at least 2. The truth values are adjusted to be in the glues as with the problem ADD.

Figure 5.17 shows an example (Brun, 2007). The tileset for the computation is shown above and is permanently present in the communication channel. An assembled message molecule with ligand "R" is shown below. This can only form when the last tile of the multiplication is present.

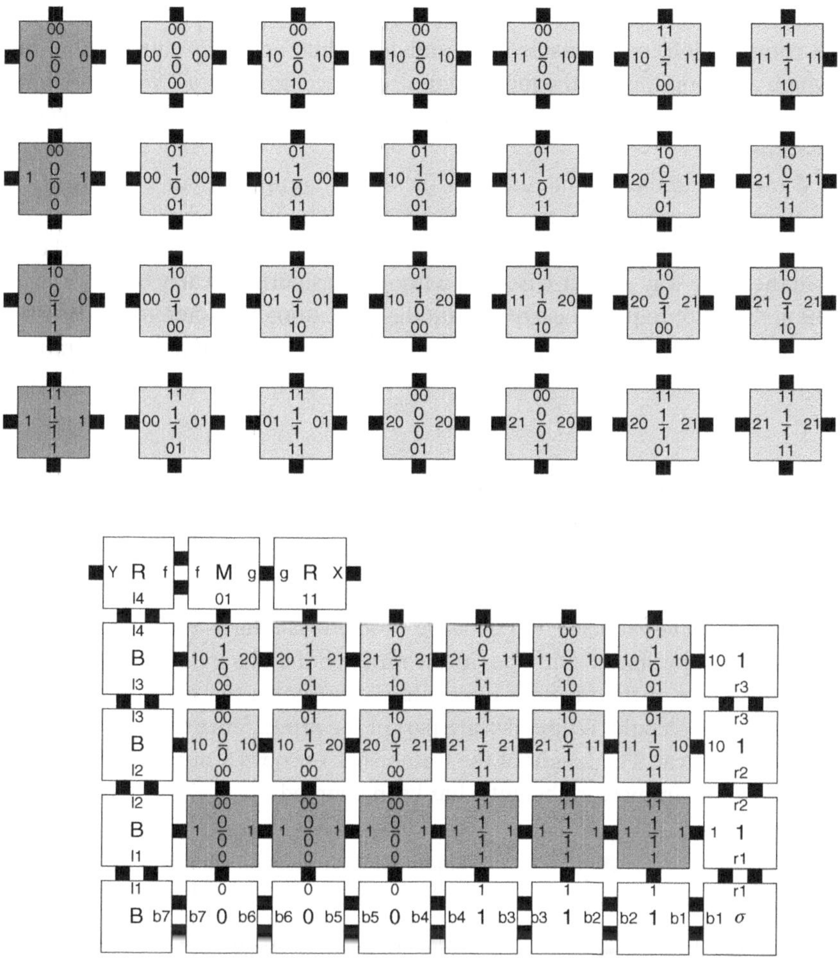

Figure 5.17 Tileset and fully assembled message molecule for the function problem MULT with the ligand. A tileset of size 28 of the algorithm is shown above, and a message molecule is shown below. Four of the tiles are only needed to start the computation. The result of the multiplication can be read on the top row of the rectangle. The upper digit of the marker shows the respective result bit. The numbers 7 and 7 are multiplied (Lau et al., 2021b). *Source*: Lau et al. (2021b)/with permission of Elsevier.

The multiplication $7 \cdot 7$ is performed in this example. The upper part of the marker on the top row corresponds to the individual bits of the result 110001_2. Different tiles "M" and "R" must exist so that a ligand can be formed for all possible results.

As with the problem ADD, the procedure corresponds to the algorithm for the written multiplication of binary numbers. However, this algorithm is inspired by an iterative addition. Each bit is computed individually by matching the other binary number. Since the result of a multiplication can be twice as long as the input, the framework of the message molecule is filled with "0" in the x direction according to the input length.

Definition 5.23 A message molecule $\mathcal{M}_{Mult}$ is a tileset that computes the function problem MULT and forms a ligand upon a successful computation.

5.6.9.6 *k*-Bit XOR

In the case of an XOR computation, nanosensors emit individual bits or binary numbers in the form of assemblies, which are computed in the message molecule according to an XOR operation. The truth value of a tile is expressed by the label/glues since "0" and "1" cannot be encoded by the presence of tiles at the same time.

Figure 5.18 shows a message molecule including a tileset for computing an exclusive OR. The algorithm must always be present in the communication channel and is displayed in white. The ligand and input are grey, and the input structure is labeled with binary markers. The procedure works in the same way as a bit-wise ADD without considering a carry bit.

While the computation does not require a carry bit, another "meaningless" glue to the right is necessary to ensure that the assembly order happens from right to left. Furthermore, a temperature 2 system could not ensure this behavior as there is no way of preventing premature bindings directly next to the ligand.

Definition 5.24 A message molecule $\mathcal{M}_{Xor}$ is a tileset, which computes the decision problem XOR and forms a ligand if the computation is successful.

5.6.9.7 *k*-Bit-COUNT

The counting problem COUNT can be modeled as a decision problem as long as the input is unary encoded. Unary encoding stores the size of a number as the number of digits in its representation.

Figure 5.19 shows the structure of the message molecule. Tiles *R* form the receptor, and tiles "F" form the framework of the molecule. White tiles are part of the input. Since this is a counting problem, all white tiles are identical. The height of the framework encodes a number *k* in unary. A ligand is only formed when the framework height *k* has been counted.

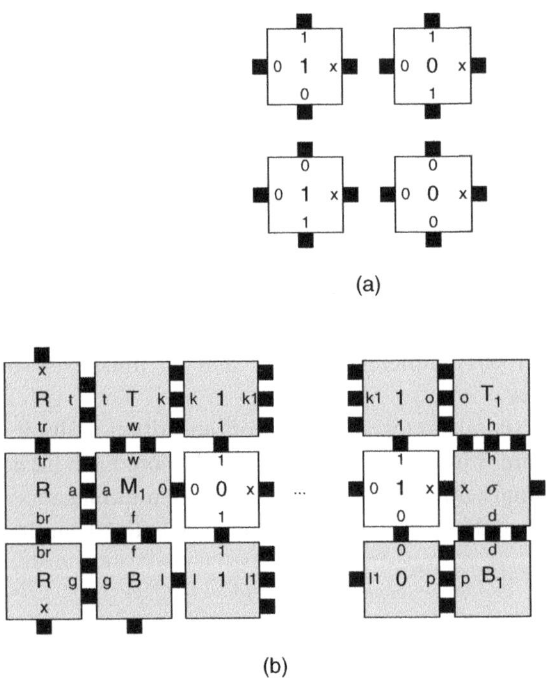

Figure 5.18 Tileset (a) and partially assembled message molecule (b) for the function problem XOR with the ligand. The XOR computation is an addition without a carry bit, but an additional x glue is necessary to meet the temperature 3 requirement. The left glue of the conditional tiles is necessary to allow for the attachment of a ligand upon finishing the computation. (a) Relevant parts of the tileset (Lau et al., 2021b). (b) Terminal XOR molecule (Lau et al., 2021b). *Source:* Florian-Lennert A. Lau.

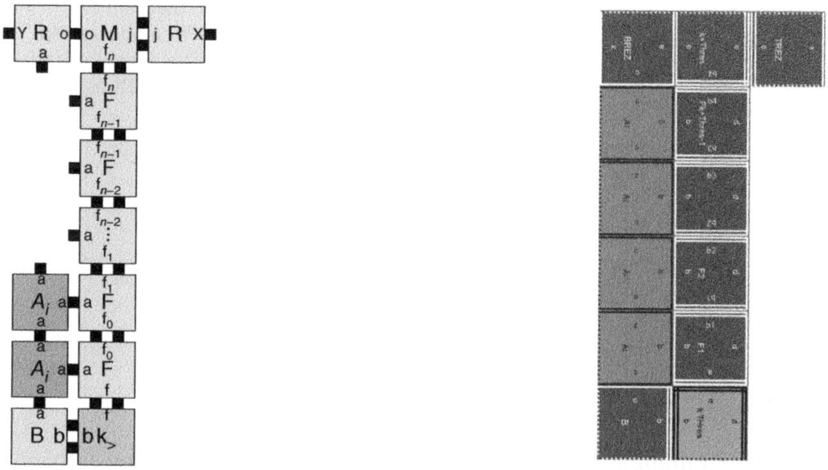

(a) (b)

Figure 5.19 (a) Assembly of a message molecule for the counting problem with a ligand. The Frame structure of tiles F demands up to k fitting tiles to bind before the ligand can be completed. (b) Result of the simulation of the same message molecule. (a) Conceptualization (Lau et al., 2022b). (b) Simulation result (Lau et al., 2022b). *Source*: Florian-Lennert A. Lau.

Definition 5.25 A message molecule $\mathcal{M}_{Count}$ is a tileset that computes the decision problem Count and forms a ligand if the computation is successful.

5.7 Finding Programs for Nanodevices

Now that we have a broad overview of various computational models at the nanoscale, we still face the problem of *finding* a suitable program for devices or especially networks of devices. While different models vary in fundamental aspects, they are usually equally computationally powerful. If we find a single program, it is likely that it can be transformed into a program for another computational model. Thus, we are more interested in initially finding such programs for nanonetworks, which is a surprisingly difficult problem due to the large number of potential participants of low complexity.

In Chapter 4, we introduced DecPOMDPcoms to accurately model nanodevices and nanonetworks. *Solving* a DecPOMDPcom means that a program for each and every network participant is generated that satisfies the specified constraints. Right now, many approaches use reinforcement learning with or without ANNs to deal with the inherent complexity of the problem. The potential interactions between potentially millions of nanodevices lead to a program search space that is infeasible to fully explore. Thus, we have to either use such heuristics or make the problem more approachable due to other "tricks." As explained in Section 5.5.4, we face numerous problems with ANNs at the nanoscale; thus, we focus on solving DecPOMDPcoms by different means.

The complexity of solving MDPs and other variations of the problem introduced in Chapter 4 can be seen in Figure 5.20. As long as only a single nanodevice is involved, the problem is comparably easy and in P. As soon as the decentralized problem variations are considered, solving the problem requires exponential amounts of space and time. This is neither feasible in nanonetworks nor on classical computers.

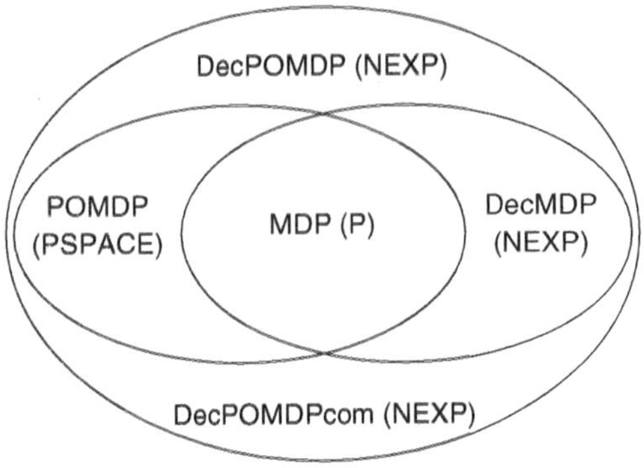

Figure 5.20 Complexity inclusion diagram of various different MDP classes. MDPs can be solved in ¶and POMDPs require PSPACE. The decentralized variations DecMDP and DecPOMDp as well as DecPOMDPcom are all in NEXP. *Source:* Florian-Lennert A. Lau.

5.7.1 Solving DecPOMDPcoms

This section formalizes the programming of nanonetworks as an offline decision-making problem, specifically, as lifted decentralized partially observable Markov decision process (DecPOMDP) (Braun et al., 2021; Lau et al., 2022a). The results can then be applied to DecPOMDPcoms, by extending the state space of DecPOMDP by messages.

In nanonetworks, there are potentially thousands if not millions of collaborating nanodevices that contribute toward the same goal. As finding programs for DecPOMDPs scale exponentially with the number of devices, the standard approaches for solving them fail. However, it might be still possible to find solutions in some cases as there are only a few different *types* of nanodevices. While there might be thousands of devices of a type, all of them are virtually indistinguishable and do not even have a unique identifier.

By *lifting*, we algorithmically focus on the number of device types instead of the number of devices (Poole, 2003). While the overall runtime remains exponential, there usually are only a small, constant number of device types or *partitions* of nanodevices. This approach is similar to *fixed-parameter tractability* (FPT) approaches, where algorithm designers find approximate solutions for NP problems by identifying the "truly hard part" of problems (Downey and Fellows, 1995).

There are numerous examples where lifting has been successfully used in both offline (Van den Broeck et al., 2011; Ahmadi et al., 2013; Braun and Möller, 2018; Holtzen et al., 2019) and online settings (Nath and Domingos, 2009; Apsel and Brafman, 2011; Gehrke et al., 2019a, 2019b). We still represent the environment by a non-lifted joint distribution as medical nanonetworks are too diverse for more structured models.

5.7.1.1 Lifting

Modeling a nanonetwork as a DecPOMDP requires a large set of nanodevices $\boldsymbol{I}$ consisting of a small number of partitions using the same setup regarding available actions and possible observations. To reduce the computational complexity, we apply lifting to $\boldsymbol{I}$.

Given a medical nanonetwork, we look at networks of indistinguishable nanodevices whose size cannot be determined precisely. The computational power of each nanodevice is limited, and all devices collaborate toward a joint goal. Therefore, DecPOMDPs that are lifted with indefinite group sizes are an excellent fit. We now explain what constitutes a *liftable* nanonetwork scenario.

Definition 5.26 A *liftable DecPOMDP* is a DecPOMDP in which the set of nanodevices $\boldsymbol{I}$ is partitioned into K sets $\mathfrak{I}$, i.e. $\boldsymbol{I} = \bigcup_{k=1}^{K} \mathfrak{I}_k$ and $\forall k, l \in \{1, \dots, K\} : \mathfrak{I}_k \cap \mathfrak{I}_l = \emptyset$, where in each partition

$\mathfrak{I}_k$, it holds that $\forall i, j \in \mathfrak{I}_k$:

$$\text{ran}(A_i) = \text{ran}(A_j) \wedge \text{ran}(O_i) = \text{ran}(O_j). \tag{5.1}$$

As a result, we may use only K variables A_k and O_k that apply to all nanodevices in a partition. With a partitioning of the nanonetwork, the joint observation $\boldsymbol{o} = (o_1, \dots, o_N)$ consists of K *partition observations* $\boldsymbol{o}_k = (o_{k,1}, \dots, o_{k,|\mathfrak{I}_k|})$.

An example of the lifting idea can be seen in Figure 5.21. The basic scenario is the same as in Chapter 4. However, the nanodevices are now colored differently, which signifies their functionality and partition. They now represent a *set* of nanodevices with identical properties. As a result, the complexity of finding a program for such nanonetworks might be shifted from the number of nanodevices to the number of partitions as we will soon see.

There are $r_k = |\text{ran}(O_k)|$ different possible observations per partition $\mathfrak{I}_k$, meaning, $\boldsymbol{o}_k$ can be encoded using a histogram,

$$\{(o_l, n_l)\}_{l=1}^{r_k}, o_l \in \text{ran}(O_k), n_l \in \mathbb{N}^0, \sum_l n_l = |\mathfrak{I}_k|, \tag{5.2}$$

or $[n_1, \dots, n_{r_k}]$ for short.

The entire idea behind lifting can indeed be motivated using histograms. When measuring, e.g., a concentration, it does not matter which individual device observed a value – what matters is the total number. We thus introduce a *counting random variable* (CRV) for every partition (Milch et al., 2008).

Based on this, the joint observation over a partitioned nanodevice set turns into

$$\boldsymbol{o} = (h_1, \dots, h_K), h_k \in \text{ran}(\#_{X_k}[O_k(X_k)]). \tag{5.3}$$

The size of $\boldsymbol{o}$ is now only $K \cdot \max_{k \in \{1,\dots,K\}} |\text{ran}(O_k)|$. Essentially, only a single variable is necessary per set/observation.

A similar principle applies to actions. A joint action $\boldsymbol{a}$ can be modeled by partition actions A_k using a CRV $\#_{Y_k}[A_k(X_k)]$. The same reduction in size as with the observations applies. Using a

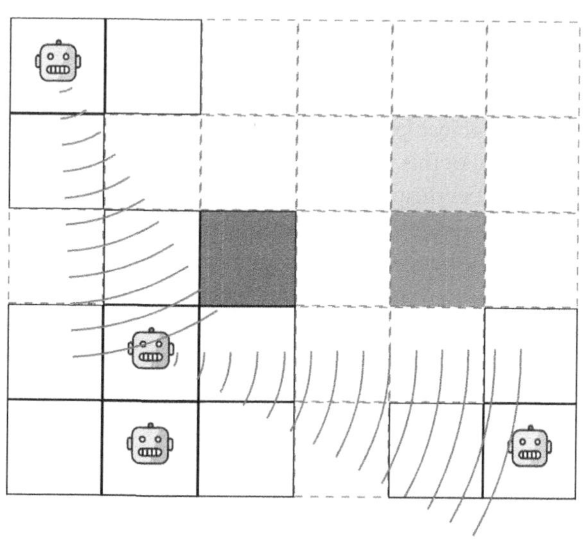

Figure 5.21 Lifted DecPOMDPcom with two types of indistinguishable nanodevices that can communicate. The top two nanobots might be of a different type compared to the bottom two. *Source*: Florian-Lennert A. Lau.

histogram allows for combining all the input mappings to the same probability into one input, reducing the size of T and Ω.

Based on the previous observation, we can now formalize lifted DecPOMDPs as follows.

Definition 5.27 A *lifted DecPOMDP* is a tuple $(\{\mathfrak{I}_k\}_k, S, \{\#_{X_k}[A_k(X_k)]\}_k, T, R, \{\#_{X_k}[O_k(X_k)]\}_k, \Omega)$, $k = 1$ to K, with

- $\{\mathfrak{I}_k\}_{k=1}^K$ being a *partitioning* of an nanodevice set $\boldsymbol{I}$,
- S being a random variable with a set of states as range,
- $\#_{X_k}[A_k(X_k)]$ being a CRV with a set of histograms as in Equation 5.2 as range, with $\boldsymbol{A} = \times_{k=1}^K ran(\#_{X_k}[A_k(X_k)])$ being the set of joint actions,
- $T(S', S, \boldsymbol{A}) = P(S' \mid S, \mathbf{A})$ being a transition model,
- $R(S)$ being a reward function,
- $\#_{X_k}[O_k(X_k)]$ being a CRV with a set of histograms as in Equation 5.2 as range, with $\boldsymbol{O} = \times_{k=1}^K \mathrm{ran}(\#_{X_k}[O_k(X_k)])$ being the set of joint observations, and
- $\Omega(\boldsymbol{O}, S) = P(\boldsymbol{O} \mid S)$ being a sensor model.

Each partition $\mathfrak{I}_k$ has a local policy π_k. The solution to a DecPOMDP is a joint policy $\boldsymbol{\pi} = (\pi_k)_{k=1}^K$.

As demonstrated in the work of Lau et al. (2022a), the worst-case space requirement is no longer exponential but polynomial in N. Given the lifted representation, we now have a chance to find optimal policies for nanonetworks. Without lifting, the problem is computationally infeasible. In the simplest case, we could now use value iteration to create conditional policies for the partitions.

To give you a better understanding, we now model a very simple DNA-based nanonetwork introduced in Section 5.6.5 as a lifted DecPOMDP. The nanosensors and nanobots each form a partition with $K = 5$. The $\kappa = 4$ types of nanosensors may observe the same biomarker. Unlike with the nanosensors, there is only a single type of nanobot $\iota = 1$ that releases medication or fluorescence markers and reacts to messages. The number of nanodevices per partition might be around 64,000, as demonstrated in the work of (Stelzner and Traupe, 2019). In total, we have $(\kappa + \iota) \cdot 64{,}000 = 320{,}000$ nanodevices.

In this simple scenario, there is only a single associated action per nanobot and a single possible observation per nanosensor. Formally speaking, there is also a "do nothing" or "sense nothing" possibility.

The state of the environment can be modeled by the presence/absence of biomarkers and finished message molecules. This leads to a state space of $2^\kappa \cdot 2^\iota$.

The goal of this medical nanonetwork is the release of medication if there is high enough confidence that corresponding biomarkers are present. This must be encoded in the reward function.

For the sensor model Ω, additional empirical information is necessary as there are a number of possible influences. As an example, nanodevices might falsely observe a marker/message since a wrong marker/message or the message itself is corrupt. Additionally, it is possible that the nanodevices do not detect any event even though it is happening.

Hence, in summary, we have 320 000 nanodevices sorted into 5 different partitions with an environmental state space of size 32. As a result, the model sizes of T are

$$S_T^{\mathrm{dec}} \in O(32 \cdot 32 \cdot 2^{320\,000}), \tag{5.4}$$

$$S_T^{\mathrm{lif}} \in O(32 \cdot 32 \cdot (64\,000^2)^5), \tag{5.5}$$

and of Ω are

$$S_{\Omega}^{\text{dec}} \in O(32 \cdot 2^{320\,000}), \tag{5.6}$$

$$S_{\Omega}^{\text{lif}} \in O(32 \cdot (64\,000^2)^5), \tag{5.7}$$

in the worst case. As you can see, without lifting, the problem is computationally intractable. With lifting, it is possible to find an optimal policy in polynomial time.

5.7.2 Value Iteration

Based on this, we could use simple value iteration over the number of partitions to find suitable programs for all involved nanodevices (Bellman, 1957). It is subject to future work to find better-suited approaches that allow for more efficient solutions for DecPOMDPcoms.

Value iteration is also sometimes called *backward induction*. As the name suggests, the algorithm is repeatedly executed until a previously defined minimum improvement to the policy is no longer met. In each step, all possible states or partitions are considered. Consecutively, for each possible state/partition, all possible actions that can be performed in that given state are evaluated.

If we consider the previous example from Figure 4.10, the entire grid might be initialized with zeroes that indicate that no reward is associated with a given state. As in the respective chapter replace the color by (consistent) east/west fill patterns. We now consider the given probabilities that certain actions are performed and update the expected rewards for each cell. In the given scenario, we multiply the probability of transitioning to a neighboring cell with the expected reward of that cell. In the next iteration, the previously defined discount rate is applied and the procedure is repeated. After a while, the algorithm terminates, and a policy of a predefined quality is found.

The policies generated from such an approach can then be implemented in an arbitrary computational model that suits the environment and satisfies all constraints. In the simplest case, a policy could be an IF/Else tree. While there are other possible algorithms to find solutions for MDPs, a single one is enough to convey the general idea. The interested reader is encouraged to have a closer look at policy iteration algorithms or prioritized sweeping methods.

5.7.3 Genetic/Evolutionary Algorithms

As the problem of finding programs is extremely difficult unless the problem difficulty is reduced by, e.g., lifting, only a few, often nature-inspired solutions are available. Those include ANNs and genetic algorithms. Another way of searching all possible programs for those that fulfill specific criteria is genetic algorithms.

This section introduces the basics for understanding genetic algorithms and is based on the work of Buss (2017). We use the established symbols that are used to describe genetic algorithms even if they are used otherwise in the remainder of the book as genetic algorithms are not used in other parts of this book.

Genetic algorithms are "meta algorithms" that have other algorithms as their output. They are designed to find "good enough" solutions to a given problem optimization problem. They are inspired by observing the genetic processes in nature and thus utilize mutation, crossover, and selection (Fraser, 1957).

The process of evolution has been described by Darwin as early as 1859 (Darwin, 1859). Evolution is the process that describes the change of species over time given their physical and social surroundings. The principles that encode those rules of change are as follows:

1. *Natural selection* occurs within a population because there is always some variation between its members. This variation can lead to certain individuals producing more/fitter offspring. For example, an individual who is better at finding food will have a greater chance of living long enough to have offspring, making it more fit.
2. In sexual reproduction, two parents *(re-)combine* their genetic material to produce offspring that bear those genes. The parents selected in the previous step, through natural selection, pass on advantageous traits to their offspring, thereby giving them an advantage in reproduction.
3. Genetic material can also change due to *mutations/errors* during replication. Although most mutations are harmful, they increase the variation within a population. If a mutation is advantageous, it can enable a species to evolve in new ways.

Evolutionary algorithms are a general procedure that allows us to find solutions for optimization problems. Oftentimes, finding an optimal solution can be difficult or even impossible due to the complexity of the problem. In such cases, it may be necessary to use a non-optimal solution, which can be obtained through an approximation strategy. Next to ANN, genetic algorithms offer such an approach. They involve starting with a set of solutions and selecting a certain number of good ones to combine and modify in order to generate new and better solutions.

Unlike classical approximation algorithms, which provide a guarantee of the quality of the solutions, genetic algorithms do not offer such guarantees. Instead, whether a given solution is considered "good enough" for a particular problem instance depends exclusively on the context and must be determined by the programmer. The following paragraphs provide a more detailed description of how genetic algorithms work.

Definition 5.28 Given an optimization problem (I, P, m, w), a possible solution $p \in P$ is called a *phenotype* or *individual*. The set of all possible solutions P is also called the *phenotype space*.

A phenotype is considered to have a higher fitness than another phenotype if it represents a better solution to an optimization problem. However, it can be unclear how to define recombination or mutation in solutions to such a problem, or what it means for one solution to have better fitness than another. Genetic algorithms address this issue by assigning a genetic representation to each phenotype. This allows for the use of standard genetic operators such as recombination and mutation, which can be used to modify and combine solutions in order to generate new and potentially better solutions. By manipulating the genetic representation of a phenotype, genetic algorithms can effectively explore the search space of phenotypes and find solutions that have higher fitness than the initial ones. This works based on an assumed "structure" of the phenotype/genotype space and the consistency of their relationship.

Definition 5.29 Given an element of a phenotype space P, a genetic representation is called the *genotype*. The set of possible genotypes is called the *genotype space* G. The process through which a genotype is generated from a phenotype is called *encoding* and is a partial, injective function $e : P \rightarrow G$. The inverse operation d is called *decoding* and is also a partial, injective function.

It is important to ensure that the encoding and decoding functions used in genetic algorithms are total. They can represent any valid solution and every possible representation is valid. Without totality, it may be possible that an optimal solution has no genotype. Similarly, if the decoding is partial, the algorithm may generate invalid genotypes that waste computation time.

The actual search for solutions takes place in the genotype space. Representations in this space need to be easily accessible, modifiable, and combinable by a computer, typically through the use of arrays or trees. Although the term "element" is used loosely to refer to any distinct part of a genotype that can be individually accessed, genetic algorithms work best if individual genes can be accessed via some index. The specific definition of an "element" depends on the particular genotype being used.

We only consider cases where the encoding and decoding functions are bijective and where all possible solutions can be represented.

Definition 5.30 Given a genotype g with elements g_i with $i \in I$ for some index set I, the different places i in g are called *genes*. The actual values g_i in those places are called *alleles*.

An example could be a tuple $g = (2, 4, 1, 6)$, where the indices $I = 0, 1, 2, 3$ represent the genes and the values 2, 4, 1, and 6 are the alleles.

In an optimization problem (I, P, m, w), the goal is to minimize or maximize a function m that measures the quality of a solution for a specific problem instance $x \in I$. While m is typically defined on the phenotype space, it can be easier to define the fitness of an individual on its genetic representation.

Definition 5.31 Let (I, P, m, w) be an optimization problem and $e : P \rightarrow G$ be an encoding. Given the genotype space G, the *fitness function* is a function $f : I \times G \rightarrow R$ such that $m(x, p) = f(x, e(p))$ for all $x \in I, p \in P$.

The fitness function of a genetic algorithm does not necessarily need to output real numbers. In most cases, it is sufficient for the codomain R to be totally ordered. This section considers both real-valued and non-real-valued fitness functions.

Typically, R is a set of floating-point numbers with a specific number of bits. The fitness function encodes the requirements for a good solution to a given problem instance x. The genetic algorithm selects prospective parents based on the fitness of individuals, similar to natural selection.

In contrast to evolution in biology, the goal of a genetic algorithm may not always be to maximize fitness. The input w to the algorithm specifies whether the objective is to maximize or minimize the fitness function, similar to the optimization problem instance (I, P, m, w).

Definition 5.32 Let P be a phenotype space. A *population* is a multiset of phenotypes in P.

A multiset is a collection that allows for repeated elements. In the context of genetic algorithms, the population refers to the set of candidate solutions at a given time. The population is the main focus of the genetic algorithm, which uses selection and modification operations to transform the current population into a new one in the next iteration.

Definition 5.33 Let P_0 be an initial population. Given a population P_i, $i \in \mathbb{N}$, a series of operations leads to a population P_{i+1}. This series of operations is called an *evolution cycle*. The index i of P_i is called the *generation*.

The process of evolving from one population P_i to the next P_{i+1} cannot be represented as a function due to the presence of random operations. The initial population P_0 depends on the application and the wishes of the programmer. It typically involves a random generation of individuals from the genotype space and decoding them into phenotypes.

Example 5.1 A simple example of a genetic algorithm is counting bits (Wilhelmstötter, 2021). The relevant parameters of the evolutionary algorithm are as follows:

- the phenotype space is $P = [0, 255] \subset \mathbb{N}$,
- the genotype space is $G = \{0, 1\}^8$,
- the genes are [0, 7], the placeholders for the values,
- encoding $e(p)$ of $p \in P$ is the 8-bit representation of p,
- the fitness of $g = (g_0, g_1, \dots, g_7)$ is $f(g) = \sum_{i=0}^{7} g_i$, i.e. the bit count of g,
- the goal of the algorithm is the maximization of f, and
- P_0 is initialized by generating random 8-bit numbers.

If we take a phenotype p with a value of 18, its binary encoding would be $g = e(p) = (0, 0, 0, 1, 0, 0, 1, 0)$ with fitness $f(g) = 2$. In this case, the fitness function is considered only as a function on G and not on $I \times G$.

This simple example illustrates the importance of carefully considering the definition of the problem. In this case, a more suitable choice for the phenotype space would have been $P = 0, 1^8$ as the fitness is only defined on the binary representations of the numbers and not the numbers themselves. This would eliminate the need for encoding and decoding operations.

The example also highlights the distinction between phenotype and genotype spaces. Without the binary representation, the exploration of possible solutions would be more difficult.

There are several different ways to alter the genetic code in each generation. Possible examples that are inspired by nature are natural selection, recombination, and mutation. While working with more structured data, many other operations are possible, which introduce a small and sometimes bigger variation to the genome.

5.8 Summary

The problem of finding a suitable computational model for the nanoscale is a central question in the nanonetworking community. While there is an array of potential candidates, not all of them are equally powerful or suited for the expected use cases. To properly evaluate the different models, we have introduced the basics of complexity theory and reductions as well as the Landau notation.

Based on that, we have analyzed the potential tasks that nanonetworks are envisioned to perform and extracted mathematical problems. We sorted those problems into complexity classes and derived somewhat realistic machine models that are fit for the task.

Among the proposed candidates, self-assembly systems seem to be a promising candidate. Realizing such systems using DNA tiles is only one possibility out of many. The advantage of such systems is that they use materials that are native to the human body and effects that are proven to work at the nanoscale. In fact, several wet-lab experiments have demonstrated that the technology can work.

Finally, we analyzed the problem of finding programs for specific nanodevices, which is surprisingly difficult and generally in the (infeasible) complexity class NEXP. Based on the "lifting" method, it is possible to find optimal solutions for some classes of scenarios nonetheless. In the end, those programs can be found using the value iteration algorithm. A feasible alternative would be approximate approaches like ANNs or genetic algorithms.

6

Simple Communication

This chapter analyzes and explains various paradigms for communication in nanonetworks. Unlike macroscopic communication, nano communication is subject to immense diversity in potential environments and it is unlikely that old paradigms can be applied. This chapter first gives a short overview of the history of long-range communication technologies ranging from telegraphs to modern 6G technologies. Next, communication is defined in its most basic form and promising communication parameters are listed. Then, we successively analyze electromagnetic, molecular, ultrasonic, and quantum communication in the context of nanonetworks. Based on this, a comparison under different assumptions is possible and potential areas of application are listed. Afterward, we try to understand how addressing is possible at the nanoscale and how to create simple routing protocols. Finally, we analyze several nanoscale networking simulators and summarize the chapter.

6.1 A Brief History of Communication

We start this chapter with a short overview of communication in general and discuss the gradual development of communication technologies.

Before we discuss any specific topic in detail, it is necessary to get a rough understanding of what communication is at its very core.

Definition 6.1 In the most general sense, a communication process is the *intentional transfer of information* from sender A to receiver B through a possibly *faulty* communication channel. The *sender* manipulates an *environmental parameter* that is capable of representing the information and the receiver B *measures* the manipulated parameter to restore the information.

If we apply this definition to human communication, we have different people who talk with each other with the intention of exchanging information. In that context, sender A and receiver B are the two people using vocal cords and ears, the channel is a volume of air, and the intention is not specified.

Human communication and communication among all animals are based on *sensor/actuator organs* that are able to manipulate and/or measure the same environmental parameter (Shannon, 1948). As a result, creatures are able to communicate via sounds, signs, smells, touch, and in a limited sense even via the sense of taste. Sometimes, living creatures (like trees) can also communicate via the exchange of pheromones that are similar but not identical to the sense of smell.

Nanonetworks: The Future of Communication and Computation, First Edition. Florian-Lennert A. Lau.

Communication among living creatures is the result of a long evolutionary process. Having the ability to sense and manipulate the environment is an advantage in a competitive environment, and those individuals with better skills were more likely to pass on their genes to the next generation. Based on this, the sensor and actuator organs were sufficiently developed at some point to exceed their original purpose. Likely by accident, creatures developed group strategies that increased their chances of survival and required some degree of communication and coordination. A group of deer is, for example, much harder to catch than an individual as many more eyes and ears could be aware of possible dangers. A single deer can then alarm the entire group, which is as a whole harder to catch as many potential targets are much harder to keep track of. In short, communication is an immense evolutionary advantage that enables group/herd strategies.

To take things a bit further into the human realm, communication allows us to form civilizations and even cultures. If every individual had to start from scratch, science and culture would be impossible. Basically, all of humanity's achievements are based on *transmitting* information from one person or group to another. Thus, the immense time investment of initially discovering or inventing something may be drastically reduced. Communication is an *enabler* and a *catalyst* for learning processes. Even reading a book is technically a communication process with high latency. In short, without communication, there is little room for complex behaviors, culture, science, or even civilization as a whole.

The first paradigm shift in the nature of communication was the emergence of mail services. Before that, letters and packages had been taken care of individually and it often took months before they reached the recipient. It was not rare that letters had a "promise of reward" assigned to them. Letters were sold from traveler to traveler until they finally reached their destination.

While there were other small systems like carrier pigeons, they were comparably insignificant. After all, there was a long process of training involved and the birds had to be shipped to a new destination only to find their way back to the old one with a small message attached to them.

All of this changed with the introduction of structured postal services. While the medium of transportation was still the same, namely stagecoaches or similar horse-powered services, sending letters became accessible to the masses due to the increase in demand. By having a dedicated service that focused on nothing else, countless letters and packages could be "bundled" and the price of a single delivery fell to affordable levels. In parts of the world, stamps even became something akin to a currency as they had a "real value" assigned to them.

The next major paradigm shift was the invention of the telegraph in 1774 by Georges-Louis Le Sage. The word *telegraph* is derived from the Greek words "tele" (distant) and "graph" (writing). Unlike prior postal services, the transmission of messages has been shifted from physical letters to electrical signals. A message could now be encoded in Morse code and transmitted closer to the speed of light.

The following paradigm shift was the patent of the telephone by Graham Bell in 1876. The word *telephone* is again derived from the Greek words "tele" and "phone" (sound). At that time, the telegraph system was already widely established and more than 19 000 km of telegraph line already spanned the United States. For the first time, a microphone (invented roughly 20 years earlier) has been used to efficiently capture a human voice and transform it into an electrical signal. This signal could now be transferred through a wire. While the principle had already been demonstrated in 1861 by the German inventor Johann Philipp Reis, the technology was not yet ready to bridge large distances. After the invention of the telephone, advancement calmed down a bit as a large network of telephone lines emerged all around the globe, even connecting different continents.

The next major innovative milestone happened roughly between 1965 and 1969 when the US Air Force started the *Advanced Research Projects Agency Network* (ARPANET). As the name and the involved agencies indicated, this network was specifically designed to accelerate the information exchange between universities and military institutions. In the beginning, the "network" only connected four universities. Nonetheless, the project offered valuable insights into digital network communication and package-based networks in general. One could even argue that the ARPANET was the predecessor of the internet as we know it today. For the first time in history, it was possible to efficiently exchange written information with nearly the speed of light. The first electronic mail ever was sent in 1971 by Ray Tomlinson and contained meaningless symbols like "QWERTYIOP" according to the author.

In 1990, the ARPANET was finally shut down and replaced by the early versions of the modern internet. At that time, the commercial phase of the internet with widespread access to the public began. In the beginning, the internet was mainly designed to exchange textual information via, for example, the *hypertext transfer protocol* (HTTP). The exchange of information mainly happened via the already-existing telecommunication network. The network at that time consisted of a backbone structure of high-bandwidth cables that connected important cities. From there, other cables connected all the less-populated areas. At the lowest level, individual households are connected to a single hub. Hence, overall, the internet could be regarded as a backbone from which recursive star graphs sprout.

Due to the already high bandwidth, the internet in that state had the potential for various novel applications that were unimaginable before. As early as 1995, people already used the internet to play digital games with remote parties or to exchange pictures and other media.

The next major step was the integration of wireless technologies into the internet. While the first wireless internet access was already available in 1991 as part of the *second generation* (2G), the technology developed significantly slower. The overall quality of the mobile internet only really changed based on the success of the wired internet. Suddenly, people wanted to have the advantages of the internet in a mobile setting. In 2001 and 2006, the 3G and 4G technologies became available, which allowed for much more complex mobile applications. As an example, 2G started with a maximum bandwidth of up to 9.6 kbit/s, while 3G offers between 0.4 and 16 mbit/s. The 4G technology already allowed for data rates of 5.8–1000 mbit/s. The high end of the spectrum can only be achieved in a stationary or slow moving scenario. The 5G technology that was released in 2018 is capable of bandwidths between 300 and 25 000 mbit/s.

In the mobile context, a heterogeneous mix of technologies is used and the problem is more difficult to solve than it seems. There is a network of satellites, cell towers, mobile devices of all kinds, and wired technologies involved. Furthermore, novel and much more complex communication protocols are necessary to handle the potential movements of devices between cell tower zones.

After this phase, we enter the territory of technologies that are still in development. Among those technologies are *wireless sensor networks* (WSNs). They often consist of a number of battery-powered, wireless devices that are designed to measure, aggregate, and communicate environmental information. Up to 95% or more of the required energy is used for wireless communication alone. While they existed as early as the Cold War, they only really gained scientific popularity in the 1990s as computer hardware became more and more powerful. In 2006, Akyildiz et al., already published extensive bodies of work on the recent and possible future developments of WSNs, even though most parts of the public have not heard about the technology yet (Akyildiz et al., 2006).

Another very related technology is the *Internet of things* (IoT). The idea for such a network seems to date back as early as 1985 when Peter T. Lewis first mentioned the term. As estimated by Cisco System, the first real-world IoT systems were realized between 2008 and 2009. In 2022, IoT devices became the new normal, and people regularly use devices like "Alexa."

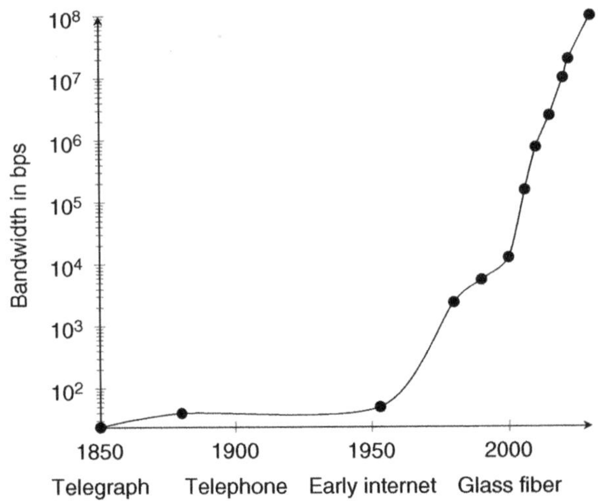

Figure 6.1 Average data rates between 1850 and 2030. While early wire-based technologies could have achieved higher theoretical data rates, other factors limited the actual speed. *Source*: Florian-Lennert A. Lau.

Figure 6.1 summarized the developments of telecommunication technologies since 1850. In the beginning, the bandwidth was extremely low but sufficient for the applications of the time. Especially since 1950, there has been an exponential increase in communication bandwidth that lasts until today.

From here on, we can only peek into the future of network technologies, which is precisely what this book is all about. Based on the observed increases in efficiency and the continued reduction in device size, it is only natural to assume that future network devices will be even smaller than today. Hence, Akyildiz et al., introduced the term "nanonetwork" in 2008 (Akyildiz et al., 2008). The nanonetwork paradigm includes all kinds of technologies that include nanoscale components.

Networks or the overarching infrastructure consists of a variety of possible devices on all scales. The novel aspect lies in the communication at the nanoscale, which fundamentally differs from macroscopic technologies. It is likely not possible to simply use slight variations of already existing technologies to solve the various new challenges at the nanoscale.

The entire nanonetworking area is a growing area of research where the scope of future applications and developments has not been fully explored yet. The remainder of this section focuses on providing an intuition on the most fundamental problems and possible solutions for networking challenges at the nanoscale. We mainly discuss everything related to point-to-point communication and only rudimentary network protocols. Actual network architectures are the topic of Chapter 12, which relief on all the previous chapters.

6.2 Definitions

With all of this in mind, we can now define communication at the nanoscale. The fundamental Definition 6.1 still applies but can be specified further. As a result, machine communication can be defined as follows:

Definition 6.2 *Machine-to-machine communication* is the transfer of coded information from one *machine model* to another machine model in the form of *words* using a priori information for interpretation in the form of *communication protocols*. To do this, an actuator manipulates an environmental parameter, which is measured by a sensor at a remote location.

The same principles as in regular computers apply at the nanoscale but less resource-intense technologies are required. Before any information exchange can take place, it is necessary to encode the information in some way or another. Afterward, the encoded information has to be *serialized*. Serialization describes the process of transforming information into a format that can be transmitted over a medium. At the destination, the serialized message has to be transformed back into a format that is suitable for the respective machine model.

Thus far, we mainly defined point-to-point communication. While this is certainly necessary, nanonetworks likely have a limited range of communication and an entire network of devices is necessary to transmit information all the way to a destination. To model this, we have to introduce several additional components.

To model entire communication networks, we typically use *graphs*.

Definition 6.3 A graph is a tuple $G = (V, E)$ from a set of vertices V and a set of edges $E \in V \times V$.

Each of the vertices $v \in V$ represents one communication endpoint. This could be a repeater that just amplifies a message, another (nano-)node, or any other device with the ability to communicate and the necessary a priori knowledge on how to understand the exchanged messages. In the context of this book, all vertices $v \in V$ represent at least nanonodes or more powerful devices. For detailed explanations of the different devices, you can go back to Figure 4.8.

That said, each nanonode is defined based on a *machine model* that specifies the general capabilities. Among them are the computational capabilities, the amount of memory, and the otherwise available components. Some of them are defined in more detail in the upcoming chapters. In short, the model fully specifies a nanodevice.

The edges $e \in E$ represent possible connections between nanonodes. Those connections are usually modeled in an abstract way to allow for all the possible implementations like wired, electromagnetic waves, or even molecules. That said, the edges themselves only model a single property of a possible connection.

Edges can also have values or *weights* assigned to them that model different properties of the connection. In the simplest case, they can be the distance, the bandwidth, or the latency. If more than one of these properties is considered at the same time, routing algorithms usually get quite complex and oftentimes infeasible.

In addition to such simple properties, edges can also be represented by a complex *channel model* that represents the physical environment in detail. In the most simple case, a channel model usually just considers a single edge and is represented by a cuboid. More complex channel models might consider different media and a variety of physical properties. There are entire physics simulations with the sole purpose of modeling edges at the nanoscale. Among the most popular software is the COMSOL multiphysics simulator (Multiphysics, 1998). The interested reader is referred to that manual for much more detailed information on the topic that is part of its own area of research.

The last necessary component for networks of devices is communication protocols.

Definition 6.4 In their most abstract form, *protocols* are programs that control the syntax, semantics, and synchronization of a communication process.

Thus, protocols are special programs that get *assigned* to all the nodes in a network. Without such programs, the nanonodes would not possess the necessary knowledge to interpret the messages they receive. After all, messages are basically just long binary numbers with no apparent structure. Interpreting the various different parts of a message requires the information specified in the protocol.

Based on all those components, we can now define nanonetworks as a whole.

Definition 6.5 A *nanonetwork* is a directed ad hoc graph $G = (V, E)$, where V is a set of nanonodes $\mathcal{N}_N$ operating in an environment S. Nanonodes are nanodevices that also have the necessary communication component $C \in K_{\text{mand}}$. E is a set of edges $e \in V \times V$.

Definition 6.6 A *medical nanonetwork* is a nanonetwork in which the participating nanonodes take the environmental constraints S that can occur in the human body into account.

6.2.1 Gateways

Yet another network component that is worth mentioning is a *gateway*. Gateways are devices that connect different networks with each other. In the context of WSNs, they have a special meaning as they are often more powerful than regular nodes. Thus, gateways usually *gather* the collected information of an entire network, perform necessary computations, and send the modified data to an external device.

In the context of nanonetworks, gateways are especially interesting as nanonodes possess next to no resources that would enable them to communicate with the internet. As a result, gateways can "bridge" that gap and transmit messages from the nanoscale to either the microscale or the macroscale. If every node had to send the gathered data directly to the internet, the overall energy requirement would be greatly increased.

An example can be seen in Figure 6.2. Many nanosensors communicate inside the human body, collect data, and transmit it to the outside. On the skin, there is a more powerful gateway that aggregates the data and communicates it to, for example, a smartphone.

While it seems likely that gateways will be used for nanonetworks in one way or another, they are the topic of ongoing research. Especially the possible combination of different communication paradigms provides unique novel challenges for network designers. It is by no means obvious how a molecular signal can be properly measured, quantified, and turned into a digital signal. The problem becomes even more difficult if multiple devices or several communication types are involved (Zafar et al., 2021).

6.2.2 Communication Parameter Overview

While we mainly use wire connections and Wi-Fi to communicate in regular computer networks, things are not as simple at the nanoscale. It would be very welcome if the already established

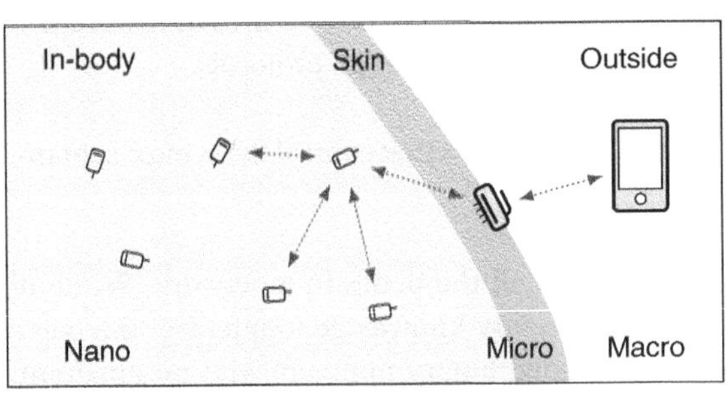

Figure 6.2 Gateway device that connects an in-body nanonetwork and less resource-constrained networks on the outside.
Source: Büther et al. (2018)/with permission of Association for Computing Machinery.

technologies could be used but that is by no means certain. Thus, we analyze different possible communication technologies at the nanoscale.

Overall, there are four major candidates:

- Terahertz communication with electromagnetic waves as a communication parameter.
- Molecular communication with molecules as a communication parameter.
- Acoustic communication with sound waves as a communication parameter.
- Quantum entanglement with unknown parameter.

We now discuss each of them in the following sections in some level of detail and list challenges as well as opportunities.

However, certain requirements must be met. No matter what parameter we choose, it should be capable of transmitting some information over a significant distance without introducing too many errors. In addition, it would be desirable if the communication latency is as small as possible. Nanonodes are often considered to be mobile devices, and the entire network topology might have shifted significantly over the course of a communication. Furthermore, the bandwidth of the medium must be high enough to meet the respective application's requirements.

Furthermore, information is usually encoded in binary. Whatever communication medium we use, it would be good if the states "1" and "0" could be distinguished properly from each other. To give you an intuitive understanding, the following list shows several possible examples of information encoding:

- On a wire, using high/low voltage through thresholds,
- via molecule concentrations and thresholds,
- via distinguishable particle types,
- via particle order,
- via the number of particles, or
- via signal amplitudes in wave-based communication.

While there are many more possibilities, these are the most frequently mentioned methods. We start by analyzing electromagnetic communication in detail.

6.3 Electromagnetic Communication

Electromagnetic communication at the nanoscale is based on the same principle as electromagnetic communication via Wi-Fi routers. These waves propagate circularly through the communication medium. In the case of either a body of air or the body of a living creature, different communication ranges are possible. The higher the frequency of the electromagnetic waves, the shorter the possible communication range. In that sense, bandwidth and range are a tradeoff.

Electromagnetic communication is one of the most often mentioned kinds of communication in the nanonetworking community, and the appeal is obvious. First, it would be possible to use variations of identical technologies in a new environment. Thus, there is a smaller barrier of entrance for many researchers who originate from the wireless communication field.

Second, nanonetworks are often envisioned to function inside difficult environments where wired connections are very unlikely if not outright impossible. In a dynamic environment like the human circulatory system, wired connections cannot be implemented. Yet, a considerable amount of proposed use cases utilize exactly that environment.

Third, communication via terahertz waves might have a low communication range, but they have an immense bandwidth. As we are limited to nanoscale structures, antennas cannot exceed that size as well. In general, antennas work best when the wavelength of the signals is about four times the length of the antenna. Even if we wanted to, it would physically be impossible to use electromagnetic waves of higher wavelengths. As a result, electromagnetic communication at the nanoscale is likely limited to not more than 2 mm. This is mainly due to molecular absorption through hydrogen.

6.3.1 History and Driving Forces

Before we analyze other properties and limitations of electromagnetic communication, we list the most important stepping stones and driving forces in wireless communication. Based on the observations, it might be possible to predict future developments toward utilizing the terahertz band.

One of the key milestones in the history of terahertz waves was the prediction of their existence by James Clerk Maxwell in the 19th century. In his mathematical equations describing electromagnetic waves, Maxwell predicted the existence of waves with frequencies between those of microwaves and infrared light. However, at the time, there was no technology available to generate or detect these waves, so they remained purely theoretical.

It was not until the late 20th century that the technology was developed to generate and detect terahertz waves. This was made possible by advances in lasers, high-speed electronics, and other technologies. In the 1990s, researchers at the University of California, Berkeley, were able to generate terahertz waves using a laser and detect them using a high-speed detector.

Since then, terahertz waves have been used for a variety of applications, including imaging, spectroscopy, and communication. In medical imaging, for example, terahertz waves can be used to create images of tissues and other materials in the body, allowing doctors to diagnose diseases and other health conditions. In spectroscopy, terahertz waves are used to study the properties of materials, such as their chemical composition and structure. In communication, terahertz waves are being explored as a potential future technology for high-speed wireless communication.

While those are rather general applications, some companies use terahertz waves for specific products. As an example, they have been explored for use in security scanners. These waves can penetrate many materials that are opaque to other forms of electromagnetic radiation, such as clothing, paper, and some types of plastic. This makes them potentially useful for detecting hidden objects, such as weapons or other contraband, on a person's body. There are several different types of security scanners that use THz waves. One type is a handheld scanner that can be passed over a person's body to detect hidden objects. Another type is a fixed scanner that a person can walk through, similar to a metal detector. THz-wave security scanners have several potential advantages over other types of scanners. They can detect a wider range of materials than X-ray scanners, and they do not use ionizing radiation, so they are safer for the person being scanned.

Terahertz waves have also been explored for use in protein recognition. Protein recognition describes the process of identifying specific proteins based on properties such as their shape and chemical composition. This is an important tool in many fields, including medical research and drug development. Terahertz waves are useful for protein recognition because they can interact with the specific arrangement of atoms in a protein, known as its secondary structure. By studying the way that THz waves are absorbed or scattered by a protein, researchers can learn more about its structure and identify it among other proteins.

Several different methods have been developed to use THz waves for protein recognition. One method is terahertz time-domain spectroscopy, which involves measuring the way that THz waves are absorbed by a protein over time. Another method is terahertz reflection spectroscopy, which involves measuring the way that THz waves are reflected off of a protein.

While there are many more applications, we focus on THz waves for communication purposes for the remainder of the book.

6.3.2 5G and 6G

Before we come to the current state of the art in THz communication, it is beneficial to understand the *fifth generation* (5G) of wireless communication. 5G provides faster and more reliable connections than previous generations of wireless technology and also supports a wider range of applications. Those include IoT applications, remote surgery, and autonomous vehicles. 5G technology uses a range of different frequency bands, including low-band, mid-band, and high-band frequencies. The low-band frequencies have a higher range, but a lower bandwidth, and the high-band frequencies have a higher bandwidth, but a limited range.

5G supports the following modes of operation:

1. Low: 600–700 MHz – 30–250 MBit/s
2. Med: 2.5–3.7 GHz – 100–900 MBit/s
3. High: 25–39 GHz – 1–3 GBit/s

Low-band frequencies, such as those in the 600 MHz range, can provide good coverage over long distances, but may not be able to support the high speeds that are possible with 5G. Mid-band frequencies, such as those in the 2.5 GHz range, can provide a balance of coverage and speed, while high-band frequencies, such as those in the millimeter wave range, can support very high speeds but may not have as good coverage as lower frequencies.

While 5G is still in development and will be steadily improved for years to come, research on the *sixths generation* (6G) has already started in 2017 and targets bandwidths of around 400 GBit/s. Especially for nanonetworks, 6G is more or less mandatory as antennas can likely only be a few nanometers in length. As a result, only very small wavelengths may be detected or emitted by them. That said, some of the often-stated concerns in classical mobile communication might not at all occur in nanonetworks, while other new challenges will likely arise.

6.3.3 Channel Models

While wet-lab experiments with THz waves are already possible, the experimental settings are usually very different from the proposed practical applications in nanonetworks. As a result, one of the biggest challenges is the accurate simulation of the environmental conditions in which THz waves might be used. For an overview of THz channel models, refer to the work of (Lemic et al., 2019).

To model the channel for THz communication in nanonetworks, several factors must be taken into account. They include the physical properties of the environment, such as the presence of obstacles and materials that can absorb or reflect THz waves, as well as the characteristics of the nanoscale devices themselves, such as their size and orientation. Additionally, the channel model must account for the effects of multipath propagation, which occurs when THz waves are reflected off of objects in the environment and can cause signal fading and interference.

Several approaches can be used to model the channel for THz communication in nanonetworks. One common approach is to use computational models that simulate the physical properties of the environment and the characteristics of the devices. Some of those models are based on physical laws, such as Maxwell's equations, and can be used to predict the strength and direction of the THz waves at different positions in space.

Another approach is to use measurement-based models, which are developed by collecting real-world data on the channel using specialized equipment. This data can be used to create statistical models of the channel. The data may then be used to predict the performance of the communication system.

In both cases, the goal of the channel model is to accurately predict the behavior of the channel so that it can be taken into account when designing and optimizing the communication system. This is particularly important for THz communication in nanonetworks, where the limited size and power of the devices can make the channel a major limiting factor.

6.3.4 Information Representation

There are several ways in which information can be encoded in THz communication in nanonetworks (Lemic et al., 2019). One common approach is to use digital modulation, which involves modulating the amplitude, phase, or frequency of the THz carrier wave to transmit data. There are several modulation schemes that can be used in the electromagnetic nanonetwork, and they are as follows:

- In *amplitude shift keying* (ASK), the amplitude of the information carrier wave is varied to represent different values.
- In *phase shift keying* (PSK), the phase of the wave is varied to represent different values.
- In *frequency shift keying* (FSK), the frequency of the wave is varied to represent different values.
- In *quadrature amplitude modulation* (QAM), both the amplitude and phase of the wave are manipulated to transmit multiple values simultaneously.

Due to the possibly severe energy restrictions as well as the restraints in computational power, it might not be feasible to communicate via *continuous waves* (CWs). As a result, it might be better to use *impulse radio* (IR), where a transmitter only sends data for a fraction of time and idles otherwise. Many publications propose a simple *On-Off keying* (OOK) scheme for nanonetworks. This is a low-cost method of transmitting data, where a digital "1" is represented by a high amplitude and a digital "0" is represented by a low amplitude or no carrier wave.

In OOK systems, the carrier wave is typically a sinusoidal signal at a fixed frequency. The amplitude of the carrier wave is modulated by switching it on and off, or "keying" it, to transmit the data. The on and off states correspond to the digital "1" and "0" values, respectively.

OOK is a useful modulation scheme for low-data rate applications. Prime examples are WSNs or nanonetworks, where a simple and robust method of transmitting data is desirable.

While the fundamental principle is always the same, several variations of OOK have been proposed. A detailed explanation of all the schemes is too much for such a general book and we focus on the method introduced in the work of Jornet and Akyildiz (2014). *Time-Spread On-Off-Keying* (TS-OOK) is specifically designed for use in nanonetworks. TS-OOK is a simple method for transmitting data that is well suited for nanonetworks due to its low power requirements and simple implementation.

In TS-OOK, bits are encoded by the presence or absence of a signal within a time period. The signal is referred to as "on" when it is present and "off" when it is absent. If the time delay between signals is known in advance, there is no need for synchronization except for the very beginning of a message. TS-OOK is a form of time-spread amplitude modulation, where the duration of the signal pulse is used to represent the data.

TS-OOK has several advantages over traditional OOK. First, it allows for a higher data rate to be transmitted as more data can be encoded in a given time. Second, it is more robust against errors as the longer duration of the signal pulse allows for better error correction. Finally, TS-OOK is more energy efficient as it allows for lower power consumption by the transmitters and receivers.

6.4 Molecular Communication

Molecular communication is an alternative approach to electromagnetic communication via terahertz waves at the nanoscale. This type of communication is inspired by naturally occurring communication processes via hormones or pheromones. The first hormone was isolated as early as 1900 by John Jacob Abel (Müller-Jahncke et al., 2005). This hormone was initially called epinephrine and is better known as adrenaline today. It serves the purpose of natural *in-body* communication and prepares the entire organism of mammals for fight or flight by suppressing pain and releasing energy from the cells into the blood. By adopting this principle, it should be possible to implement long-range but low-data rate communication inside living organisms.

A similar observation has been made by Adolf Butenandt who observed a molecular communication process between different members of the same species (Karlson and Butenandt, 1959). The process is very similar to communication via hormones; only the communication channel is different. Instead of in a liquid medium, pheromones are usually transmitted through the air. The process can further be compared to the sense of smell that also detects particles. However, most pheromones cannot be consciously detected but influence the behavior of creatures nonetheless.

6.4.1 Classical Molecular Communication

Molecular communication in nanonetworks is a form of communication that uses molecules to transmit information between nanodevices. These devices can be biological or man-made, and they are unfathomably smaller than traditional communication devices, such as smartphones or laptops.

In molecular communication, information may be encoded in a variety of different ways. The most obvious methods would be either the concentration or the type of the molecules in question. The work of (Farsad et al., 2016) offers an overview of the most popular encoding schemes. They are based on the following underlying principles:

- concentration,
- molecule type,
- molecule ratios,
- time of emission,
- molecule order,
- DNA words, or
- a combination of the above.

Based on these building blocks, more complex molecular communication protocols may be devised. Oftentimes, these are motivated by techniques like OOK from WSN.

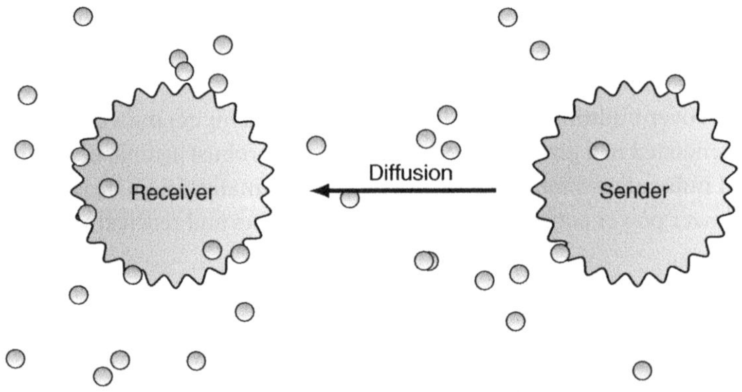

Figure 6.3 Simple molecular communication channel model. *Source*: Florian-Lennert A. Lau.

Molecules may be detected by other devices, which interpret the message based on the characteristics of the molecules that were received. The overall idea is depicted in Figure 6.3. The sender to the right emits certain molecules that travel through the communication channel, and the concentration of molecules can be detected by the receiver to the left. The process that carries the molecules from one place to another is usually assumed to be diffusion. Diffusion describes the tendency of particles to achieve an equal distribution throughout the entire medium given enough time.

Figure 6.4 shows the basic principle of detection of certain molecules. The orange octagons represent molecules with specific sizes, shapes, and chemical/electrical properties. Based on these, a binding interaction between the receptors and the molecules is possible. The molecules attach to the receptor, and a signal cascade is triggered that communicates the successful detection.

There are many possible particles and techniques that allow for information exchange between nanodevices (Bi et al., 2021). As early as 1987, researchers already analyzed the speed at which different particles move. As an example, nowadays calcium ions are often mentioned as a possible way to implement molecular communication (Donahue and Abercrombie, 1987). They are mainly subject to diffusion as a means of transportation and can achieve speeds of up to $5.3 \cdot 10^{-10}\ \mathrm{m^2/s}$.

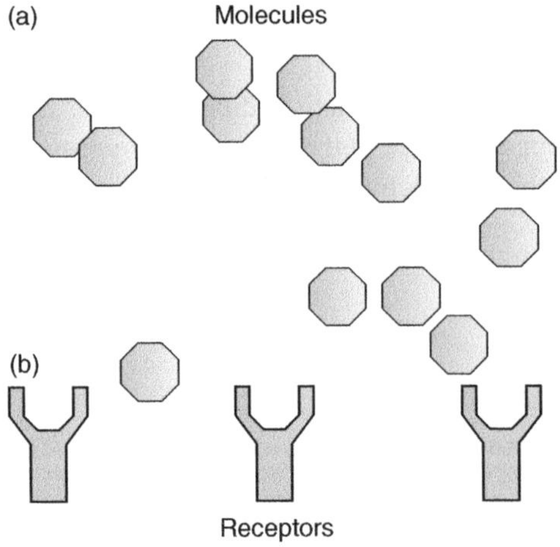

Figure 6.4 Interaction between a simple receptor (b) and fitting molecules (a). *Source*: Florian-Lennert A. Lau.

Next to calcium ions, there is an entire class of particles like the aforementioned pheromones and hormones that propagate using diffusion. Depending on their size, they move at different velocities. As a rule of thumb, the smaller a particle is, the faster it can move.

Yet another class of molecular communication is based on the *advection–diffusion* process (Fagrell et al., 1977). Advection–diffusion is a process used in mathematics and physics to describe the movement of material or energy within a medium. This movement can be caused by advection, which is the movement of the material through the medium, and diffusion, which is the spreading of the material due to differences in concentration.

An example of the advection–diffusion process is the spread of pollutants in a river. The pollutants are carried downstream by the flow of the river (advection), but they also spread out due to differences in the concentration of pollutants in the river (diffusion). The advection–diffusion equation describes how the concentration of pollutants in the river changes over time and how it spreads out in different areas of the river.

The same principle can be applied to the human bloodstream and the exchange of particles between capillaries and cells. Both forces combined, as well as some degree of Brownian motion, model the movement of particles sufficiently well. Due to the much stronger advection effect, particles may move at speeds of up to 3.5^{-4}–$9.5 \cdot 10^{-4}$ m/s.

The last naturally occurring effect worth mentioning in this context is kinesin propagation. Kinesins are a class of the so-called *motor proteins* that are capable of transporting particles in a targeted manner through cells. They usually use a rail system of *microtubule* to move large molecules that cannot be transported using diffusion alone. As an intuition, an often extremely big piece of cargo is attached to the kinesin which drags it to the desired location along the predefined railway system.

Finally, the emitted molecules have to be measured at some point. A variety of different effects can be used to achieve that. A few examples are antibody-antigen interaction, gap junction (holes in cells), or simple ligand–receptor interaction to count the concentration of molecules or to identify their type. A special type of detection might be implementable using DNA.

6.4.2 DNA

While DNA is most well known for its naturally occurring properties of encoding genes, the molecule can be modified in many different ways. DNA is one of the most well-researched molecules in existence, and it can be manufactured in almost arbitrary shapes and sizes in the laboratory. As early as 1982, Seeman already suggested using DNA as a material for construction, deviating from the naturally intended purpose for the first time (Seeman, 1982).

Similarly, DNA may also be repurposed as a molecule for communication. Especially the DNA tiles mentioned in Chapter 4 are prime candidates for communication purposes. First of all, they are comparably small and thus still subject to diffusion as a means of transportation. Depending on the manufacturing details, tiles can be as small as 10 nm in diameter.

Additionally, tiles have the advantage that they come with a built-in addressing mechanism that can be encoded in the open ends. They behave like normal molecules during any diffusion process but may only bind to their respective inverse base sequences. As a result, it should be much easier to implement point-to-point communication in nanonetworks. As of right now, most experiments focus on the exchange of information between exactly two parties. That limitation might be removed by using simple DNA tiles or other DNA-based molecules as a means of communication.

While a single type of DNA tile is already a powerful candidate for molecular communication, the main benefit of tiles lies in the potential for interaction with other tiles or DNA-based technologies.

As explained in Chapter 5, if many intelligently designed tiles are combined in a communication channel, they start interacting. That interaction usually leads to the assembly of a DNA-based crystal. However, the immense potential lies in the fact that this assembly takes place under certain conditions that are encoded in the open ends of the tiles. Thus, those DNA crystals can be designed to only fully assemble once a computation successfully finishes. The resulting structure is called a message molecule $\mathcal{M}_\Phi$, where Φ is the Boolean formula that the molecule computes.

An example can be seen in Figure 6.5. The right tiles with a dashed outline represented the part of the message molecule that performs a computation. The left part represents the *ligand* of the molecule that can only form once the remaining molecule is already assembled sufficiently. Furthermore, the ligand may *only* form in case of a "successful" computation. Message molecules mainly solve *decision problem*. As a result, a finished message molecule always also represents a "yes" decision. As an example, a finished molecule could mean that there is high confidence that a certain disease is present.

The finished molecule has at least two unbound glues at the left that can interact with other DNA-based devices. An example can be seen in Figure 6.6. The two glues and the form of the molecule allow for a stable binding with the receptor to the left. The binding might trigger another mechanism. This might be the opening of a DNA box that either directly treats a disease, amplifies the signal by releasing other already complete message molecules, or simply communicates the successful computation via fluorescence markers to the outside.

Finally, DNA-based technologies come with the advantage that they are per definition biocompatible as they naturally occur in all living organisms. However, tiles also interact with their physical surroundings and many unwanted side effects are to be expected in realistic environments. It is more than likely that all kinds of proteins and molecules in the human body will "coat" any DNA molecule and thus change its intended properties. Furthermore, the "waste material" will likely also interact with the human body in one way or another. At least parts of the tiles that are no longer used for communication will likely be transported into cells and might cause harm there. While we have filter organs that can dissolve DNA, unwanted side effects cannot be properly predicted and thus cannot really be avoided. While those effects are not unique

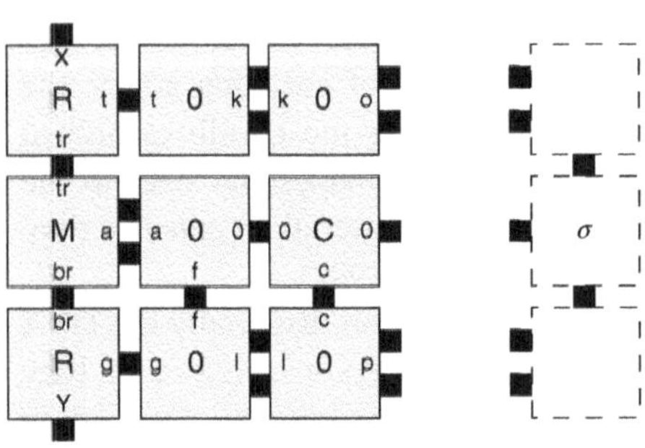

Figure 6.5 Ligand of a message molecule at temperature 2. The tiles to the left with marker "R" can only attach to the assembly if tile C in the middle has a right neighbor. The white tiles with a dashed border to the right represent an adjacent message molecule that has to be finalized before the ligand can form (Lau, 2020). *Source*: Florian-Lennert A. Lau.

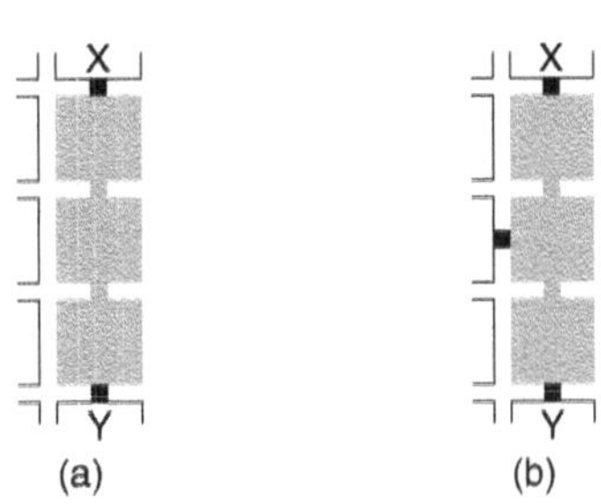

Figure 6.6 Receptors, to which message molecules can stably bind (a) at temperature 2 (b) at temperature 3. The binding condition is encoded in the open glues with markers "X" and "Y". *Source*: Lau et al. (2019)/with permission of Elsevier.

to DNA-based devices and will likely occur more drastically with other types of devices, it is still necessary to mention them.

6.4.3 Channel Models

Molecular communication channel models describe the factors that may affect the transmission of information via molecules. The information itself may be encoded in the type and concentration of molecules used, their size and shape of the molecules and nanodevices, and the presence of obstacles that can interfere with the transmission. Channel models may also consider other chemical, physical, and biological factors into account if they have a potential influence on communication.

Different types of molecules have different properties that can affect how they are transmitted and received. For example, some molecules may be more stable or reactive than others, which can affect their movement and interaction with the environment. The concentration of molecules can also affect the transmission of information as higher concentrations may lead to faster or more reliable transmission. However, too high a concentration may lead to interference or saturation of the system. In addition, temperature, humidity, and other factors in the environment can also affect the properties of the molecules.

In general, most channel models are defined using the mathematical tools of information theory. In the most simple sense, a communication channel can be described by a mathematical function that receives an input signal and transforms it into an output signal. The function may also consider an arbitrary set of other influence factors named above that may affect the signal.

As computations slow down drastically if more parameters are considered at once, most publications focus on a small number of influences. A good overview of different channel models in the work of Farsad et al. (2016). Most of them use diffusion as a means of transportation for molecules, while the earliest works focused on Brownian motion (Eckford, 2007). More complex models also consider the blood flow in addition to diffusion processes.

However, the vast majority of models do not take the complex chemical environment of, for example, a human body into account. As of right now, it is simply not feasible to simulate realistic conditions in a meaningful way. Especially advanced and precise bottom-up physics simulation engines like COMSOL multiphysics reach their limits very fast (Multiphysics, 1998). In most cases, only a handful of molecules may be simulated at once, while the entire process might already take several hours to finish. As a result, the complex task of simulating biological communication channels is modularized into many smaller pieces in a top-down approach. Some information likely gets lost in the modularization process, and it is unclear if the sum of the partial solutions offers a solution for the initial problem. Nonetheless, modularization at least offers some means of getting insights into this problem even if it is not perfect.

6.5 Acoustic Communication

Another alternative for communication at the nanoscale is *acoustic communication* (Hogg and Freitas Jr., 2012). Acoustic communication works based on sound waves instead of electromagnetic waves but behaves similarly otherwise. The communication signal is transmitted by a mechanical wave, and the speed of the communication is limited by the communication medium. As an example, sound waves travel at roughly 343 m/s through the air. When acoustic signals are transmitted through water or watery solutions, the speed can be as high as 1481 m/s.

Among the earliest motivations for an alternative to electromagnetic communication are submarines. Submarines have to communicate with their surroundings but are usually below a thick layer of water for prolonged periods of time. Water absorbs any type of electromagnetic wave and limits the signal propagation distance drastically. To still be able to send signals to submarines, extremely long antennas and low-frequency signals were necessary. Submarines usually pulled long cables behind them that acted as antennas, while huge antenna towers were necessary on land. As an example, US submarines communicated via 76 Hz signals during the Cold War. This limits the data rate to just a handful of bits per second.

Due to the severe limitations in both the bitrate and communication distance at the nanoscale, researchers stumbled upon acoustic communication as an alternative. One could argue that this type of communication is inspired by whales or other sea-dwelling creatures who naturally communicate via sounds over sometimes 2000 km distance.

As a result, underwater robots often communicate and map their surroundings via sounds (Steinmetz and Renner, 2022). For this, the so-called *underwater acoustic modems* are used. Depending on the desired range and data rates, modern underwater modems can achieve up to 62.5 kbit/s using a 120–180 kHz frequency range over 300 m (Evologics, 2000). In good environmental conditions with little channel noise, distances of up to 8000 m or more can be achieved while maintaining a data rate of up to 6.9 kbit/s.

Another already-established application in medicine is ultrasound in the 1–70 MHz range. Ultrasound is mainly used as a diagnostic tool to visualize certain areas of the human body. The most well-known application would be fetal imaging, but the technology is also used to analyze cancer tumors. While ultrasound imaging has many advantages, the resolution of the technology is rather bad and it is often difficult to correctly visualize structures below 1 cm in size.

6.5.1 Nanoscale Acoustic Communication

At the nanoscale, the environmental constraints share many more similarities with medical ultrasound than submarines. Oftentimes, possible applications for nanonetworks target the human body or parts thereof. As a result, many of the macroscopic problems with acoustic underwater communication do not necessarily apply. However, the environmental diversity of the human body and bloodstream might pose additional problems that are already difficult to deal with in underwater robotics.

As the human body mainly consists of water, the speed of sound can be estimated to be around 1500 m/s at an ambient temperature of around 37 °C. Furthermore, the human body is not as homogenous as a body of seawater – it consists of many different types of tissue that often rapidly change from location to location. As a result, acoustic waves that carry desired information will likely be reflected or scattered by different tissue types, leading to signal inference. As a result, parts of the signal may be canceled out, amplified, or modified, leading to different kinds of communication errors. That said, at the proposed communication frequencies of 10–100 MHz and the resulting wavelength of 15 μm, little interaction with surrounding tissue is to be expected (Hogg and Freitas Jr., 2012). Furthermore, due to the attenuation of the acoustic waves, communication ranges of more than 100 μm are unlikely.

6.5.2 Medical Constraints

While the proposed range of 100 μm sounds a bit less promising than the 2 mm of electromagnetic communication, there are some additional limitations when it comes to acoustic communication.

As an example, applying sound waves of too high an intensity can result in tissue damage. It is not rare that visitors to concerts who stand too close to speakers end up with mechanical organ damage due to pressure variations. Similarly, continuous exposure to vibrations can also heat up tissue but is considered safe at energies of around 2000–6000 W/m^2.

These dangers might not be too realistic as nanonetworks are severely energy restrained and likely will not be able to output as much energy to cause lasting harm. As an example, acoustic nanorobots will likely communicate pulse-based and thus only transmit data a portion of the time. The maximum amount of energy a robot of around 4000 nm in diameter might output is around 100 pW, which could generate 1000 Pa pressure variations. In comparison, cell membranes are believed to rupture at around 100 000–1 000 000 Pa. In general, the smaller the nanorobot, the less energy it has at its disposal. The nanodevices analyzed in this book come nowhere near the required energy levels to damage a cell membrane.

While the results from the work of Hogg and Freitas Jr. (2012) sound promising, there is much more work to be done. The effect of environmental noise needs to be studied as this is one of the major problems with underwater robots too. Unlike electromagnetic waves, mechanical waves naturally occur at all times and in most frequency bands. Thus, some inference is always to be expected. As an example, it can be surprisingly difficult to maneuver a macroscopic aquatic robot using just acoustic waves to map its surroundings. The same task is likely even more difficult given the limited resources and generally less powerful hardware at a very small scale, let alone the nanoscale. Nonetheless, more research is necessary and will shed light on the remaining open questions.

6.6 Quantum Communication

In 1935, Einstein, Podolsky, and Rosen published a paper intended to criticize quantum mechanics (Einstein et al., 1935). The authors suggested a thought experiment that led to a contradiction with the then-accepted limitations of physics. By entangling two photons (this would also be a special case of electromagnetic communication) and later measuring their spin, the information had to be transmitted faster than the speed of light, namely instantaneously. While the authors viewed that as a contradiction at their time, it proved to be true. This "spooky action in the distance" is still an active research subject and is often proposed as a technology for communication. In 2021, researchers proved that it is possible to bridge distances of around 4600 km in an instant (Chen et al., 2021).

Quantum entanglement is a phenomenon that occurs when two or more particles become connected in such a way that the state of one particle can affect the state of the other, even if they are separated by large distances. This connection between the particles is called an *entanglement*, and it can be thought of as a sort of "link" between them.

One way to create an entangled pair of particles is to start with a single particle and split it into two smaller particles, such as by shooting a photon through a beam splitter. The resulting two particles will be entangled in a way that cannot be explained by classical physics. When the state of one of the particles is measured, the state of the other particles will be the exact opposite, instantaneously transmitting the information of the measurement. At least in theory, this works over arbitrary distances.

However, very specific environmental conditions are necessary to maintain the state of entanglement (Homeister, 2008). As soon as it is even *possible to know* the state of any of the entangled particles, the wave function collapses and both particles assume one of the two discreet states. In a sense, the "possibility of knowing" the state of a particle already counts as a measurement.

In order for quantum communication to work, it is necessary to isolate the particles from the environment around them. This is because any interaction with the environment can cause the quantum state of the particles to change, which can corrupt the information being transmitted. As a rule of thumb, bigger particles or even molecules are much harder to isolate than small particles. There are several ways to isolate quantum particles from the environment, such as using specialized shielding to block external influences or cooling the particles to extremely low temperatures to reduce thermal noise. It is also possible to use error correction techniques to mitigate the effects of environmental noise on the quantum state of the particles.

While it is already possible to use this effect for macroscale communication, researchers also propose the use of quantum communication at the nanoscale (Cacciapuoti et al., 2022). The idea is surely enticing as an instantaneous transmission of information without channel noise would get rid of a majority of the challenges in nanonetworks. However, there are several challenges when it comes to using quantum communication at the nanoscale. First, the particles that are suggested for communication do indeed fulfill the size requirements as fundamental particles are more or less "points" without any measurable size.

The devices that are necessary to isolate particles from their environment and the devices that are necessary to measure the state are much bigger. In fact, they are so much bigger and more complex that there is no realistic way of establishing quantum communication channels using nanodevices with currently available technologies. The use of quantum communication in nanonetworks is an active area of research and development. There has been significant progress in recent years in demonstrating the feasibility of this approach. However, it is still an emerging field, and there are many challenges that need to be overcome in order to fully realize the potential of quantum communication in nanonetworks. Thus, this book will not analyze quantum communication for nanonetworks in more detail.

6.7 FRET

Förster resonance energy transfer (FRET) is a phenomenon that occurs when the energy of an excited state molecule (the donor) is transferred to another molecule (the acceptor) through non-radiative energy transfer (Kulakowski et al., 2017). This process can occur between molecules that are in close proximity to each other and is important in many biological processes, including protein folding, DNA replication, and cell signaling. When a molecule absorbs light at a specific wavelength, it enters an excited state. As this state is unstable, the molecule will release the absorbed energy sooner or later.

The efficiency of FRET depends on the distance between the donor and acceptor molecules, as well as the relative orientation of their electronic transitions. When the distance between the molecules is too big or the orientation is not favorable, the energy transfer is inefficient and the donor molecule will fluoresce instead. Typical Förster distances are between 3 and 9 nm, which would limit the transmission range to around 10 nm. If more than one donor is used at the same time, the distance might be increased, but the interaction efficiency decreases with the sixth power of the molecule distance. If the conditions are favorable, the delay of the energy transfer is often less than 20 ns.

Based on this, it is possible to measure the energy intensity to determine the distance between the sender and the receiver. By attaching fluorescent molecules to different parts of a nanonetwork and measuring the energy transfer between them, it might be possible to better understand the interactions between the molecules and their physical positions.

Additionally, FRET is sometimes proposed to be directly used as a means of communication in nanonetworks. Given the harsh restrictions concerning communication distance, the technique is only really feasible for ultra-dense networks or as a means of contact-based information exchange. Furthermore, the technology is still at a very early stage of development and many unsolved problems exist. Among these are the inability to store the received information and problems with nanodevice movement due to the ultra-low communication range. As a result, we will not analyze FRET in more detail in the following chapters.

6.8 Nanophotonics

The last means of communication we analyze in this book is based on *light amplification by stimulated emission of radiation* (laser). Lasers are among the only available tools to accurately and precisely influence matter at the nanoscale. While lasers are technically electromagnetic waves, laser light spreads beam-like and is not spherical. As a result, lasers can achieve incredible levels of accuracy and precision.

Lasers are also useful tools for nanonetworks. The field of *nanophotonics* involves the manipulation and control of light at the nanoscale. Lasers are among the most developed technologies to enable interaction between the macroscale and the nanoscale.

Many proposed applications from Chapter 3 are based on this technique. Oftentimes, a container that includes medicine has to be opened at the right place. This could, for example, be a liposome (Enzian et al., 2020) that contains a drug that is intended to only affect a small area inside the human body or a single cell. Other applications involve the targeted heating of nanoparticles to stimulate metallic nanoparticles so that the resulting heat may kill adjacent cancer cells. In fact, nearly all wet-lab experiments that require high levels of accuracy use some kind of laser as either a trigger mechanism or as another integral part of the experiment.

While lasers are well suited for bridging the gap between the macroscale and the nanoscale, this link of communication only properly works in one direction. Lasers are currently only suitable for macro–macro, macro–micro, and macro–nano communication. Communication between microscale or nanoscale devices or communication from a very small scale to a much bigger scale is currently not really feasible and is subject to intense research in other areas too. While lasers are certainly used as a medium of communication at the macroscale, there is a long way to go to effectively and efficiently use lasers as a classical communication technology at the nanoscale. As a result, lasers are much more interesting as actuators or triggers and will be discussed in that context in more detail.

6.9 Comparison

This section compares the different communication paradigms at the nanoscale. We discuss the advantages and disadvantages of the respective technologies at various scales but will not discuss the quantum entanglement or the Coulomb forces used for QCA in detail. Quantum entanglement is simply too immature to be of near- or midterm benefit. Either the Coulomb interaction between electrons on the other hand is wire-bound or the communication distance is too short and prone

Table 6.1 Comparison between electromagnetic, acoustic, molecular, FRET, and laser communication at various size levels.

	Size	THz	Sound	Particles	FRET	Laser
Top-down	Macro–micro	✓	✓	✓	✓	✓
	Macro–nano	✓	✓	✓	✓	✓
	Micro–nano	✓	✓	✓	?	✓
Same	Macro–macro	✓	✓	✓	✓	✓
	Micro–micro	✓	✓	✓	✓	✓
	Nano–nano	✓	✓	✓	✓	x
Bottom-up	Nano–micro	✓	✓	✓	✓	x
	Nano–macro	x	x	x	x	x
	Micro–macro	✓	✓	?	x	?

Source: Florian-Lennert A. Lau.

to errors and signal inference. As a result, this section focuses on THz wave, acoustic, molecular, laser, and FRET-based communication.

We start with a short overview of the suitability of the given communication technology at the macroscale, microscale, and nanoscale. Table 6.1 compares the different technologies at different sizes and takes communication between different size levels into account.

Top-down communication works reasonably well for most of the proposed technologies. Given the available technologies, it is possible to get information from the macroscale to any other scale using all of the proposed technologies. The only outlier is FRET-based communication where it is unclear if information can be propagated from the microscale to the nanoscale given available technologies.

Communication between devices of roughly the same size is almost always possible. The only exception would be lasers, where it is not clear if there is sufficient energy to power a laser at the nanoscale.

The most interesting direction of communication would be bottom-up. This type is especially interesting for sensing nanonetworks or single sensing devices. The main observation here is that it should be possible to bridge a single magnitude in size but rarely two of them at once. Except for lasers, all the technologies should be able to transmit information from the nanoscale to the microscale, while none of them can skip the microscale to directly reach macroscale devices. The micro–macro link is a bit more interesting as it might be necessary to switch the communication technology to electromagnetic waves of lower frequency.

In summary, most proposed technologies can get the job done if they are combined with, for example, gateway technologies.

Next, we want to compare the same communication technologies using different metrics. Table 6.2 shows an overview that lists the communication medium, speed, bit rate, range, and energy consumption.

For many of the technologies, it is either not possible or not useful to state explicit numbers. Even the displayed numbers should be taken with a grain of salt as they simply represent the current "best-case scenario." Some of them have been subject to change every year.

We now compare the technologies according to the displayed metrics one by one.

As THz, FRET, and lasers/photonics all work based on light, the exchange of information happens with the speed of light and much faster than both acoustic signals (1500 m/s) and molecular

Table 6.2 Comparison between electromagnetic, acoustic, molecular, FRET, and laser communication types in nanonetworks.

Type	THz	Acoustic	Molecular	FRET	Lasers
Medium	THz EMW	Mechanic	Molecular	Photons	EMW
Speed	Light	1500 m/s	4.9–19 cm/s	Light	Light
Bit Rate	Very high	10^4 bit/s	Very low	Low	High
Range	<2 mm	100 μm	Arbitrary	10 nm	?
Energy	Very High	10 pW	0 pW	low	High

Quantum communication has been omitted as there is no nanoscale proof of concept yet (Hogg and Freitas Jr., 2012; Scharringhausen, 2018).
Source: Florian-Lennert A. Lau.

signals in the bloodstream (4.9–19 cm/s). From that perspective, light-based communication channels are more suitable. However, the ultimate goal is feasibility and it requires a combination of different factors. That said, in an ideal (if unlikely) scenario, it *would be* great if we could just continue to use electromagnetic waves.

Similar to the speed, the bit rate per second is heavily influenced by the communication medium. THz waves may achieve bit rates that far exceed any possible use case at the nanoscale and even top the current data rates at the macroscale. Similar rates may be achieved using lasers. Acoustic signals may also achieve considerable data rates at the macroscale but are limited to roughly 10 000 bits/s at the nanoscale which should be more than enough for most proposed applications (Hogg and Freitas Jr., 2012). Molecular signals will likely achieve the lowest data rates of all the presented technologies. It might even take a few seconds or even minutes for a single bit to be transmitted. That said, many of the proposed nanoscale applications may tolerate such a low bandwidth. Even though FRET transmits information at high speed, there is often a significant delay involved at the donor/receiver.

The communication range differs significantly from the achievable bit rate in terms of its quality. Due to molecular absorption and signal attenuation, THz waves may be limited to just 2 mm communication ranges. Acoustic signals may achieve even shorter distances of just 100 mm. The lowest range might be FRET with just 10 nm, which renders the technology dysfunctional for most applications. Molecular communication, however, may achieve nearly arbitrary ranges in a closed system like the bloodstream and thus overcomes one of the most severe challenges of nanoscale communication. While lasers may achieve high ranges, they likely will not be able to achieve the same at the nanoscale.

The last metric we analyze is energy consumption. While there are many techniques to reduce the required energy in THz networks, they still require a lot of energy and each device might only be able to send a message once every few minutes or even hours. Acoustic communication might require a bit less energy but that comes at the expense of an even lower communication range. Molecular communication requires no additional energy at all as molecules only use passive movement types like diffusion or blood flow. For FRET and lasers, the energy requirements are difficult to estimate but will likely be somewhat high.

In summary, there is no clear favorite as no technology can achieve all the desired qualities right now. The main tradeoff exists between the speed/bit rate and range/energy. Thus, THz networks might be well suited if energy is not an issue and low communication ranges may be tolerated due to a high node density. Molecular communication may be suitable for all applications that require high ranges, and low energy, and can tolerate very low data rates. Acoustic communication might be suitable for applications that lie somewhere in between. FRET and lasers are likely of little use as of right now, but that might change in the future.

Finally, we want to compare different kinds of molecular communication as this is likely the most diverse technology. It is likely the slowest type of communication but naturally occurs in many biological systems already, which proves the feasibility of the underlying principle. The speed of information exchange further varies significantly according to ambient temperature, as well as the type and size of the molecule. As an example, DNA molecules of different sizes move between 0.81 and 53 mm/s (Farsad et al., 2016). The hormone insulin has a diffusion coefficient of roughly 150, and water molecules move at around 2100 mm/s. That said, the isolated speed of diffusion is only interesting in some scenarios as molecules are often affected by a mix of Brownian motion, diffusion, and flow.

Table 6.3 shows a list of molecular communication concepts and some associated properties based on the work of Farsad et al. (2016). The types of communication are categorized based on their underlying concepts.

The table makes it clear that most communication mechanisms spread information on the basis of diffusion or flow. Only bacteria-based techniques deviate from this as they sometimes use chemotaxis (Balasubramaniam and Lio', 2013). *Chemotaxis* describes the forces of attraction and repulsion used by bacteria.

Particle-based methods mostly use molecules that naturally occur in the environment. If the environment is the human organism, this can affect the body's own communication mechanisms. Tiles and bacteria on the other hand have little potential for conflict. Nevertheless, it is difficult to compare the different types of communication with each other in that regard. This is especially true for tile-based communication since message molecules are first assembled from tiles, which in turn can be used in communication like conventional molecules. The hierarchical process of message composition has hardly been studied in terms of networks.

Nonetheless, Table 6.3 shows that tile-based methods can be assessed as positive in comparison since they are less bioinvasive and the underlying DNA technology allows for a lot of information being sent per particle and they come with a built-in addressing mechanism. The low level of bio-invasiveness is due to the fact that DNA can occur in practically any place in the human body, and, therefore, there are hardly any antibodies. It can therefore be assumed that DNA tiles have less of an impact on existing communication mechanisms in the human body than other

Table 6.3 Comparison of different molecular communication methods from the work of Farsad et al. (2016).

Parameter	Propagation	Invasiveness	Bits/Particle
Tiles (DNA)	Diffusion/flow	Very low	$1/k$
Particles	Diffusion/flow	High	$1/k$
Concentration	Diffusion/flow	High	$1/k$
Type	Diffusion/flow	High	$1/\log_2 n$
Order	Diffusion/flow	High	$2/n$
Bacteria (DNA) (Balasubramaniam and Lio', 2013)	Chemotaxis	Low–high	several
Viruses (DNA) (Walsh and Balasubramaniam, 2013)	Diffusion/flow	Low–high	several

The variable k describes the particles per message, and n is the number of particle types (Lau, 2020).
Source: Florian-Lennert A. Lau.

methods. Concentration-based methods convey comparatively few bits per particle since only the total concentration of particles in a medium can be measured, which often has to be very high in order to be able to meaningfully distinguish between a logical "1" and a logical "0."

Overall, no definitive answer may be given about which type of molecule or information encoding is best. As of right now, many mechanisms work properly and have to be tested in complex environments like the human body. Only wet-lab experiments and actual tests may show if the theoretical results may be transferred into real-world applications.

6.10 Multi-hop Communication

Now that we have explored multiple methods for communication between a single sender and receiver, we can start to explore simple strategies for communication in nanonetworks. As Table 6.2 indicates, the communication range of each of those techniques does not allow for simple point-to-point communication and some message forwarding through a network of devices will be necessary. The only exception to that is likely molecular communication, where molecules may reach any desired area on their own. Thus, message forwarding or routing is not necessary, strictly speaking. However, it might be necessary to use repeaters to re-amplify a message when the concentration is too low. We now discuss, we first discuss different addressing schemes and later simple routing protocols that function with next to no available resources. Finally, we discuss the impact of multiple more powerful gateway devices that control and aggregate the information flow in a nanonetwork.

6.10.1 Addressing

The first topic we need to analyze is addressing. An address is an *identifier* (ID) for either a single device or a group of devices. Without such an ID, it is impossible to forward any relevant information to the desired location. Addresses enable the identification of specific sources and destinations of multi-hop networks.

A typical device in a regular network of devices has either an IPv4 or IPv6 address that serves as a unique ID for communication. Such an address is either 32 or 128 bits long, which might create problems as typical transistors nowadays are around 20 nm in size and may only store a single bit worth of information. Marketing names like "7.7 nanometers technology" are often misleading in that regard. As a result, storing a single identifier might already require devices or at least a few hundred nanometers in size. More generally, a unique identifier further requires $O(\log n)$ much memory, where n is the number of devices in the network. Simply "counting them" and thus assigning them an ID require that much memory, and the possible requirements of routing tables only make things worse. As a result, both IPv4 and IPv6 are likely not suitable for nanoscale applications, even though the number of possible addresses in IPv6 might be more than sufficient.

Definition 6.7 An *address* is a unique number that identifies a device unambiguously.

If a unique identifier is infeasible, it might be better to use address prefixes or address classes, where only a set number of bits are allowed and the uniqueness as a condition is dropped. As a result, only groups of devices might be targeted given a specific message. However, this might not be a problem at all as devices likely have to cooperate to achieve a given goal due to their minuscule size and limited impact on their own.

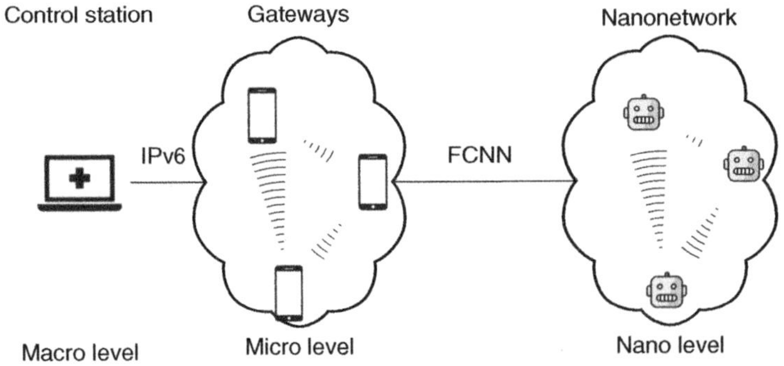

Figure 6.7 Function-centric networking on different levels of size. *Source*: Adapted from Stelzner et al. (2016a).

Definition 6.8 An *address class* is a number that uniquely identifies a set of devices with similar capabilities.

This idea is further illustrated in Figure 6.7. At the macro level, it is still possible to use regular devices and the already available protocols. As a result, the control stations likely communicate with gateway devices via IPv6. Those gateways might be able to use the same technologies or simplified versions thereof. The communication with the nanoscale has to deviate from that scheme due to address constraints.

Instead of using lengthy addresses, the number of different nanodevice types and the interesting locations are categorized in advance (Stelzner et al., 2016a). As a result, a number of necessary distinguishable messages can be derived. These are now used to address, for example, parts of the human body in which specific nanodevices may perform actions if the address class fits.

In the simplest case, these address classes could be implemented using regular expressions that may be implemented using *deterministic finite automata* (DFA). They are among the weakest classes of computational devices that still can do something, namely recognize if a number is part of an accepting list of words. This only requires a constant amount of memory and very simple mathematical operations.

In the unique case of molecular communication, things are a bit different as messages usually cannot even represent an address properly. Oftentimes, only a few bits are transmitted over a lengthy period of time. Thus, addressing is difficult by conventional means.

An exception to this is DNA-based molecules like tiles. Those naturally encode conditions in their open DNA strands that already serve as addresses. Those might also be used over longer distances and are only able to properly attach to specific other devices that express the respective complementary DNA strands. If additional repeaters are necessary, they can also express such inverse DNA sequences or an array thereof to simply increase the concentration of DNA molecules again. As a result, DNA molecules might be an excellent candidate for communication in molecular communication networks that come with the additional capability for computing mathematical functions as we have learned in Chapter 5. With all of this in mind, we can now start to analyze simple routing protocols at the nanoscale.

6.10.2 Routing Protocols

At the macroscale, the problem of digital communication between two or more parties has long been solved. Even at the microscale, potential solutions are available or at least would be feasible given sufficient demand. However, things are more complicated at the nanoscale. The following itemizations show several peculiarities that nanoscale communicating devices have to consider. All of them apply for communication both over a single link and over entire networks of devices.

- Volatile communication channel
- Unstable channel memory
- Biocompatibility
- Garbage management
- Safety/security
- Low range
- Little to no memory
- Not Turing-complete
- No unique identity
- Frequent device failure/malfunction
- High energy per bit
- Slow energy harvesting
- Passive movement
- No known topology
- Only approximate localization
- High number of participants
- Possible redundancies
- Long time to live
- High bandwidth
- High communication speed
- Only specific scenarios necessary

While there are many potential problems or constraints, there are also some unique aspects that might exceed the capabilities of regular networks. Most of the problems result from the small size of the devices. As a result, there is little available energy, little memory, a short communication range, and subjectivity to all three types of passive movement. Consequently, nanodevices have little information about the ever-changing local network topology, let alone the global one. Thus, the environment in which nanonetworks have to function is much more challenging than any type of macroscale WSN.

Yet, there are also several positive aspects. Nanonetworks likely consist of thousands if not millions of devices which allows for a high degree of redundancy and possible new protocols. It might even be possible to use the communication channel itself as a type of outsourced memory that the devices themselves cannot offer due to the limited size. Furthermore, a single nanonetwork likely only fulfills a single function and can be optimized for that. The more the environmental parameters are known in advance, the less complex the behavioral programs of the individual devices can be.

Given all these constraints and possibilities, we can now introduce a simple but effective first strategy for routing at the nanoscale.

6.10.3 Hop-count Routing

Before specifying how an algorithm or protocol works, it is necessary to precisely define the capabilities of the involved (nano-)devices. In this case, we try to solve the general communication problem in nanonetworks using as few resources as possible to properly reflect somewhat realistic assumptions.

A reasonable assumption for the amount of available memory is $O(1)$. Nanodevices are capable of storing exactly a single (short) integer. Concerning the necessary operations, the devices require the ability to compare integers and to increase as well as reset a number in the memory. Apart from that, nanodevices have no additional capabilities and are not even able to store a unique identifier that most high-level routing protocols rely on. However, it is still possible to create a rudimentary network topology and route messages to a gateway given the listed constraints (Büther et al., 2018).

Once the capabilities of individual devices are specified, it is necessary to make some assumptions about their positions in space. Figure 6.8 shows several possibilities.

The image to the left shows a fully randomized placement of devices. The center image shows a smart placement on a grid in, for example, metamaterials, and the right image shows a random placement of nanodevices in grid cells. This kind of device placement considers possible repulsions, and this is likely the most realistic device placement.

Finally, we have to make a few simplifying assumptions about the scenario. For the sake of simplicity, we assume that all involved devices do not require energy to send and are otherwise not prone to errors. Nanodevices are further assumed to be (quasi-)static and do not move. All those simplifying assumptions can likely be considered in the following algorithm, but it is easier to understand given fewer variables.

The hop-count routing protocol is divided into two general phases: a propagation phase, where the network establishes a short-term topology and determines the connectivity between devices, and a retrieval phase, where nanodevices may forward gathered information to a gateway. The following listing shows the program that governs the propagation phase:

```
onReceiveMessage (Message m) {
  if (m.type == Type && this.hopCount > m.hopCount + 1) {
    this.hopCount = m.hopCount + 1;
    sendMessage (Type, this.hopCount );
  }
}
```

As Figure 6.9 indicates, all nanodevices in the network initiate the variable that stores their individual hop count with the reserved value ∞. This value may not be reached by any other means but by resetting the network topology.

Once the propagation phase starts, the gateway (displayed in grey) starts sending propagation messages to all other nanodevices within reach. The device that initiates the process sets its own

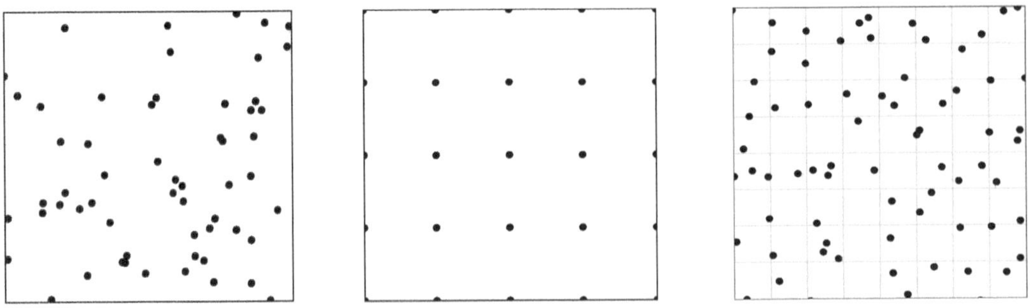

Figure 6.8 Different types of device distributions in a communication channel. *Source*: Adapted from Büther et al. (2018).

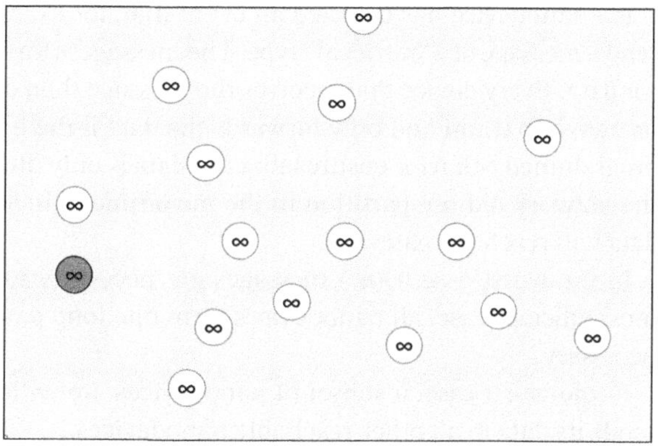

Figure 6.9 Initial state of the network at the start of the propagation phase. *Source*: Büther et al. (2018)/with permission of Association for Computing Machinery.

hop count to 0 and attaches it to the propagation message. Once a nanodevice receives a message, it compares the hop count in the message with its own hop count. If the own hop count is higher than the one in the message, the receiving nanodevice sets its own hop count to the received hop count + 1.

Each of the receiving nanodevices then repeats the process that has been started by the gateway and sends a new propagation message to all adjacent devices with the adjusted hop-count value. After some time, all nanodevices within reach will have an assigned hop-count value. Furthermore, a device should receive multiple propagation messages, and it will always only keep the lowest value. Thus, the stored hop count always represents the shortest known distance to the gateway as the "number of hops."

Figure 6.10 shows the result of the propagation phase after a finite amount of time. The grey circles represent the communication range of the nanodevices, and each number represents the number of hops that are necessary to reach the gateway node. The dashed lines indicated the shortest distances to the gateway and form a *minimal spanning tree* (MST).

Once each connected nanodevice has received a hop count, the network can enter the second phase in which the non-gateway devices can measure and forward data to the gateway. The nanodevice behavior in the retrieval phase is defined in the following listing:

```
onReceiveMessage (Message m) {
  if (m.type == SensorType && m.hopCount > this.hopCount ) {
    sendMessage (SensorType , this.hopCount , m.data );
  }
}
```

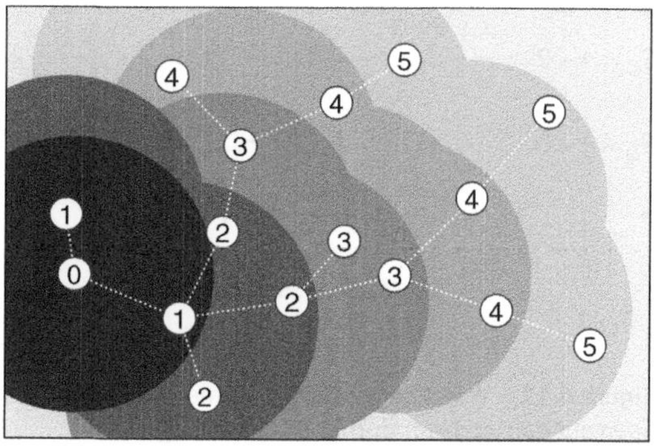

Figure 6.10 Resulting internal hop-count values of all nanodevices once the propagation phase is finished. The connectivity is indicated by a *minimal spanning tree* (MST). *Source*: Adapted from Büther et al. (2018).

If a nanodevice has detected an event that, for example, exceeds a predefined threshold, it may send a message of a "retrieval" type. The message is forwarded to all neighbors and contains the sensor data. Every device that receives the message then compares the hop count in the message with its own hop count and only forwards the data if the hop count of the message is bigger. Thus, the programmed behavior ensures that the data is only directed closer to the gateway node. Given that the network did not partition in the meantime (which we excluded as a precondition), all sensor data will reach the gateway.

In the worst case, $O(n^2)$ messages are necessary to assign a hop count to each device. In the most efficient case, all nanodevices form one long path, and only $n - 1$ propagation messages are necessary.

In the worst case, a subset of nanodevices are within each other's reach and each nanodevice sends its data to all other reachable nanodevices.

Figure 6.11 describes the problem in more detail. Each of the nanodevices may send a message to all other nanodevices with a lower hop count. If the messages are delayed unfavorably, all receiving devices also forward the message to all other devices. As a result, the number of retrieval messages grows quadratically with $O(n^2)$.

Figure 6.12 shows the result of a simulation given the above-mentioned network topology. As we can see, a linear number of messages can be expected given an average distribution of devices.

The retrieval phase is a bit more difficult to analyze as the condition to send no additional messages is less strict. As a result, the worst-case number of messages can grow exponentially in the hop distance between the gateway and the nanodevice that detected an event. In the worst case, $O(n^k)$ messages are necessary, where k is the hop count diameter of the nanonetwork. This problem is known as a *broadcast storm* in WSN and is harder to avoid with identity-free networks.

Figure 6.13 shows simulation results that prove the analytical results. Without any additional effort, the number of messages grows exponentially.

If a *destructive retrieval* approach is used, the number of messages may be reduced significantly. Each of the involved nanodevices will reset its own hop count upon successfully forwarding a message. As a result, it is impossible to send more than a linear number of messages. However, this comes at the expense of a destroyed network topology and phase 1 of the algorithms has to be repeated. Yet, this might be necessary anyway as nanodevices are generally assumed to be mobile and the once-calculated distances are likely to change rapidly. We discuss this approach in more detail in Chapter 9 in the context of saving energy.

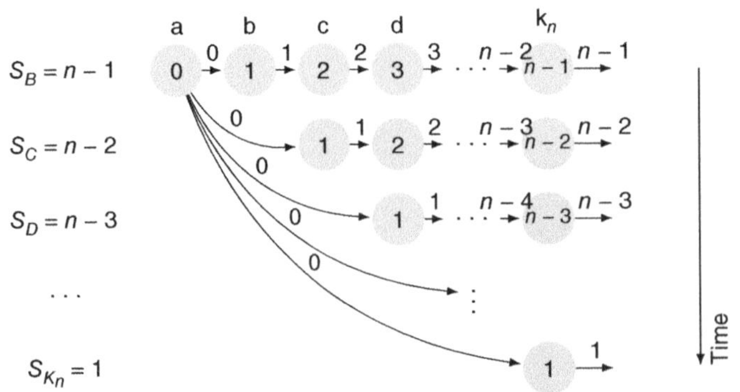

Figure 6.11 Worst case number of messages that are necessary to retrieve sensor data from the network. *Source*: Adapted from Büther et al. (2018).

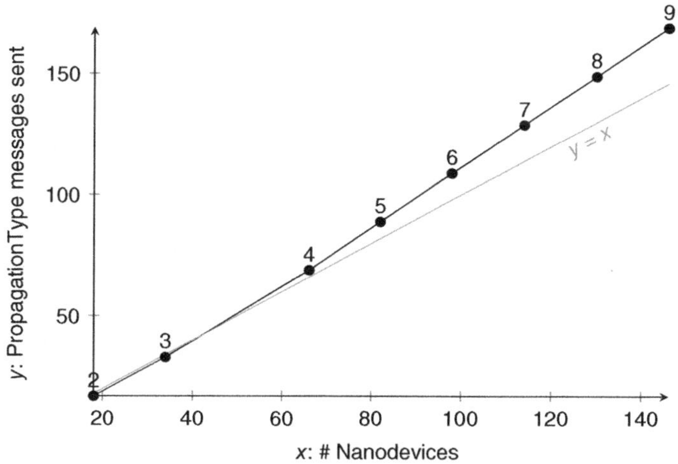

Figure 6.12 Number of propagation messages sent. *Source*: Büther et al. (2018)/with permission of Association for Computing Machinery.

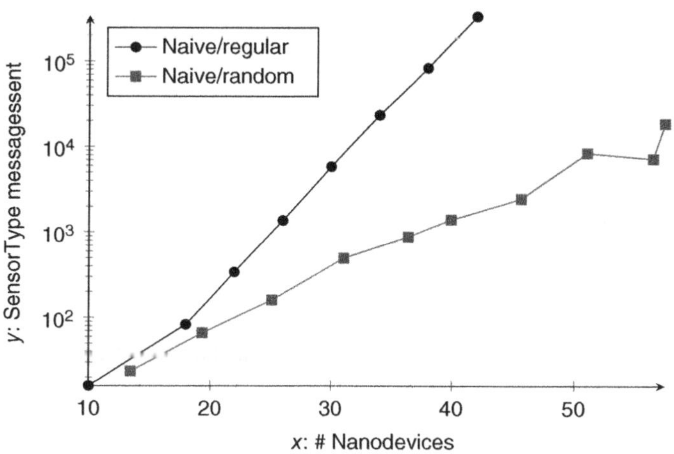

Figure 6.13 Messages sent during the retrieval phase using destructive retrieval as opposed to the native approaches. *Source*: Büther et al. (2018)/with permission of Association for Computing Machinery.

6.11 Communication and Network Simulators

Since networks are usually complex systems, theoretical analyses often fail to provide satisfying results. As a result, most scientists in the field use simulators to test their initial hypotheses. A detailed analysis of existing simulation tools was carried out in the work of Nguyen (2017). Based on these results, we examine several simulation software.

Most simulation tools for nanoscale networks refer to *Body-Area-Networks* (BANs) or *In-Body-Nanonetworks* (IBNs). These model macroscopic components such as communication protocols and channel models as well as some environmental constraints. Yet, in order to correspond to the accuracy of in-vitro or in-vivo experiments, it is necessary to map the environment very precisely.

With IBNs, we have to distinguish between a large number of models. Above all, these include characteristics of communication as such. For example, there are approaches to simulating electromagnetic communication in the terahertz band, chemical communication, or communication through special molecules (Guney et al., 2012; Jornet and Akyildiz, 2013; Galluccio et al., 2015; Pierobon and Akyildiz, 2010; Bicen and Akyildiz, 2014; Kuran et al., 2012). In the work of Galluccio et al. (2015), the models presented have been expanded to also include acoustic communication.

The IEEE P1906.1/Draft 1.0 standard recommends general practices for molecular communication at the nanoscale without providing an interface for more complex types of molecular communication (Bush et al., 2015).

The first nanoscale network simulator was NanoNS (Gul et al., 2010). It is based on the popular network simulator ns-2, which has been expanded to include modules that allow various new types of nanoscale communication.

Another example is the BiNS2 simulator (Felicetti et al., 2012). Due to its modular architecture, other types of communication can also be carried out in the same framework. The truthfulness of the BiNS2 simulator has already been confirmed in comparison with in vitro experiments (Felicetti et al., 2013).

The well-known network simulator ns-3 has been expanded to include modules that allow simulations at the nanoscale. This also includes molecular communication. Two well-known components are nanoNS3 (Jian et al., 2016) for electromagnetic communication and Nano-Sim (Piro et al., 2013) for molecular communication.

Older simulators like Nano-Sim usually extend ns3 in some way or another and often leave out crucial parts of the communication process, like payload and propagation delays, as well as bit collisions. The same is true for more realistic large-scale network simulators like Vouivre (Boillot et al., 2015). Scenarios that include more than a few hundred nodes are usually impossible to simulate. This is even more true for physical simulations using COMSOL (Multiphysics, 1998).

Another well-known and precise tool for simulating electromagnetic nanonetworks is the *Bit simulator* (BitSim) (Dhoutaut et al., 2018). It takes most of the aforementioned weaknesses into account and specializes in nanonetworks with a large number of participants. For realistic simulations in electromagnetic nanonetworks, a simulator like BitSim is likely necessary.

Finally, we have a look at simulation tools for DNA-based nanonetworks. Since DNA-based nanonetworks communicate via molecules that have to self-assemble before they may be used, they require more specific simulation software called *tile-assembly simulators*. The first among those was Xgrow by Eric Winfree (Xgr, 2003). The tool is written in the programming language C and is only available for Windows. However, the software is rather old, cannot be executed on many modern systems, and does not allow for much networking support.

An alternative is the ISU TAS software by Patitz Patitz (2009). This powerful tool provides three different simulation models: aTAM, kTAM, and 2HAM. In combination, they allow for a decently precise prediction of DNA-based nanonetworks. However, the software is no longer developed and is intended to be replaced by the WebTAS software (Patitz, 2022).

The most suitable software for the simulation of DNA-based nanonetwork applications is the NetTAS (Kaussow, 2022). Unlike the WebTAS in its current state, the NetTAS offers support for all previously mentioned models and may be executed in either the Chrome or Firefox browsers. Furthermore, the NetTAS offers the new KTHAM model that combines the advantages of the kTAM and 2HAM models and offers several comfort features like statistical assembly analysis and TikZ export for images.

6.12 Summary

In summary, the problem of communication in nanoscale networks is extremely complex and multiple influence factors exist. As a result, it is yet unclear which physical properties might be used in the future to represent information. The three main contenders are electromagnetic, molecular, and acoustic communication. Each of them comes with its individual strengths and weaknesses and countless models exist already.

As most questions of practicality and feasibility cannot be answered yet, it is necessary to use simulation tools to gain insights into the potentially novel effects at the nanoscale. It might not be the most precise method available, but simulation is a suitable low-cost alternative for wet-lab experiments. Additionally, it is sometimes not even possible to manipulate matter in the desired ways to test the hypotheses generated by the nanonetworking community.

Based on the presented protocols and communication technologies, many of the other problems in nanonetworks like energy consumption or localization can be addressed. They are discussed in the upcoming chapters in more detail.

7

Movement and Localization

This chapter explains the important topics of locomotion and localization at the nanoscale. No matter how nanodevices or nanonetworks are physically realized in the end, they will always be subject to certain kinds of passive movement mechanisms. At the nanoscale, warmth equals movement and nanodevices will always be subject to differences in particle concentration, Brownian motion, and different flow effects either in air or in liquids.

As a result, the positions of nanodevices cannot be assumed as static and must be determined on the fly. This problem is known as "localization", and it is unclear if the established mechanisms at the macroscale may be applicable at the nanoscale.

This chapter first discusses the three types of passive transportation mechanisms nanodevices are always subjected to. Afterward, we analyze different kinds of active movement that are based on a variety of physical effects. Among them are bacteria, molecular motors, and nanojets/nanoswimmers, and completely artificial approaches.

We then discuss possible techniques that can be used to estimate or determine the positions of nanodevices in their environment. For that purpose, gateway trilateration, proteome fingerprinting, or machine learning of "age of information" might be applicable.

Finally, we discuss two simulation approaches that allow for the estimation of particle position in the human bloodstream. We also analyze a possible modularization approach that separates the human body into different layers of complexity for which certain important constants might be generated. They may then be used as an approximation for low-level effects in abstract parts of the model.

7.1 Definition

Over the past 15 years, researchers suggested many different kinds of both active and passive movement mechanisms at the nanoscale. As of right now, it is completely unclear how nanodevices will actually manipulate their position in space and it is thus necessary to use general definitions that encompass all those ideas. The following itemization shows several examples that are explained in more detail later in the chapter. The top 6 entries are active forms of movement while the last 5 are passive.

- Molecular motors,
- bacterial flagella,
- chemotaxis,
- external magnetic fields,

Nanonetworks: The Future of Communication and Computation, First Edition. Florian-Lennert A. Lau.

- enzyme-based locomotion,
- nanojets,
- Brownian motion,
- diffusion,
- blood flow,
- wind,
- and many more.

At the most general level, movement and locomotion are ways of influencing the own position in space. As a result, the locomotion component L could be regarded as a special actuator A. As locomotion is sufficiently different from other types of actuation and generally regarded as fundamental, we define it apart from simple actuation as follows:

Definition 7.1 The locomotion component L describes the ability of a nanodevice to change its location. Locomotion is divided into two categories: *active* and *passive* movement. Passively mobile nanodevices do not require any internal influences.

An example would be nanorobots, which move through an organism together with the blood flow. Similar to communication, actively mobile nanodevices use a special actuator component. The environment is further influenced with the intention of locomotion.

7.2 Passive Movement

The first class of movement at the nanoscale is of a passive nature. Passive types of movement or locomotion include all effects that affect the position of a nanodevice or nanonetwork via natural effects. It does not matter if the effect the devices are subject to is voluntarily or involuntarily used. This is especially true for nanoscale entities without an active means of locomotion like certain cells, viruses, and molecules. In the following paragraphs, we analyze Brownian motion, diffusion, and flow in detail as these three effects dominate most proposed scenarios.

7.2.1 Brownian Motion

The most fundamental type of passive movement at the nanoscale is Brownian motion. This type of movement was first discovered and documented by the botanist Robert Brown in 1827. During his study, he examined the pollen of certain plants in water under a microscope. The pollen moved in a completely chaotic, random-seeming pattern. Almost 80 years later, Einstein found a convincing source of this type of movement in water molecules randomly colliding with each other and the pollen.

The higher the temperature of the medium is, the faster the molecules move and the harder it is to follow the individual paths of molecules. As a result, ambient temperature is always equivalent to a high degree of movement at the nanoscale.

While individual molecules display the behavior of a random walk, it is not immediately visible from the outside. As a result, it is computationally infeasible to simulate Brownian motion and only stochastic models exist. This is especially important as all nanoscale structures are always *also* subject to Brownian motion, no matter what other locomotion component they are equipped with. As a result, any type of simulation that tries to accurately determine the positions of individual nanodevices will always be flawed to some degree. However, the same is also true for measuring the positions of individual devices due to Heisenberg's uncertainty principle.

7.2.2 Diffusion

Diffusion is the process by which particles move from an area of higher concentration to an area of lower concentration. Diffusion is fueled by a gradient in chemical potential and affects all sufficiently small things like atoms, ions, or molecules. As a rule of thumb, diffusion scales with the size of the involved particles in Dalton. Bigger particles move much slower than smaller particles.

As opposed to Brownian motion, diffusion leads to a concentration balance due to random thermal movement. It can occur in gases, liquids, and solids and is also the result of the random motion of the particles, albeit on a more general level as Brownian motion. The diffusion of particles can be observed as the spreading out of a substance, and it is the net result of many random collisions between the particles. The discovery of diffusion can be attributed to Thomas Graham, a Scottish chemist, and inventor, who first described the process in 1829 in his paper "On the Laws of Diffusion of Gases" (Graham, 1833).

Overall, diffusion can be further categorized into processes that follow Fick's law and processes that do not (Fick, 1855). The former kind is also called *normal diffusion* and can be described by the following formula:

$$\sqrt{2nDt}.$$

In this case, n describes the dimension of the medium, t is the time, and D is the *diffusion coefficient*. In the simplest case of a constant concentration source in one dimension, the formula simplifies to $2\sqrt{Dt}$. In simple wet-lab experiments using molecular communication, this simplified law is often sufficient to describe the observed behavior.

7.2.3 Blood Stream and Bulk Flow

The next type of passive movement is called *bulk flow* or *advection* in its most general form (Hundsdorfer et al., 2003). Bulk flow describes the macroscopic movement of large quantities of particles in roughly the same direction. An example would be a river, particles carried by the wind, or the blood flow in the human body. Another example that is very related to the human blood flow is the absorption of liquid and nutrients by plants through their root systems. Whenever such effects occur in an environment, all nanodevices or nanonetworks in that context are equally subject to it.

It is important to keep in mind that neither of those flow effects can be described by either diffusion or Brownian motion. Additionally, flow is often the dominant force in such systems. At the same time, all particles in, for example, the bloodstream are also subject to all the other passive movement effects at once. In the research literature, this combined effect is sometimes called *advection–diffusion*, where the diffusion component is due to temperature differences.

Especially in the context of medical nanonetworks, the blood flow in the human body might be the most interesting advection–diffusion effect. The flow itself originates from pressure differences created by the heartbeat. Especially close to the heart, the blood behaves in a turbulent manner and it is not feasible to simulate particle positions accurately (Csanady, 1973). Apart from that, most of the flow effects in the body can be described by a *laminar flow*, which is much easier to model and simulate.

The simulation of flow effects is of interest to multiple institutions. It is necessary to understand the behaviors of groundwater flows, rivers, and, for example, pollutants spread in them. In the human blood flow, doctors want to know how, for example, the distribution of MRI contrast agents that help visualize difficult parts of the body. Multiple commercial solutions are available for this problem. However, there are still unsolved problems in the area of fluid dynamics and simulation. As an example, the Navier–Stokes equations are used in most mathematical models that try to predict turbulent flow. However, it is currently unknown if there is always a solution for those equations in the three-dimensional case. The problem is even listed among the six remaining millennium prize problems for which a bounty of one million dollars each exists (Jaffe, 2006).

7.3 Active Movement

The second overarching type of locomotion is an active manipulation of the environment to move from point A to point B. This can either happen through adapting naturally occurring processes or via introducing artificial means of transportation. The active type of movement can take many forms, and we discuss several options in the following sections.

7.3.1 Chemotaxis

The first and intuitively best-known type of active movement at the nanoscale is utilized by bacteria. Most people in the Western world have examined bacteria under a microscope at some point or another, and it is plainly evident that they can move on their own. An example of such a bacteriophage can be seen in Figure 7.1.

The organism has certain sensory components around its outer hull that allow it to recognize the presence of certain chemicals. According to its genetic code, some of those substances are

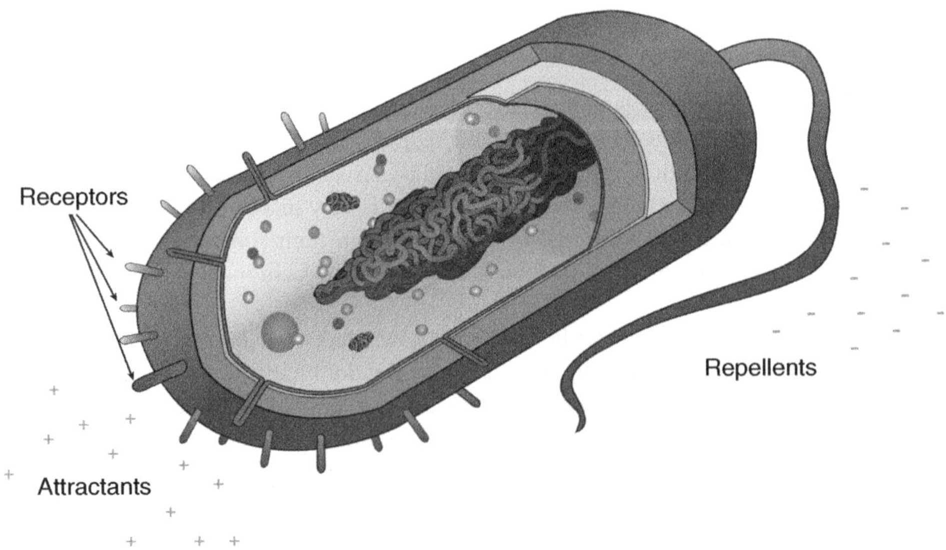

Figure 7.1 Bacteriophage with receptors that moves from an area of repellents to an area of attractants. *Source*: Florian-Lennert A. Lau.

deemed as good and some as bad. The entire process is based on the receptor–ligand interaction we introduced in Chapter 6.

As bacteria do not possess the same sensory organs that higher forms of life have, they need another mechanism that guides them toward beneficial substances and helps them avoid danger. The process behind this is called *chemotaxis* and originates from Latin, where it means "ordered march." To achieve this, bacteria need some degree of control over their environment that allows them to change their position in space.

For this, bacteria use the so-called *flagella* (Gregori et al., 2011). A flagellum is a long, whip-like structure that some organisms use for movement. It is a type of appendage that extends from the cell body and is composed of a protein called flagellin. Some single-celled organisms, such as bacteria, have one or more flagella that they use to swim through liquids. Additionally, some eukaryotic cells, such as sperm cells, also have flagella to aid in movement.

While bacteria have some degree of control over their position, they are still subject to all the aforementioned passive types of movement. However, as flagella emerged from an evolutionary process, they clearly provided a sufficient advantage to the species. Otherwise, they would not have stood the test of time. Only those things that provide enough benefit can establish themselves in the genes.

As a result, many researchers proposed nanonetworks that function based on the targeted movement of bacteria (Gregori and Akyildiz, 2010). These cells or bacteria serve as both a vessel and a means of transportation. In this case, the transmitted information is DNA that is stored inside of the cell which can be modified to a certain degree. While such a design is certainly promising, the actual human contribution to the process is very limited and the natural processes are more or less kept as they are with only slight modifications to the DNA. As a result, it might not be fitting to call such bacteria-based nanonetworks the first implementations of artificial nanonetworks. Yet, they are still feasible and may transmit plasmid-encoded information over long distances requiring a relatively long time.

7.3.2 Other Motor Proteins

Another type of nature-inspired locomotion functions using *motor proteins* (Moore et al., 2006). Motor proteins are special biomolecules that utilize energy from the hydrolysis of ATP to generate mechanical force/movement. They can be divided into two types: those that move along a filamentous track, such as microtubules or actin filaments, and those that rotate a cilium or flagellum.

Examples of the former kind of motor proteins include myosin, kinesin, and dynein. These proteins are involved in several cellular processes such as cell division, muscle contraction, and the transport of nutrients inside of cells. A schematic example of the process can be seen in Figure 7.2.

The track represents a filamentous track made from microtubules. Directly attached to it is a motor protein that drags cargo along the railway system inside a cell. The cargo can be many times bigger than the motor protein itself, and ATP is necessary to power the process.

There are many types of motor proteins that all function based on the same overarching principle. This type of locomotion is best suited to transportation inside cells and thus works on an even smaller scale than other types of locomotion or molecular communication. It is unclear if such a network of microtubules may be created outside of cells to create a type of wire-based communication or transportation system in the human body.

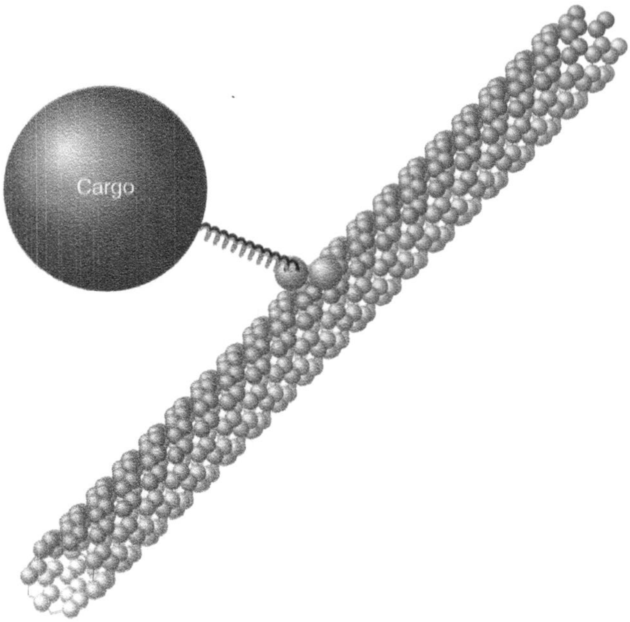

Figure 7.2 Motor protein moving cargo along a filamentous railway system of microtubules (Pan et al., 2017; Moore et al., 2009). *Source*: Florian-Lennert A. Lau.

7.3.3 Artificial Movement

The last type of locomotion at the nanoscale is completely artificial and does not utilize naturally occurring phenomena. In the following, we roughly distinguish between three broad paradigms:

- chemical self-propulsion,
- magnetic field manipulation, and
- movement via electrical fields.

Of those three, chemical propulsion is likely the easiest to understand. An example of this type of movement is the so-called *bubble propulsion*, which was first proposed in 2009 (Gibbs and Zhao, 2009). Nanodevices equipped with just a means of locomotion are sometimes called *nanomotors* and may move in specific chemical environments which contain hydrogen peroxide. Such nanomotors likely use a similar physical effect compared to a rocket.

An example of such a nanomotor can be seen in Figure 7.3. In the simplest case, such a nanomotor can be modeled as a silica sphere or circle. The right half of the sphere gets coated with a platinum catalyst that speeds up a chemical reaction with the peroxide in the environment.

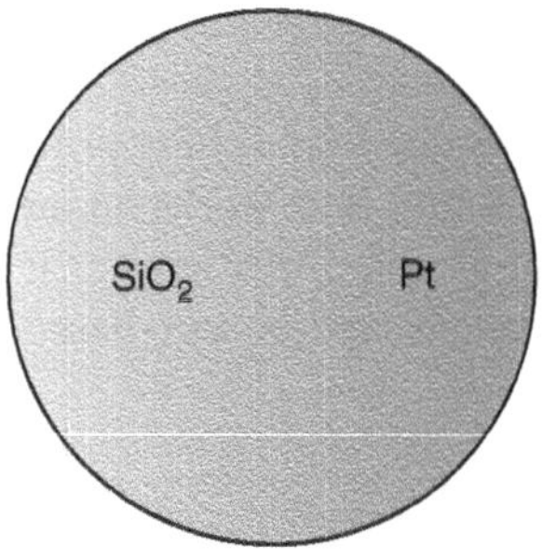

Figure 7.3 Bubble propulsion of a spherical nanomotor. The half of the silica microsphere-based nanomotor is coated in platinum (pt). The platinum reacts with the peroxide in the environment and creates oxygen bubbles that propel the nanomotor. *Source*: Florian-Lennert A. Lau.

The result is an asymmetrical chemical reaction that produces oxygen bubbles on just one side of the nanomotor. As a result, a driving force away from the platinum coating is created.

The remaining two types of active movements have both a macroscale and a nanoscale component. We start with the nanodevices that generate directional movement through the application of magnetic fields (Yu et al., 2018). There are two common types of such motors: rotating magnetic field and oscillating magnetic field propulsion.

The nanodevices that utilize rotation via magnetic fields are inspired by bacterial flagella and utilize the same basic mechanism. Oftentimes, they are equipped with an artificial flagellum that reacts to externally applied magnetic fields. Due to the oscillation, a rotating force is applied that propels the nanodevice forward. However, the movement direction cannot be precisely controlled.

The second type of movement via magnetic fields also relies on the oscillating magnetic field but uses them in a different way. Instead of generating a flagellum-like rotation, nanodevices with multiple joints made from magnetic materials may generate force by moving a fin up and down. It is possible to incorporate more than one joint into the movement, but this usually leads to an increase in movement complexity and generally unpredictable behavior due to the chaotic nature of such systems.

The last type of locomotion at the nanoscale exploits the electric properties of nanostructures through the application of an electric field. In principle, this type of movement uses the same effect as the PCR, namely electrophoresis. In essence, charged particles tend to move relative to a liquid or gel they are in when a uniform electric field is applied. Nanodevices may also use this effect for directed movement. This is especially true when they are capable of manipulating their own electrical properties or surface structure.

Apart from that, there are numerous other concepts for the generation of movement at the nanoscale. Among them are fish-like nanoswimmers, scallop swimmers, and Janus microdimer surface walkers. The interested reader is encouraged to read the work of (Yu et al., 2018) for an in-depth survey.

7.3.4 Comparison of Locomotion Types

It is difficult to compare the different types of locomotion at the nanoscale as they are so different in nature. While some of them cannot be avoided and must always be accounted for, others are options to select from. We try to evaluate the different methods regarding feasibility and compatibility with other technologies.

The first thing to mention is that it is completely unclear what type of locomotion component future nanodevices will use. All we can do is give a rough estimate based on the current state of the art and similar developments in other areas.

We start with the passive types of movement as those are easiest to judge. Flow, diffusion, and Brownian motion are going to affect any nanoscale device imaginable and must always be either accounted for or explicitly used as a means of transportation. Neither of the naturally occurring effects requires additional energy or changes to the environment. Furthermore, as many naturally occurring structures or organisms use those effects to their advantage, nature already provided a proof of concept. As a result, it is feasible to utilize those effects as transportation mechanisms, for example, for molecular communication or tile-assembly systems.

The active and artificial kinds of locomotion are more difficult to judge. While many working proofs of concept certainly exist, the compatibility with other nanodevices or with, e.g., the environment in the human body is unclear. It seems unlikely that techniques that require specific chemical environments that differ from the bodies of living creatures will ever work inside human bodies.

While nature-inspired approaches that modify cells or bacteria can certainly work, it is unclear how much such organisms may be modified before the cell is too stressed. As an example, cells that are used as genetic circuits may only produce a small number of additional proteins before they die off. As a result, it is unclear if a simple modification of a cell is sufficient to fulfill all the preconditions for the proposed applications. Nonetheless, as only small changes to existing organisms are necessary, some bacteria-based approaches might likely be the first proof of concept of an actively mobile nanonetwork.

The approaches that work based on external magnetic or electrical fields show potential for immediate use (Zhou et al., 2021). In this case, for example, an artificial flagellum might be attached to an existing nanodevice that reacts to the application of an external magnetic field. This relatively simple procedure seems compatible and feasible given the availability of external machinery that provides a suitable magnetic field. As the entire approach does contain crucial macroscale components for the nanonetwork to function, it is more fitting to call it a macro–nano hybrid network.

All in all, there is a surprisingly vast amount of available research when it comes to locomotion at the nanoscale. Chemists seem interested in the topic and provide valuable insights into the processes of movement at the nanoscale. While the question of compatibility with more complex devices is yet to be answered, the entire field is comparably advanced.

7.4 Localization

While locomotion in itself is a valuable skill already, it is only of practical use when specific areas may be identified. In almost all proposed applications, nanonetworks are envisioned to *only* affect specific areas. This requires the ability of nanodevices to recognize their position relative to the area of application with a high degree of accuracy. Even small errors may have non-trivial effects on the body.

In the following paragraphs, we analyze three techniques that may allow for the localization of nanodevices or specific areas inside the human body. First, we discuss an approach based on trilateration inspired by GPS systems and routers. Second, we discuss a machine learning method that tries to utilize the travel time of messages to identify their origin. Finally, we present a method that tries to identify a unique fingerprint for areas inside living creatures based on molecules and proteins that only occur in specific regions of the body.

7.4.1 Multi-gateway Hop-Count Localization

In Chapter 6, we have learned about the hop-count routing protocol, which enables us to communicate in nanonetworks with very few resources. While hop-count routing enables some simple applications, the protocol is far from perfect and leaves much to be desired Stelzner and Traupe (2019). However, a few simple changes can already improve the protocol tremendously while still keeping the memory requirements low.

One of the biggest weaknesses of hop-count routing is the inability to route packages in a targeted spatial direction. While it is possible to route messages from a node to a gateway, the whole network is involved in that process in the worst case, which requires a lot of energy.

A simple solution for this problem is the introduction of more than one hop count. Using two different gateways each with a unique hop count already allows us to create a curve-linear coordinate system where individual quadrants may be addressed. An example can be seen in

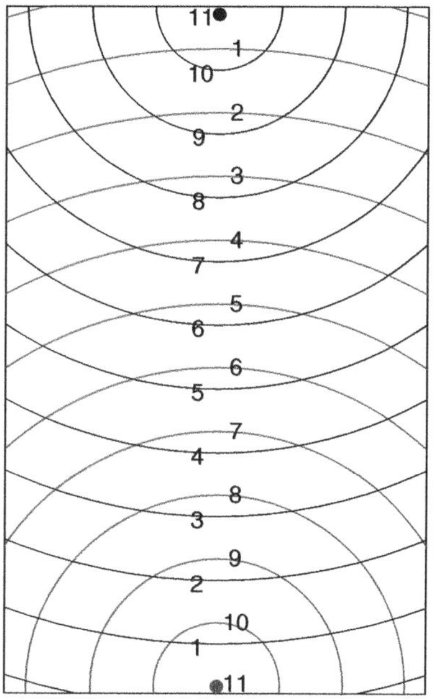

Figure 7.4 Two overlapping hop-count zones forming a curve-linear coordinate system. *Source*: Florian-Lennert A. Lau.

Figure 7.4. The concentric circles starting from the top and from the bottom gateways represent the approximate. Each overlap represents an addressable quadrant that may be used for directional routing purposes.

As a result, devices are now capable of forwarding a message "along an individual hop count." As we will see in Chapter 9, it is even possible to send messages along a quasi-straight line.

While two gateways and hop counts offer significant improvements, the absolute position in space is still undetermined and requires the introduction of a third gateway for trilateration. Figures 7.5–7.7 show the result of a simulation using three hop counts and three gateways.

Figure 7.5 displays the initial state of the network similar to using just a single gateway. The three gateways have the first, second, and third hop count of each respective nanodevice assigned to them. The underlying circles represent the communication range of the nanodevices. Initially, all hop counts start at ∞ and assume the lowest hop count they receive via any propagation message. If the network is connected, each node will receive a hop count from each gateway in a finite amount of time. These represent the fewest hops required to reach each gateway from the node's position.

Figure 7.6 shows the same network after the propagation phase has finished. The opacity of the underlying range indicators now represents the proximity to the other gateways. Furthermore, the nanodevice variables that store the hop counts are no longer set to ∞ but to the respective distances in hops. The gateways also store the distances to other gateways respectively.

Figure 7.7 shows additional information about the localization process using hop counts. The thick, colored circles represent an approximation of the distance in hops from each gateway. The intersection between the three circles may be used to address nanodevices in a specific area. As an example, node (5,4,6) may be addressed like that given the chosen distances.

Furthermore, if the distances between the gateways are known in advance, they may be used to estimate the average range of a hop of approximately – in this case 1.64 mm. In this example, the distance between green and blue is 8 hops or 12.73 mm. The distance between the bottom left and

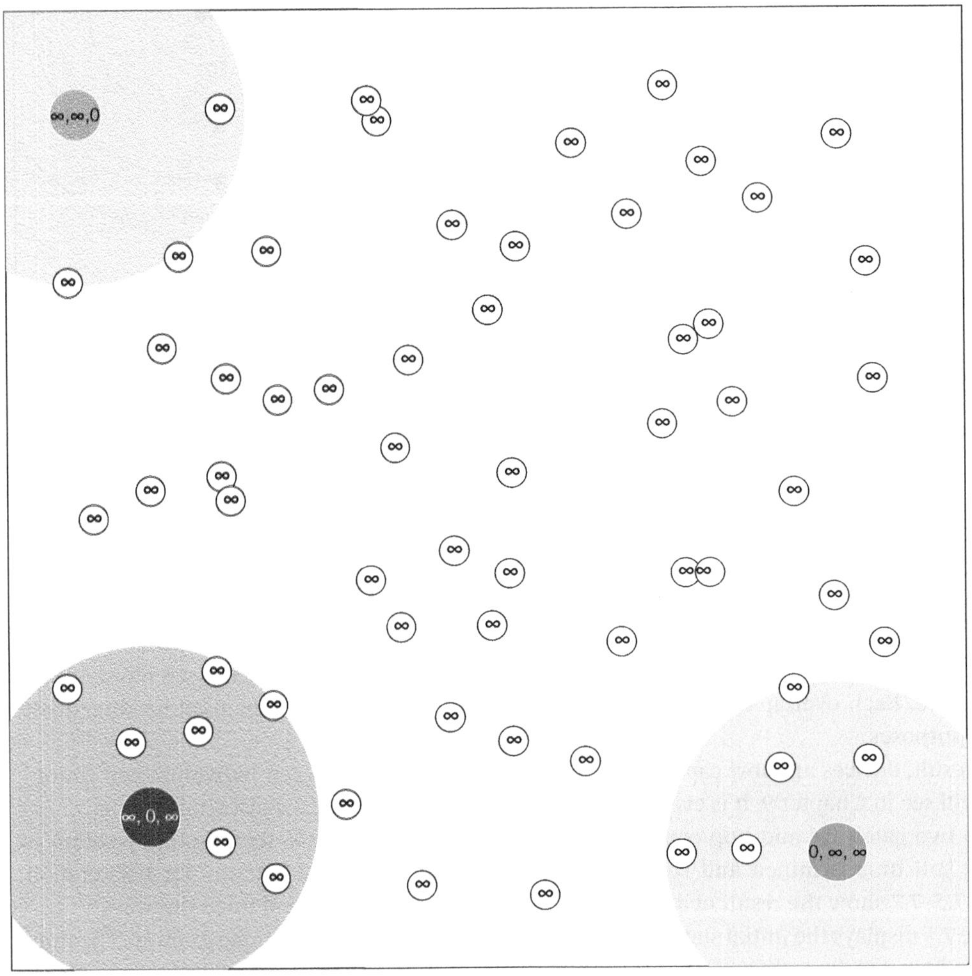

Figure 7.5 Initial network state where each hop count is set to ∞ modified and based on the work of (Stelzner and Traupe, 2019). The top left gateway has the rightmost hop count, the bottom left gateway the middle one and the bottom right gateway the right hop count. The underlying circles show the initial state of the network before a topology is established. *Source*: Adapted from Stelzner and Traupe (2019).

bottom right gateway is 5 hops or 8.28 mm, and the distance between the bottom left and bottom right gateway is 5 hops or 8.47 mm.

Figure 7.8 shows a much more advanced example that uses the same principle given a human body model generated by Geyer et al. (2018b). The three gateways are positioned at both shoulders and the right hip and the shading represents the distance to the respective gateways.

The black nodes are disconnected from the network and have no hop count assigned. The x-axis, y-axis, and z-axis display the distance in meters and roughly correspond to an average human in size.

As we can see, even using 64 000 nodes and a communication distance of roughly 10 mm is not sufficient to guarantee the hop-count network connectivity. Especially nanodevices in the legs and

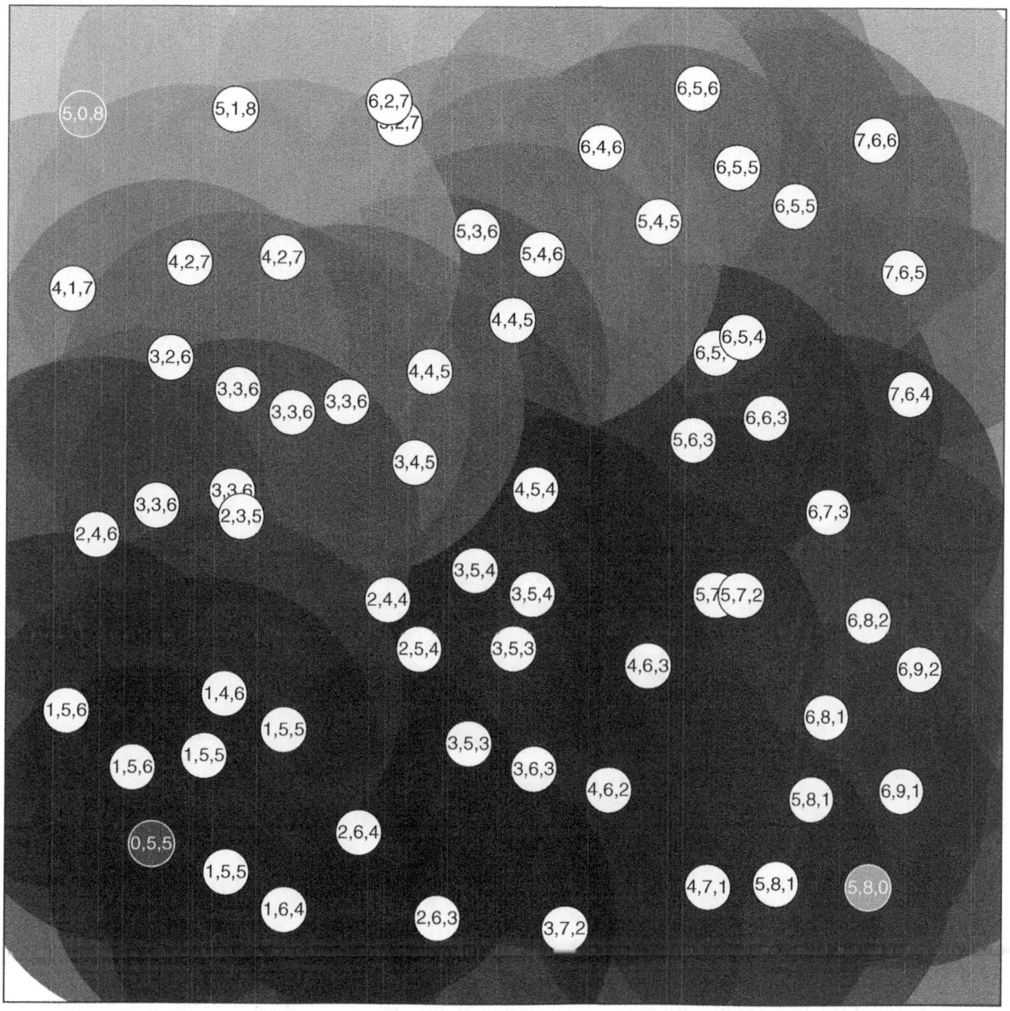

Figure 7.6 State of the hop-count network with assigned hop-count values to each and every node. The color of the zones around the nodes represents the proximity to the respectively colored gateway. *Source*: Adapted from Stelzner and Traupe (2019).

other peripheral regions of the body are often disconnected from the rest of the network. Nonetheless, the center parts of the human body are well connected and messages may be exchanged there. That said, much higher numbers of nodes might be necessary to ensure full connectivity in a majority of cases.

In summary, using three times the initial memory already allows us to establish a rudimentary location awareness of nanodevices. This does not change the memory complexity, and the protocol still operates using $O(1)$ memory. However, it is now possible to send messages to any location in the body with only a small error. In summary, it is already possible to create electromagnetic nanonetworks with very few resources available. This is surely a good starting point given the harsh resource constraints.

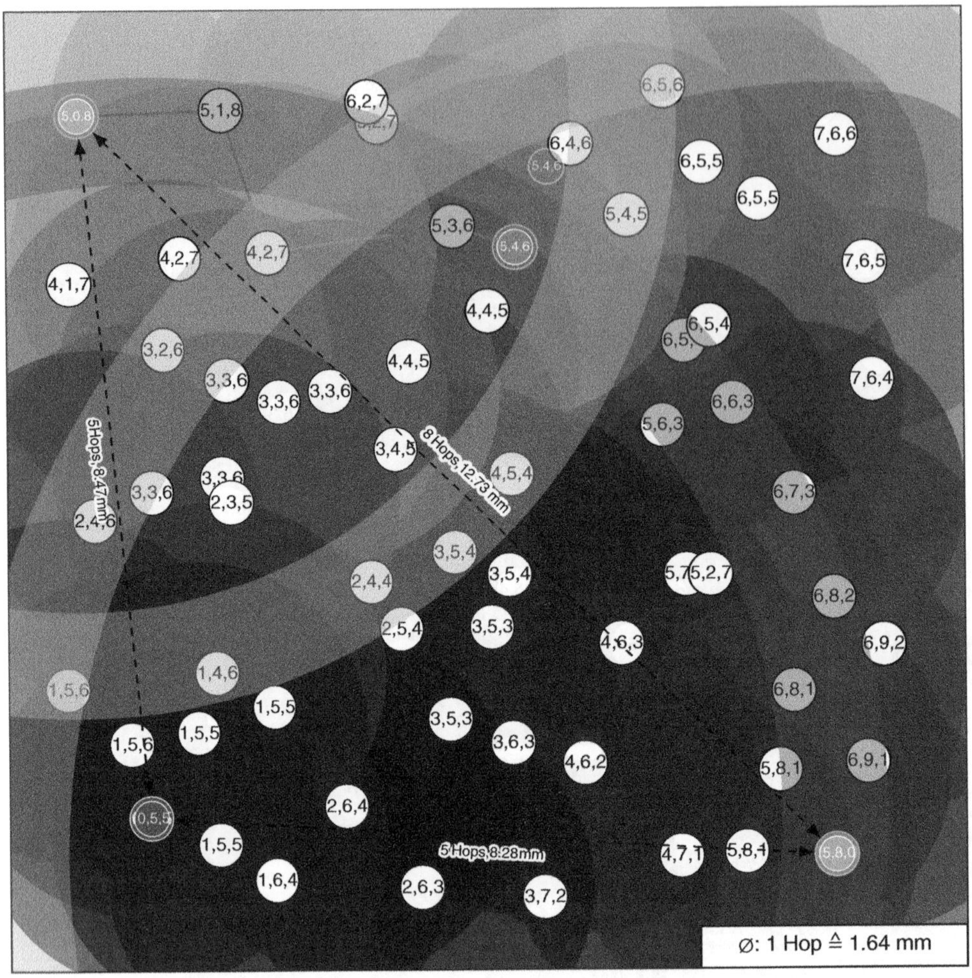

Figure 7.7 State of the hop-count network with several examples for trilateration. The absolute positions of the nodes may now be estimated through the a priori knowledge about the gateway positions and the average distance between two nanodevices. For this scenario, a communication distance of 2 mm was used. *Source*: Adapted from Stelzner and Traupe (2019).

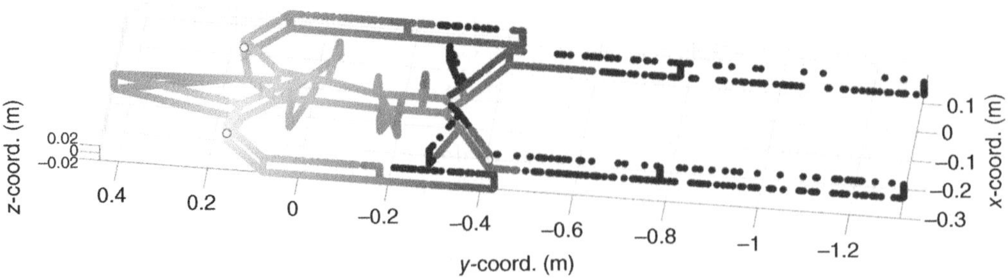

Figure 7.8 3D hop-count network after the propagation phase in a human body model. The three gateways are marked in yellow, and the saturation of the colors stands for the distance to the respective gateways. The black nodes are disconnected from the network and have no hop counts assigned. *Source*: Adapted from Stelzner and Traupe (2019).

7.4.2 Age of Information

Another approach for estimating the positions of nanodevices inside of a known environment is based on machine learning of the *age of information* (AoI) of molecular messages (Gómez et al., 2022). If each nanodevice that participates inside a nanonetwork possesses the necessary timing components (see Chapter 10), it is possible to timestamp messages. Given the required level of precision, a variable number of bits is necessary to measure the delay between sending a message and its arrival at the target destination. If the environment is sufficiently predictable like, e.g., the human bloodstream, additional conclusions might be drawn. As blood moves rather slowly, 8–16 bits might offer sufficient resolution, given that older messages lose relevance over time and might be omitted at some point.

With enough a priori information and a specific network topology (like the human circulatory system), it should be possible to isolate the areas of origin of specific messages. For this to work properly, a well-defined network architecture with few unknown parameters must be presupposed and the same message must be detected in different places. Based on that, messages might either be directly detected by a more powerful gateway device or must be collected and preprocessed by other devices. As the blood always passes by the heart, it might be a suitable location for a gateway device. This could be necessary as detecting molecules directly might require direct contact with a sensing device at the heart, which poses too big of a health risk.

Thus, instead of sending messages directly from point A to point B, special *nanocollectors* gather and preprocess all information (Szott et al., 2022). The nanocollectors themselves then move through the human body with the blood flow and transmit the gathered information to a more powerful device that executes machine learning methods on the transmitted data. Based on the concentration of the messages and the time, it is sometimes possible to relate the information to specific areas in the body. Especially in the torso, there are multiple options with similar travel times and it is not too easy to distinguish between them. However, areas that are further away are easier to unambiguously identify.

Given a rough categorization of the human body into

- head,
- shoulders,
- upper arms,
- elbows,
- hands,
- thorax,
- spleen,
- kidneys,
- liver,
- intestines,
- pelvis,
- hips,
- knees, and
- feet,

many areas can be successfully identified.

Results from the work of Szott et al. (2022) suggest that all the peripheral areas in the body can be detected without any errors. However, the internal organs are difficult to distinguish. The error for the spleen, kidneys, liver, and intestines varies between 40% and 65%. In those cases, the age of information/traveling time combined with the concentration is not sufficient for a good prediction. Nonetheless, the technique performs well overall and rightly predicts 85.91% of all data points. That said, more research is necessary to address the problems in the center area of the body and it is unclear if artificial intelligence is suitable in a medical environment.

7.4.3 Tissue Fingerprinting

The last type of localization we analyze is *tissue fingerprinting*, where we try to find ways to identify specific regions in the human body given available body parameters (Wendt et al., 2023; Lau et al., 2021a). For this to work, two general procedures might offer insights: tracking trace elements and tracking the occurrence of special proteins/genes.

The first tries to identify areas in the human body by using trace elements or gases inside the circulatory system. The O_2, CO_2, and pH levels vary according to the type and the size of the blood vessel and may be used to identify very rough regions in the body. The further away the blood is from the lungs where it gets enriched with oxygen, the lower the concentration gets. However, this approach is not suitable for distinguishing more than a few very rough regions. Furthermore, physical exertions heavily influence the oxygen levels and thus influence measurement results in an unpredictable manner (Mairbäurl, 2014).

Other possible substances that might be correlated with specific body regions are trace elements like cobalt, copper, iron, zinc, and many more (Versieck, 1985; Hassanin et al., 2021). Some of these fulfill more than 300 known functions in the human body. As a result, they occur in multiple regions but in different concentrations. Combining multiple elements might yield a suitable fingerprint with a high enough resolution. However, due to the very high variance and the relatively small number of possible trace elements, it is likely that only a small number of rough areas may be distinguished from each other.

The last and most promising method for identifying specific areas in the human body is *proteome fingerprinting*. The proteome represents the set of all proteins that may naturally occur in the human body. Related to the proteome is the *genome*, which contains the necessary genetic information to encode the aforementioned proteins. Each protein fulfills an often unique function in the body and thus only occurs in a limited number of areas. Unlike trace elements, there are thousands of possible proteins and it is likely possible to identify bodily regions with a much higher resolution.

In fact, in the protein atlas, a majority of all human proteins are mapped to specific tissues (Uhlén et al., 2015). Of the 20 000 proteins, roughly 9000 show elevated levels in specific areas of the body and may be used for localization. Table 7.1 shows several example proteins that may identify the

Table 7.1 Six gastric-specific proteins that can be used as biomarkers for uniquely identifying certain tissues.

Gene	Tissue specificity score	Tissue distribution
PGA5	2014	Detected in many
CBLIF	1859	Detected in single
GKN1	1840	Detected in many
PGA4	1281	Detected in many
PGA3	613	Detected in many
ATP4B	305	Detected in some

Source: Florian-Lennert A. Lau.

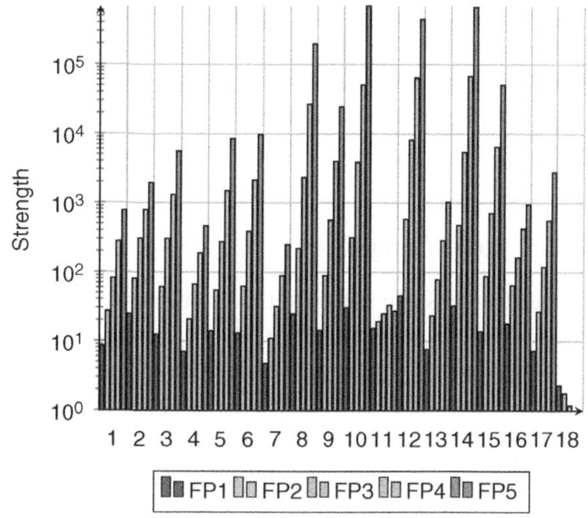

Figure 7.9 The likelihood of correctly identifying one out of 18 organs based on the 1–5 most commonly occurring proteins. *Source*: ITM (Regine/Florian-Lennert A. Lau).

human gastric system. Using more than a single protein allows us to identify a single area with almost arbitrary accuracy.

A similar table may also be created for all other organs. While there are sometimes thousands of genes that show elevated concentration in specific areas, a handful is sufficient for a high probability identification. Thus, using as little as two to five genes that are unique enough for an area constitutes a proteome fingerprint. Figure 7.9 shows examples of strengths of five-gene fingerprints for different organs. The strength represents how strongly a fingerprint identifies an organ.

Except for the bladder (18), all organs can be better identified by adding more genes to a fingerprint. However, for most tissues, two genes are more than enough to uniquely identify the organ without much error.

All in all, this method of localization looks very promising and might be utilized in many ways. It might further be combined with DNA-based self-assembly systems to decide if a sufficient number of proteins or genes are locally present. The simple architecture of an AND message molecule is a perfect match for the fingerprint requirements.

7.5 Simulation

With all the different types of locomotion and localization in mind, real experiments are still infeasible in many cases. Either the necessary tools are missing or the cost-to-reward ratio is simply not sufficiently good to justify funding for such research. As a result, researchers have to come up with other solutions like simulations.

Simulations abstract and generalize complex physical systems and try to only represent the factors that are of relevance to the task at hand while omitting everything else. This is necessary as a small number of interrelated simulation parameters or rules can already render a computational task *intractable*. While it might be possible to conceptualize an algorithm for the task, it will not terminate in an acceptable number of steps.

Consequently, good simulators are quite difficult to come by but can offer valuable insights into a topic if applied to the right problems. In areas like nanonetworks, where only a small number of primitive devices may be manufactured, they are invaluable tools for further research. Especially in safety-critical environments like the human body, there is simply no initial alternative to simulations.

As a result, simulations of parts of the human body are of special importance when it comes to nanonetworks. In the following, we discuss two simulators: BloodVoyagerS and MEHLISSA. The former software simulates the positions of nanodevices or particles in the human circulatory system. The latter modularizes the human body into four levels of precision where different physical effects dominate to generate further knowledge.

7.5.1 BloodVoyagerS

BloodVoyagerS is a software that simulates the distribution of nanodevices and other particles in the human circulatory system (Geyer et al., 2018a). This information is valuable for any type of routing protocol simulation inside the human body as any type of random distribution would be a mere estimation or assumption. In reality, countless influence factors exist that invalidate such simplifications. As a result, simulation is the only feasible option for deriving realistic network topologies under the given circumstances.

BloodVoyagerS tries to accurately model an average female body of roughly 172 cm in height. The software includes a representation of all major blood vessels and organs as well as their dimensions. However, the z-axis is only rudimentarily implemented as of right now.

The general model can be seen in Figure 7.10. To the left, 94 different points of interest are represented in a schematic representation of the human circulatory system. For more information on the different node numbers, see the work of (Geyer, 2017). The left lines represent the vessels that start at the human heart. The right lines represent all vessels that pump the blood back to the heart after passing through various organs and capillaries.

Overall, the circulatory system of all living beings is a closed system that follows Kirchhoff's first rule. The sum of the inflow of blood is equal to the outflow of blood at any given point where vessels fork or merge. Furthermore, a good intuition for the entire system for engineers is two binary trees that attach to each other on the level of leaves (capillaries). In the end, all blood must pass by the heart where the two trees reattach at a single node. Due to the forking tree structure, there is an exponential growth of the sum of the length of all vessels. As a result, the entire system is close to 100 km long and can span the entire planet Earth more than two times.

BloodVoyagerS mainly models the influence of bulk flow as it is the dominant movement factor. On average, it takes about one minute for the blood to circulate the entire system and arrive at the heart again. Just relying on diffusion, the process would take much longer.

Figure 7.11 illustrates how blood vessels themselves are usually modeled. The most relevant parameters are the spatial dimensions, contained particles and nanodevices, and the distances and sizes of all contained nanodevices.

In BloodVoyagerS, each vessel has the following attributes:

- BloodvesselID,
- bloodvesselLength,
- angle,
- velocity,
- bloodvesselType,
- startposition,
- stopPosition,
- deltaT,
- nextBloodvessel1,
- nextBloodvessel2,

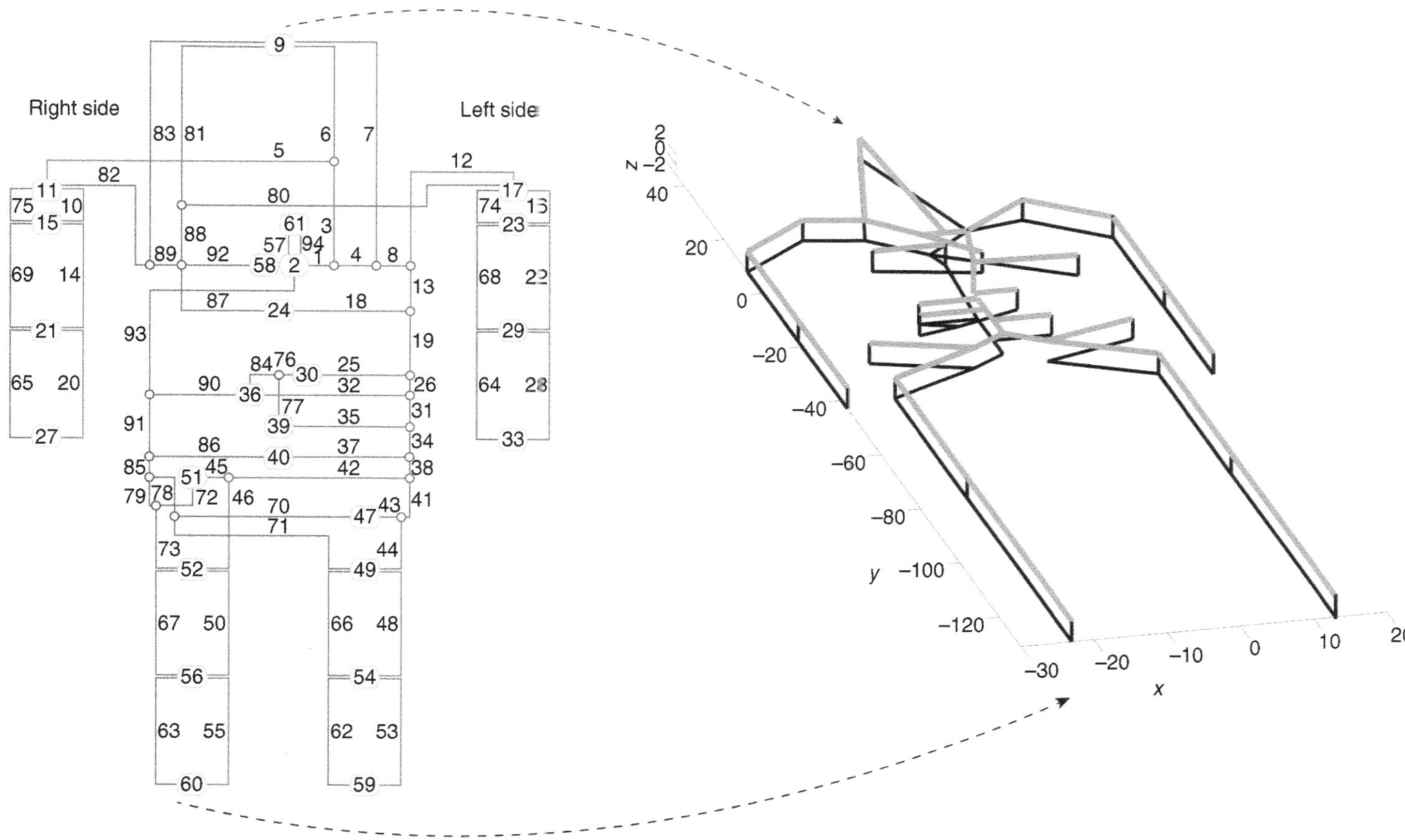

Figure 7.10 Relationship between the 94 modeled vessels and organs in a schematic view and the model used for simulation. *Source*: Geyer et al. (2017) /with permission of Association for Computing Machinery.

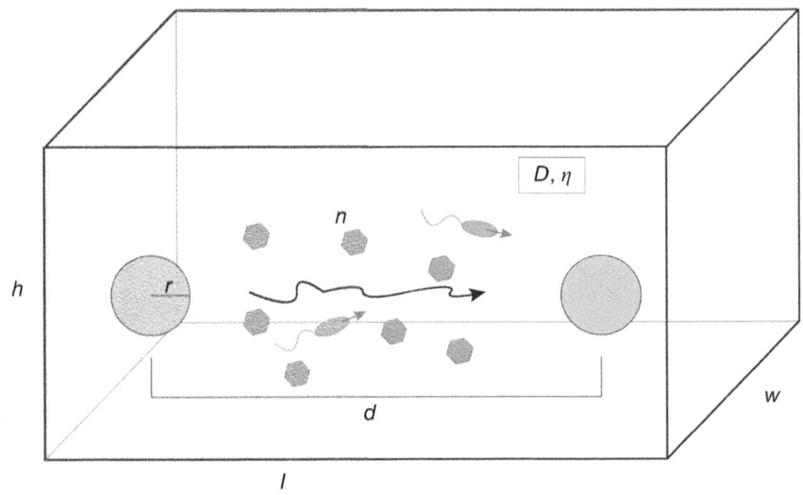

Figure 7.11 Model for all blood vessels in nanonetworks. It includes a height h, a length l, a width w, and information that is relevant to nanobots of a certain size. *Source*: Adapted from Geyer (2017).

- numberOfStreams,
- streamWidth, and
- nanobots[].

While the length of each vessel is accurately represented, the y- and z-axis are modeled by an adjustable number of streams. This simplification allows for a much faster simulation of often large numbers of nanobots. It is not rare that hundreds of thousands of devices must be simulated at once, which can lead to poor performance otherwise.

In the end, all the vessels are collected in a blood circuit class and form a holistic model. The simulation then continues to run in a game loop for a set number of steps or until manually terminated. All the relevant information is written in a CSV file whenever necessary and may be downloaded. The software is still updated as of October 2023 and might contain several refinements that are not reflected in this section.

7.5.2 MEHLISSA

While simulating the positions of nanodevices or particles in the human circulatory system is of considerable interest, it is not sufficient for all nanonetwork applications. On different scales, devices will likely be subjected to different physical effects or other emergent phenomena. To tackle this question, (Wendt and Fischer, 2020) proposed a simulation framework based on modularizing the human body into different levels of detail.

An overview of this idea can be seen in Figure 7.12. The entire human body is broken down into four levels of detail that likely differ sufficiently from each other. The framework intends to isolate aspects of the body that are sufficiently different and simulate them on their own. In the process, constants are generated that serve as input for lower or higher layers. This is necessary as simulations of such a complex environment with multiple, possible interconnected parameters are infeasible otherwise. Once generated, the generated constant and simulation results may hopefully be assembled into a solution for the "whole."

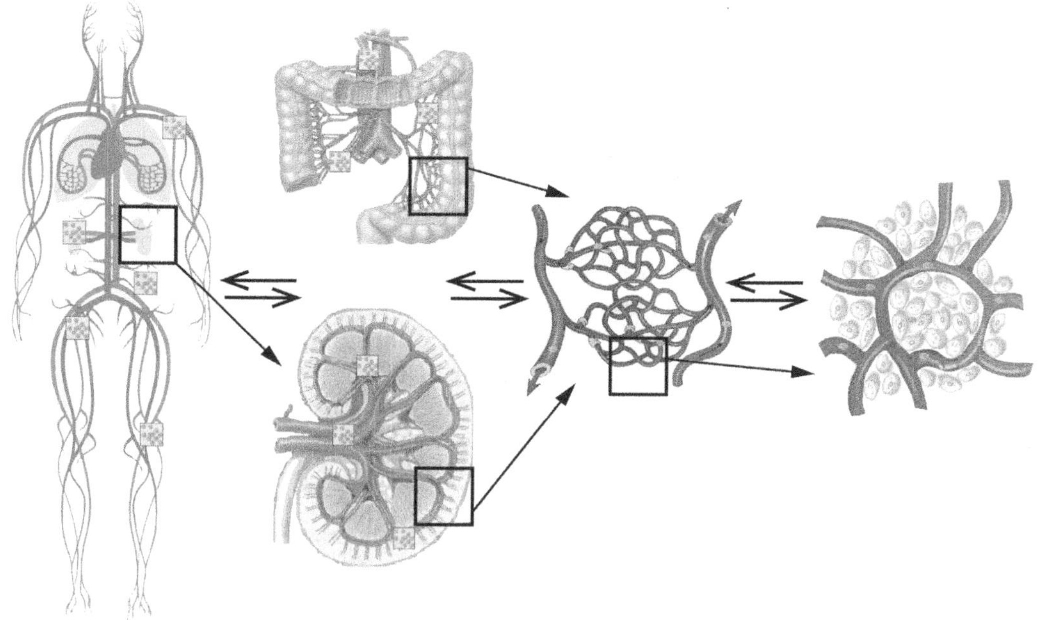

Figure 7.12 Different modules and relationships of the MEHLISSA framework. There is a body layer, an organ layer, a capillary layer, and a cell layer. *Source*: Wendt et al. (2020)/with permission of Association for Computing Machinery.

We now present the four different models in sufficient detail to give the reader an intuition of the expected problems and proposed solutions. That said, this framework is still in development and thus incomplete, yet the overarching idea is likely the only possible way.

7.5.2.1 Body Module

We start with the body module, which has already been addressed to some degree in Section 7.5.1. This module simulates the global position of nanodevices in the body and accounts for the main type of passive movement: blood flow. The blood flow dominates all types of macroscopic bulk movement, and smaller effects like diffusion fade in comparison but might become relevant at smaller levels of size.

Apart from that, BloodVoyagerS answers the questions of the body module layer in sufficient detail. Of all the proposed layers of the framework, this one is best researched. That said, future developments also consider the varying movement speed of particles of different sizes.

However, as with all simulations, empirical tests are necessary to either confirm or invalidate the hypotheses generated by the simulation. On their own, simulation results are only of limited value and it is important to caution against giving them too much significance. That said, the only real wet-lab results of particle distribution in the human circulatory system originate from angiograms that track the distribution of a contrast agent inside the bloodstream.

7.5.2.2 Organ Module

The next level of detail is the organ module. The organs inside the human body vary significantly in structure, size, function, and metabolic activity. All of them require varying amounts of macronutrients and micronutrients to function properly. As a direct result, they also produce different waste products that have to be excreted via the blood through one of the filter organs or the skin pores.

Apart from that, all the organs require different amounts of oxygen to function properly, and as a result, more or fewer vessels span the entire organ. This is only worsened by other factors like the level of physical activity. Muscles can consume up to ten times the amount of oxygen while exercising and will switch to a different mode of anaerobic energy production if the amount of energy that burning oxygen can supply is exceeded. Thus, it is not too easy to estimate the positions of nanodevices through, for example, trilateration or measuring travel times of particles.

Additionally, the size of organs differs tremendously. While an average kidney is roughly the size of an average human's fist, the colon is several meters long and fills a considerable portion of the human torso. Furthermore, there are obviously individual differences. The average human might be around 160–170 cm depending on the gender, yet the range of possible heights spans the entire range between 55 and 271 cm. In addition to that, the size of, for example, the human heart might be double for endurance athletes compared to an average individual.

As a result of all of this, individual measures are likely necessary. Especially at the nanoscale, such differences do matter and must be accounted for in both simulations and wet-lab experiments.

7.5.2.3 Capillary Module

While the total length of the combined circulatory system is roughly 100 km, most of that enormous length stems from the capillaries. Capillaries are the smallest blood vessels in the human body and only measure between 5 and 10 μm in diameter. In comparison, a red blood cell is roughly 7–8 μm in diameter but only 2.5 nm thick. They can only pass through many capillaries due to their flexibility.

The capillary layer is metabolically different from organs and big blood vessels as the transfer of nutrients and waste materials in and out of cells happens here. To achieve that, the bigger arterioles split multiple times into smaller vessels until the majority of cells are either directly adjacent to a vessel or there are at maximum two cells in between. This repeated splitting of vessels leads to an exponential growth in their total length, while the contained volume remains identical.

The total number of capillaries as well as their average density heavily depends on the involved organ/tissue as well as the activity levels of a person (Jensen et al., 2004). The human body has an energy conservation drive that attempts to minimize its energy expenditure whenever possible. On average, only as many structures as possible are created as everything else would be a waste of otherwise valuable energy. This results in considerable variance from person to person and body region to body region when it comes to capillary density.

In addition to that, some types of tissues, like cartilage, are only indirectly supplied with nutrients. There, the ring muscles of the arterioles are closed most of the time and only a very limited amount of blood may enter those areas. As a result, the required number of nanobots is vastly different if a collaborative task has to be performed.

Furthermore, it is completely unclear how and if it is possible to communicate via, for example, terahertz waves between capillaries that are technically in range but separated by other cells. Additionally, it might be interesting to know how much time a passively moved nanobot spends in such an area on average to perform, e.g., medical tasks. All in all, this layer might be one of the most interesting when it comes to the research of actual medical applications using nanonetworks.

7.5.2.4 Cell Module

The last module concerns itself with the interior of cells. In general, three overarching types of cells have to be distinguished: eukaryotes, archaea, and bacteria. The latter two are sometimes summarized as *prokaryotes*.

Prokaryotes do not possess a cell nucleus and are usually simple, single-celled organisms. Furthermore, they are usually much smaller than eukaryotes and measure just 1–2 m in diameter.

All the substances such cells require to function float freely in the cytoplasm of the cell. This includes the DNA of the cell as well as smaller DNA rings called plasmids.

Eukaryotes on the other hand are usually part of multi-celled organisms like humans, animals, fungi, and algae and possess a much more complex internal structure. Due to the additional separation of the cell nucleus, eukaryotes do not need as much DNA to prepare the cell for varying environmental parameters. In addition, eukaryote cells have other separate internal organelles like mitochondria or the Golgi apparatus. Furthermore, eukaryote cells are usually much bigger and span 10–38 μm in size.

The DNA of eukaryotes is confined to the nucleus and is stored there as chromosomes. Those unfold on demand when a gene has to be transcribed into a protein or the cell wants to divide itself.

In the following, we mainly focus on eukaryote cells as they are most interesting to medical applications. Any medicine that influences the human body does so on the level of cells. Thus, any type of nanonetwork-based medical application must also affect the human body on this level.

Due to the unique internal cell structure, it might be possible to use or exploit different kinds of locomotion and localization. It should be possible to determine the position of nanodevices in relation to cell organelles that are always similarly structured. Furthermore, many cells contain a "railway system" made from microtubules that many of the aforementioned motor proteins use to transport nutrients and waste products to their target destination. As a result, many of the otherwise proposed mechanisms will either not work or work in a fundamentally different manner inside cells.

The cell module of MEHLISSA aims at simulating the interaction between nanodevices and medication in blood vessels and cells. As many envisioned scenarios propose a targeted drug release, this step is of great importance, and time bounds as well as other side effects have to be properly researched.

7.6 Organs-on-Chips

While simulation is likely the most cost-effective way to generate early insights into novel areas of research, they are almost universally subject to errors and must be repeatedly refined. This is due to the inherent loss of information and accuracy that comes with modularization as well as generalization. There is simply no guarantee that the simulation results of partial problems offer a solution for the original, bigger problem. As a result, wet-lab experiments offer more reliable insights and should be conducted once simulations have generated interesting hypotheses.

Especially in the area of medicine where human lives are on the line, wet-lab experiments of yet untested technologies are unthinkable unless the patient is already terminally ill. However, many of the applications of nanonetworks are about prevention and early detection and thus cannot be tested under those conditions.

As a result, other alternatives are necessary. One of those alternatives is organs-on-chips, which are currently researched as tools to prototype medication without having to accept the risks of testing on living humans or animals (Huh et al., 2011). Organs-on-chips are tools to simulate the environments in a living body accurately. Instead of fully relying on computer simulations, those devices often contain multi-channel 3D microfluid cell cultures on an integrated circuit. These chips emulate the entire physiology of organs or parts of a body.

While numerous papers on the topic exist and several products are already available on the market, the entire area is still in its infancy. However, all the major organs of the human body as well as skin, bones, cartilage, and even the blood vessels have already been created (Low et al., 2021).

Especially in the context of this chapter, the vessels-on-a-chip are of special interest as they might be used to validate the results from the BloodVoyagerS software. In later stages, they might also be used to simulate the previously conceptualized organ module of MEHLISSA.

All in all, organs-on-chips offer a potentially more feasible alternative to animal testing or situations where tests on living creatures are not possible for ethical reasons. Furthermore, such tests are likely much cheaper and less regulated and thus more accessible to research institutions.

7.7 Summary

In summary, locomotion and localization are two interrelated topics of extreme importance at the nanoscale. Without sufficiently good solutions for both problems, any application that requires a high degree of precision/accuracy is impossible.

Thus, we have discussed possible solutions for the locomotion of both the passive and active kinds. While it is unclear which of these methods will ultimately be used, it is nonetheless beneficial to know the scope of possibly exploitable physical effects.

Next, we discussed different methods for localization and compared the quality of the solutions. While not all of them are directly implementable using currently available technologies, the overarching paradigms are nonetheless valuable for further research. Especially software-based modeling frameworks and simulation tools offer insights into otherwise infeasible questions.

8

Sensors and Actuators

This chapter analyzes the topics of sensing and manipulating matter at the nanoscale in a targeted manner. We discuss both the sensor and actuator components and highlight why many of the other functional building blocks of nanorobots are special instances of the two. Before that, we reiterate several important scenarios that nanonetworks and nanotechnologies in general aim to solve.

Afterward, we discuss and analyze many of the currently used techniques to identify chemicals, diseases, or other types of matter. In medicine, those are typically divided into the two classes of *qualitative* and *quantitative* procedures.

Finally, we try to understand several sensor paradigms and possible implementations as well as actuator technologies that are already available. Contrary to popular belief, creating simple sensors or actuators is not that difficult. Only the accurate and precise application as well as the compatibility with other components poses challenges.

8.1 Application Scenarios

Before we begin with a detailed analysis of sensors and actuators, we have to contextualize them given the possible use cases. We have already listed many of them in detail in Chapter 3. Now, we further analyze some of them in terms of sensor and actuator requirements.

Medical applications form the majority of proposed applications, while all the others fade in comparison. Among medical scenarios themselves, there are three categories of often-named problems:

- Monitoring of disease parameters without stationary treatment.
- Direct treatment of diseases while they still emerge.
- Local treatment without far-reaching systemic side effects.

All of the aforementioned scenarios require interaction with the environment in one way or another. Any kind of treatment can and should also be understood as a special kind of actuation, even if it is just the release of medication. As soon as the right place, time, or combination of circumstances have to be met, some degree of sensory input is necessary. In combination, none of those scenarios may work without at least primitive kinds of sensors and actuators. They form the basis of intentional interaction at the nanoscale. Many high-level applications further require at least actuators and sensors that enable communication.

Nanonetworks: The Future of Communication and Computation, First Edition. Florian-Lennert A. Lau.

8.2 Measuring Systems

Measuring both environmental data like temperature as well as health data dates back several thousand years. As far back as ancient India, primitive doctors and health-enhancing procedures were already available.

While measuring environmental data is relatively simple, biological systems are much more complex and it is thus harder to find causes for problems. In the broadest sense, all kinds of diagnostic measures for human health can be divided into four categories. They can be seen in Figure 8.1 (Scharringhausen, 2018).

There are four broad categories of diagnostic tools. Of those, anamnesis and physical examinations cannot be performed by nanodevices as global knowledge about the system and vast computational power as well as medical databases would be necessary. As of right now, it is not even possible to properly perform those tasks using machines at the macroscale, let alone at the nanoscale. That said, some of the hindrances are more political and less of a technical nature.

The remaining two categories of apparative examinations and laboratory analytical methods are of potential interest to nanonetworks. Both can be verbally defined as follows:

Definition 8.1 *Apparative procedures* use complicated equipment to examine patients. Imaging procedures include procedures that make structures inside the body visible without the need for surgical intervention. These include X-rays, computed tomography, and sonography. Electrophysiological methods use the body's electrical fields to measure the activity of the heart, muscles, nerves, and sensory organs. The known procedures are EKG and EEG.

Definition 8.2 The *laboratory analytical methods* include quantitative methods and qualitative methods. The number, concentration, or activity of selected particles is determined using quantitative methods. In contrast, qualitative methods only return "positive" or "negative" as a test result. For diagnostically relevant statements, these results are linked to pathological threshold values.

It is clear that possibly all apparative methods may not be used in nanonetworks due to the size constraints alone. None of the named devices is available at the required dimensions.

Some laboratory analytical methods however offer potential. However, most qualitative procedures are tied to predefined thresholds that vary a lot from person to person. As a result, those tests

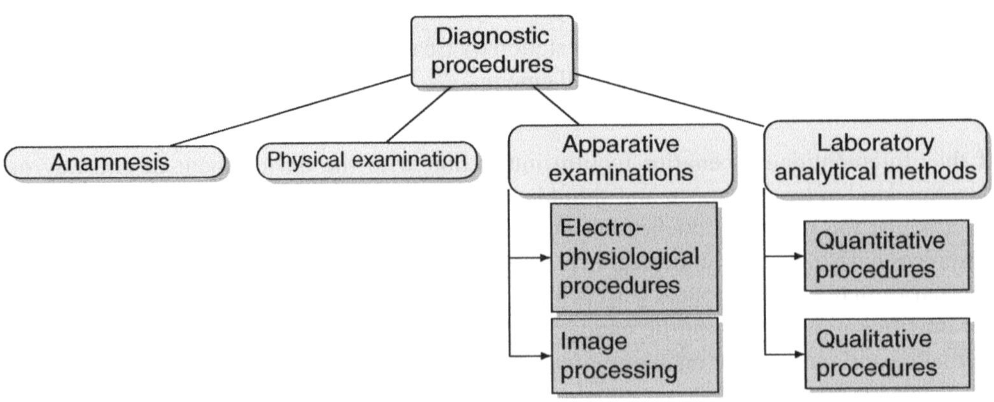

Figure 8.1 Overview of diagnostic procedures in medicine (Scharringhausen, 2018).
Source: Florian-Lennert A. Lau.

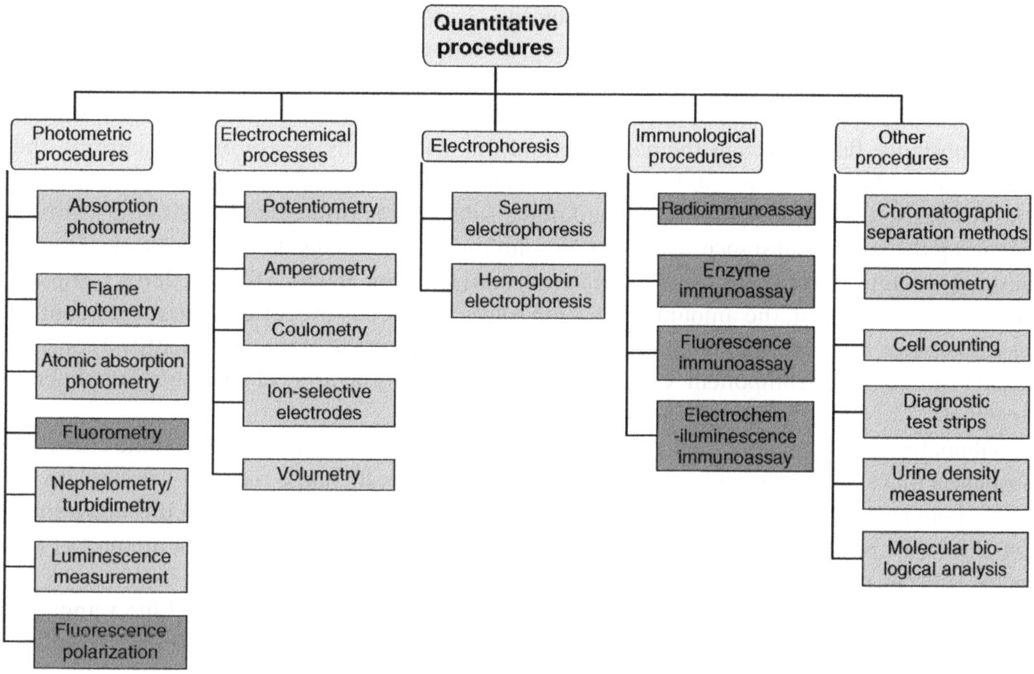

Figure 8.2 Overview of quantitative procedures of laboratory analytical methods (from Bruhn et al. (2011)). *Source*: Bruhn et al. (2011)/Schattauer.

are barely used anymore and have been replaced by, for example, test stripes that display a gradient instead of a binary value.

An overview of the more promising quantitative methods can be seen in Figure 8.2. They can further be categorized into photometric, electrochemical, electrophoresis, immunological, and other methods.

In general, all quantitative procedures have the same overarching goal of finding evidence for specific substances. The first category of photometric methods is based on using either visible or ultraviolet light to gain insights into a substance. In some way or another, a sample is hit by a beam of light, and the properties of the light are measured before and after a sample is hit to calculate a difference.

Absorption photometry measures the attenuation of the initial signal after passing through a sample. Atomic absorption photometry first vaporizes a sample and later measures the radiation. Flame photometry is based on burning a sample and analyzing the chemical composition of the resulting products of the reaction. Nephelometry and turbidimetry measure the clouding of a sample.

All of the above methods have the following in common: Relatively strong light is required, and the analysis of the reflection requires either complex machinery of vast amounts of data for comparison and/or computational power. Only luminescence measurement and fluorometry have some potential to be used by nanoscale devices as natural effects are used that might influence nanodevices.

The next group of diagnostic procedures concerns itself with electrochemical processes. Those include potentiometry, which is an analytical method in chemistry used to determine the concentration or activity of ions in a solution. A potentiometer is used to measure the potential between a reference electrode and a working/sensing electrode.

Amperometry measures a current caused by the electrolysis of the solution when a specific voltage is applied to an electrode. One electrode is used as the working or measuring electrode, and another is used as the reference or counter electrode. When a certain voltage is applied to the working electrode, a current proportional to the concentration of the electrochemically active substance flows through the electrode and the solution. By measuring this current, the concentration of the substance in the solution can be calculated.

In titrimetry, a solution containing an unknown substance is titrated with a titrating solution containing the known substance. The titrating solution is slowly added to the unknown solution, constantly monitoring the progress of the reaction, until the point at which the reaction is complete is reached. At this point, the amount of titrating solution added is proportional to the amount of the unknown in the solution. Volumetry works similarly but instead of trying to identify a concentration of an unknown component, volumetry is used to identify an array of unknown components altogether.

Coulometry is a chemistry-based analytical method used to quantitatively determine chemical substances in a solution. It is based on measuring the electrical charge required to complete an electrochemical reaction. An electrical current is used to pass a predefined charge through a solution. While the reaction takes place, the charge is measured and used to infer the amount of the desired substance in the solution.

While the entire category of methods sounds feasible, it usually requires big machinery and currents that are not feasible at the nanoscale. Thus, none of them seems applicable at the nanoscale, given the current knowledge.

The next group of diagnostic tools is based on electrophoresis. Electrophoresis is based on the different movement speeds of proteins, DNA, or RNA in a fluid under the influence of a spatially uniform electrical field. Once a sample of (macro-)molecules or proteins is added to the test device, the charged particles will move in the direction that is opposite to their own electrical charge. Depending on their size, molecules will move varying distances and it is thus possible to identify substances or even specific DNA sequences with a high degree of precision.

This procedure is likely most well known from PCR tests and has gained increased public attention due to the 2020 Sars-Cov-2 pandemic. There, the multiplied DNA is added to an agarose gel bed where the charge is applied.

While the overall procedure is extremely successful and among the most well-known and accurate tools, it requires vast amounts of energy and molecules to function properly. This is no problem at all at the macroscale but likely impossible to feasibly implement at the nanoscale with limited resources and available energy.

While there are many other diagnostic procedures available at the macroscale, none of them can likely be applied to the nanoscale as they are optimized for macroscale use and circumstances. The only exceptions are immunological procedures, which are the most promising diagnostic tools at the nanoscale.

Immunological methods work based on the interaction of antibodies and antigens. Antigens are special surface properties of substances we want to identify. Antibodies are substances that have special receptors for a fitting antigen. Both interact based on the previously mentioned receptor–ligand relationship that provably works at the nanoscale.

Overall, four different categories of immunological procedures can be distinguished. These are the radioimmunoassay, enzyme immunoassay, fluorescence immunoassay, and electrochemiluminescence immunoassay. All of these work based on the same underlying principle, but there are different ways to make the results of the antigen–antibody interaction visible. In addition to the binding itself, the interaction also produces a measurable signal as a by product.

As the principle of complementarity is involved in a vast number of biological and chemical processes, there is no reason to assume that it should not work in nanonetworks too. Of all diagnostic procedures, immunoassays are most likely implementable at the nanoscale. Figure 8.2 further shows an overview of potential candidates for further analysis marked in darker grey.

8.3 Sensors

Next to the classical diagnostic procedures, a number of other sensor paradigms exist. They may use physical effects that are not suitable for medical, macroscopic tests. Nonetheless, nanoscale sensors must still meet a number of constraints.

In general, sensors always work on a similar principle. They measure environmental parameters through a component that reacts in a predictable manner to changes in a specific parameter. An example could be a potentiometric displacement sensor which might be sensitive to physical movement. The movement itself gets converted to a voltage or an electrical resistance. If the sensor is well designed, there is a near-linear relationship between the changes in the output signal and the physical displacement that serves as an input.

In general, the following physical properties/effects might be of interest in sensor concepts (Khanna, 2016):

- mechanical,
- molecule/atom numbers/concentrations,
- thermal,
- acoustic/vibrations,
- optical/light,
- electricity/charge,
- (electro-)magnetic,
- chemical properties,
- biological, and
- antigen/antibodies.

In general, the macroscale principle also applies at the nanoscale, but the components have to be much smaller and usually autonomous. While the nanoscale only spans dimensions between 1 and 1000 nm, the prevalent definitions allow for sizes up to 4000 nm as that is the size of the smallest human blood vessel. Anything bigger is definitely not suitable for most proposed medical applications. Apart from physical limitations, sensors should be reusable more than once and must be biocompatible.

For all of the above-mentioned conversion phenomena, several sensor concepts exist. In the following, we only highlight several examples of nanosensors that are of special interest to nanonetworks. For a detailed analysis of everything related to nanosensors and the underlying physical effects, the interested reader is encouraged to have a closer look at the work of (Khanna, 2016).

8.3.1 CNT-based Sensors

While holistic designs for sensors are still theory only, many potential materials fulfill the necessary preconditions to act as a sensor. Especially promising candidates are sensors based on CNTs as these structures are often proposed as the construction material for nanodevices anyway and thus likely offer compatibility with such designs. As explained in Chapter 4, CNTs can have a variety of

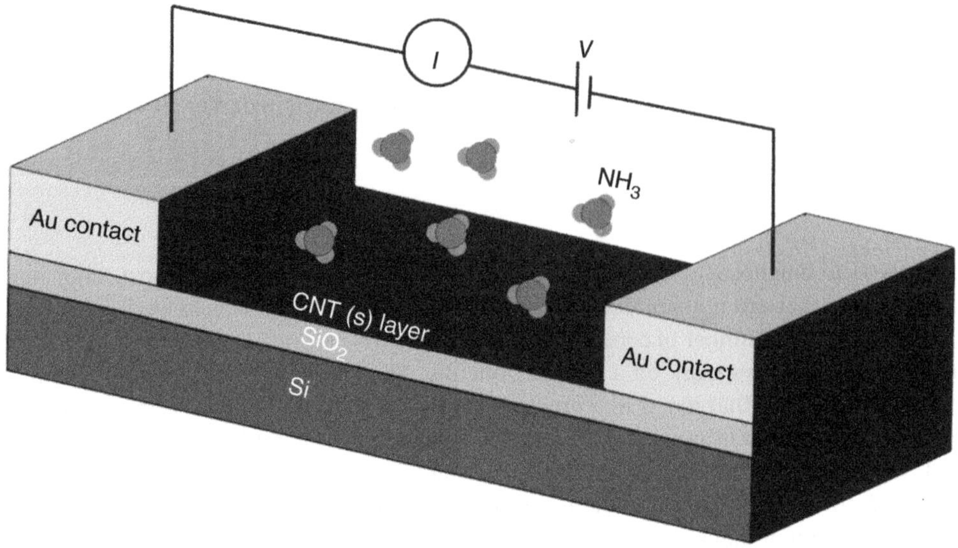

Figure 8.3 Type of CNT-based sensor that converts a physical property into a voltage. CNT-based sensors might detect differences in external force, gas concentration, or even the presence of antigens/antibodies (Wang et al., 2016). *Source*: Wang et al. (2016)/The Royal Society/CC BY 4.0.

different properties based on their internal configuration. As a result, different CNTs can be used to measure heat, electricity, and a variety of other environmental parameters.

A generic example can be seen in Figure 8.3. The idea behind most sensor concepts is always the same. A CNT is placed between a source and a drain, and several external parameters may influence its electrical properties. This change may be converted to another signal that can be interpreted/used by other components.

However, as CNTs are versatile, multiple parameters might influence them at once. Thus, it is often not too easy to identify the dominant influence factor from the "noise" of other parameters. Furthermore, the problem of precise component placement still remains and it is difficult to place a single CNT where you want, let alone ensure compatibility with other components.

8.3.2 Magnetic Sensors

Another type of sensor works based on magnetism. Magnetic sensors convert a magnetic field into an electrical signal. The sensor itself is able to detect deviations from a predefined range of values. As an example, if the source of the magnetic field is displaced, the output signal might get stronger or weaker. Given other environmental constraints and special sensor architectures, it is possible to infer what most likely happened from the change in the output signal.

At the macroscale, a variety of magnetic sensors are used for sensing electrical current or for motion-sensing tasks. At the nanoscale, similar applications might be possible.

8.3.3 Molecule Counters and Biosensors

The last type of nanosensor we discuss in this section is based on the receptor–ligand or antigen–antibody interaction.

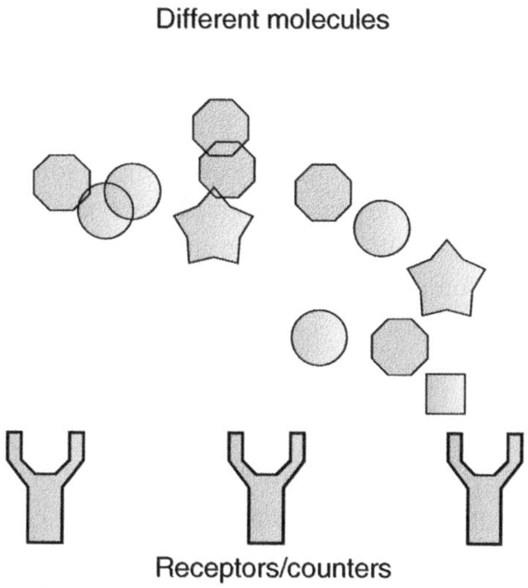

Figure 8.4 Interaction between a simple receptor/counter (bottom) and molecules. An output signal is only produced when fitting molecules (octagons) are present. This sensor could be based on enzymes, antibodies, cells, or other microorganisms (Khanna, 2016). *Source*: Florian-Lennert A. Lau.

One of the most important tasks in medical nanonetworks is the recognition of specific substances. They could be molecules, proteins, DNA, or more complex structures like viruses or cells.

The basic idea behind this type of sensor is illustrated in Figure 8.4. At the bottom, there are several receptors that are sensitive to specific substances and may only interact with a single substance that is available in the environment. In this case, only the octagons may be detected, while all other shapes are incompatible.

The receptors themselves can work based on a number of different effects. The interaction between deoxyribonucleic acids that has been discussed before is a popular possibility, but other approaches based on cells, enzymes, antigens, and many others are possible. In general, the output signal of a nanobiosensor can either be a number that represents a concentration or a simple yes/no value that indicates if a substance is present or not.

8.4 Actuators

Actuators are devices that are used to control or move a mechanical system or a physical process. They usually convert electrical signals into mechanical movement, allowing them to influence certain environmental parameters. Actuators are used in a wide range of applications, including robotics, manufacturing, automotive engineering, aerospace, and many more. Some examples of actuators include motors, hydraulic cylinders, but also dispensers for medication and even antennas or clocks.

In the context of this book, only actuators that operate at the nanoscale are of interest. They are typically used to control and manipulate nanoscale materials and devices. Nanoactuators operate using various principles such as electrostatic, piezoelectric, and magnetic forces. These principles allow for precise and controlled movements and manipulations.

Compared to conventional macroscale actuators, nanoactuators have several unique properties. For example, they can operate in very small spaces, allowing for the precise manipulation of tiny components. They have high force-to-mass ratios, allowing them to generate significant forces despite their small size. You could intuitively compare their relative strength to the strength difference between ants and humans.

Nanoactuators have many potential applications in nanorobotics, nanoscale manufacturing, and nanoscale medicine. For example, they could be used to precisely manipulate and transport nanoparticles or even medication to specific locations in the body for targeted drug delivery. As we have learned in Chapter 3, this is one of the biggest remaining problems in medicine and solving it would benefit basically everyone at some point in their lives.

In the following paragraphs, we shortly analyze the most common types of actuators at the nanoscale. We only try to understand them in a level of detail that is appropriate for engineers and computer scientists to get a good intuitive understanding of the overarching problems and possibilities.

8.4.1 Motors

The first and likely most intuitive kind of actuator at the nanoscale is a motor. We have discussed them already in some detail in Chapter 7, but will now shortly analyze them from the perspective of actuators.

Nanoscale motors are a type of nanoactuator that converts energy into rotational motion. They typically consist of a rotor and stator, similar to conventional motors, but are much smaller and operate using different principles. There are many types of nanoscale motors, including molecular motors, which use chemical reactions to power their movement, and synthetic motors, which are powered by external stimuli such as light or magnetic fields. They can be as small as a few nanometers in diameter, making them orders of magnitude smaller than conventional motors.

Unlike most macroscale motors, nanoscale motors usually have to function autonomously. As we have learned before, there is only a limited degree of achievable accuracy when trying to manually influence matter at the nanoscale. As a result, self-organization is a likely more promising paradigm for most nanonetwork applications.

To create a CNT-based motor, we typically start with a single-walled CNT that has been functionalized to create a rotor section (Kim et al., 2014). This rotor section is then attached to a stationary stator, which is typically another functionalized CNT or a graphene sheet. The motor operates by using an external energy source, such as an electric field, to induce a charge imbalance between the rotor and stator sections. This imbalance creates an electrostatic force that causes the rotor to rotate. The direction of rotation can be controlled by adjusting the polarity of the external electric field. If the electric field is applied in one direction, the rotor rotates clockwise. If the field is reversed, the rotor rotates counterclockwise.

As is most often the case, the nanoscale motor does indeed work as intended, but the precise and accurate combination with other functional components of nanorobots is still an open problem.

8.4.2 Antennas

Transceivers are a combination of transmitters and receivers and technically a combination of special actuators and special sensors. In electromagnetic communication, transceivers usually use an antenna for both detecting and sending electromagnetic waves. In the context of this section, we only analyze the sending of a signal as that is technically an act of manipulation of electromagnetic waves. Antennas for transmitting information are thus actuators in their most basic sense.

Antennas work by converting electrical energy into electromagnetic waves or electromagnetic waves into electrical energy. When an electrical current flows through an antenna, it creates an alternating electromagnetic field around it. This field propagates through space and may be received by another distant antenna. When that wave is detected by the antenna, an alternating electrical current is created. This current is usually amplified and later processed to recover the original signal.

The wavelengths that an antenna can interact with depend on its physical characteristics, including its size and shape and the material it consists of. The frequency of an electromagnetic wave is inversely proportional to its wavelength. Higher frequencies mean shorter wavelengths. As an example, a radio wave with a frequency of 100 MHz has a wavelength of approximately 3 m. In comparison, a microwave with a frequency of 10 GHz has a wavelength of approximately 3 cm.

The length of an antenna determines the wavelengths of the electromagnetic waves it can interact with. An antenna that is designed to operate at a particular frequency should have a length that is roughly half the wavelength of that frequency. As a result, nanoscale antennas have a very limited scope of signals they may send and receive.

CNT antennas are highly efficient at transmitting and receiving signals in the radio frequency range. They possess some unique electrical properties, including high conductivity, a high surface area, and low resistance.

8.4.3 Medication

The next type of actuator that is of special interest in the context of nanotechnologies is medication. While medication is rarely thought of as an actuator, it is nonetheless true in the most basic sense. Medication has the purpose of manipulating a body parameter in a targeted manner to allow the human body to function normally again.

As health is of special interest, a vast array of different medications exist that all function based on different underlying principles. At the very core, they are all actuators that sometimes have only a single use.

Optimally, medication only affects those regions of the human body that are actually affected by sickness. Everything else is best left untouched, but achieving that is among the biggest remaining problems in medicine. Nonetheless, it would be desirable if a medication works conditionally, which might be achieved using nanonetworks and localization mechanisms presented in Chapter 7.

One example of a nanoscale actuator used in medicine is the liposome. Liposomes are nanoscale vesicles that can encapsulate medication or other substances and release them when encountering specific cells or tissues. They may be equipped with antibodies that recognize and bind to specific receptors on the surface of specific cells.

Another example is the use of nanoparticles as drug carriers. Nanoparticles can be engineered to have properties that allow them to bypass biological barriers while accumulating with a high probability in predefined tissues.

There are many other examples of nanoscale actuators used in medicine, including dendrimers, CNTs, and magnetic nanoparticles, among others. However, it is important to note that not all medications are nanoscale actuators, and the majority of drugs are still delivered in traditional forms such as pills, capsules, or injections. Yet, even many of those types of medication ultimately affect the human body at the nanoscale and manipulate certain signal pathways in the body. In the broadest sense, influencing or manipulating is an "act" and the substance that initiates this effect could be viewed as a special type of single-use actuator.

8.4.4 Dispenser

Very related to medication are special actuators called *dispensers*. Dispensers are autonomously working machines that release an item or substance when triggered. In the simplest sense, this could be a soft drink from a vending machine or a container that reminds elderly or forgetful people to take their medication. In the nano context, dispensers are of special interest as they at least partially solve the problem of local treatment of diseases if it would be possible to implement them.

There are numerous approaches to implementing medical dispensers named in the literature (Aghebati-Maleki et al., 2020). To name just a few, medication could be stored inside liposomes (Wang et al., 2015), inside DNA boxes (Andersen et al., 2009), or inside dead/weakened bacteria. All of these approaches have the following in common: A certain substance is isolated from its environment and thus temporarily prevented from affecting its surroundings. Only upon releasing the stored substance at possibly the right place will it take effect.

The general principle can be seen in Figure 8.5, which is based on the work of Andersen et al. (2009). In this case, one of the localization mechanisms from Chapter 7 is implied, where certain DNA or RNA markers are only present in special regions of the human body. Once a box reaches such a place by means of passive movement, it might encounter such a key sequence. In that case, the key might form a stronger binding with the lid of the box than the bottom part can. As a result, the box opens and dispenses a part or all of its stored (medical) payload.

A similar effect may be achieved by the aforementioned nanoparticles in cancer therapy. Instead of a medication directly affecting the cancerous tissue, special nanoparticles that have been coated with certain substances may bind only to infected tissue. In the second step, the nanoparticles are "activated" via, for example, lasers to vibrate and kill nearby cells. While this is technically not achieved by releasing a substance, the underlying principle of local activation is identical.

Lasers may also be used as a means to open other transporting structures like liposomes to release medication where it is necessary (Enzian et al., 2019). In this case, the transport structure receives a special coating that allows it to enter certain cells or areas of the body that may later be hit by lasers to destroy the transport structure and release the medication. This mechanism mimics a dispenser but cannot be reused or partially opened.

In summary, nanoscale dispensers are technologies that aim at solving the problem of targeted drug delivery. A considerable research budget is invested in this area every year as increasing concerns about bacterial resistance to the medication surface (Neu, 1992). This is in part due to the systemic effects medication has on the body. In low enough doses, medication may not kill bacteria but acts as a stimulus for adaptation that breeds new bacteria that are ultimately unaffected by certain substances.

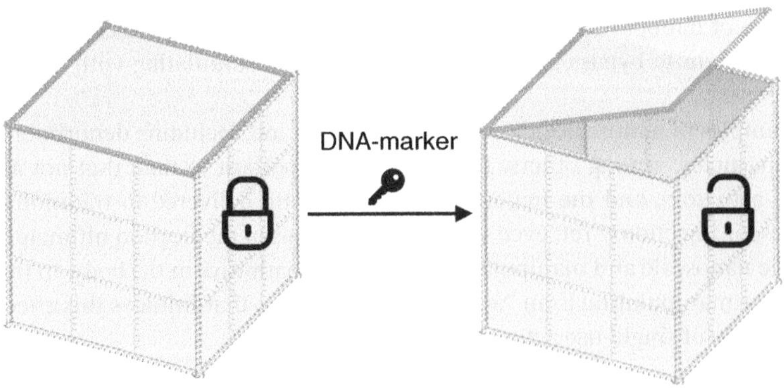

Figure 8.5 Dispenser based on a DNA-origami box that can open and close based on the presence of special DNA or RNA sequences (Lau et al., 2021a). *Source:* Lau et al. (2021a)/with permission of Association for Computing Machinery.

8.4.5 Switches

The more general form of dispensers are *switches*. In the most simple sense, switches activate or deactivate an actuator or other type of machine/system. The simplest example at the macroscale would be a light switch that closes an electric circuit to turn on an LED.

At the nanoscale, switches are a bit more interesting and versatile. This is due to the number of possible involved technologies that require precise timing, accuracy, or a simple safety mechanism. As an example, the DNA-based self-assembly systems use special seed tiles to initiate the growth of a DNA crystal and those tiles thus act like on-switches. DNA off-switches could be implemented using tiles that have just one glue in a single direction that also fits an assembly at certain points but blocks any further growth. This is a much-needed safety mechanism for otherwise self-organizing systems that might have a risk of getting out of control – even if it is small.

In a similar fashion, the special DNA or RNA sequences that may open DNA origami boxes presented in Chapter 4 are switches too. The same is true for the bindings of message molecules to nanobots that might trigger other events. A generous interpretation would even allow many types of DNA bindings to be viewed as switches.

In electrical systems, switches can be implemented using simple transistor technologies. Transistors are the fundamental components in electronics that can be used as switches or amplifiers for electronic signals. They consist of three components: a source, a drain, and a gate. The source and drain are separated by a channel through which the current can flow. The gate is an electrode that controls the current in the channel.

At the nanoscale, the channel is usually only a few nanometers wide and all involved components are in close proximity. The gate could be made of materials that form a thin layer on top of the channel, such as silicon dioxide or hafnium dioxide. When a voltage is applied to the gate, it creates an electric field that can either attract or repel electrons. If the voltage is positive, an electric field that attracts electrons toward the gate is created. This "narrows" the channel and reduces the flow of current between the source and the drain. If the voltage is negative, an electric field that repels electrons away from the gate is created. This "widens" the channel and allows more current to flow between the source and the drain.

Nanoscale transistors can be made using a variety of materials and techniques, such as nanowires, CNTs, or graphene. The only real constraint placed on the materials is the ability to efficiently control the flow of current using another signal.

While there is a variety of other possible technologies for switches, we limit ourselves to the most prominent ones presented here. In practice, there is also a vast array of chemical "switches" that start a reaction once introduced to other media. However, from an engineering and computer science point of view, transistors and DNA bindings are sufficient for most technologies.

8.4.6 Mechanical Actuators

The last type of actuators we discuss in this section are mechanical. Mechanical actuators convert electrical, chemical, or thermal energy into mechanical motion. An example of a nanoscale mechanical actuator is a piezoelectric actuator. Piezoelectric materials may generate an electric field when they are deformed. Vice versa, they change their shape when an electric field is applied to them.

Another example of a mechanical actuator is a thermal actuator. Thermal actuators use the thermal expansion/contraction of materials to create motion. By carefully choosing materials, it is possible to create a nanoscale thermal actuator that can react to changes in environmental temperature.

There are countless other physical effects that may be exploited to create mechanical nanoscale actuators. As we have seen many times before, there is no shortage of physical effects – it is the compatibility and requires precision to combine different pieces of technology into bigger systems that are still lacking. Nanoscale mechanical actuators can be created using techniques, including electron beam lithography or focused ion beam milling. However, a lot more research is necessary until many of the science-fiction-sounding scenarios like cellular surgery or the mechanical removal of blood clots may be implemented in a feasible manner using nanorobots (Li et al., 2017).

8.5 Summary

As nanonetworks are a relatively young area of research, there is still a lot of change happening on a yearly, sometimes monthly basis. As a result, it is difficult to predict or even accurately assess the different concepts or implementations for nanosensors and nanoactuators. Thus, it makes more sense to analyze the underlying principles and overarching themes and limitations to extrapolate possible future trends.

One of the biggest open questions when it comes to sensors and actuators at the nanoscale is the ever-present issue of compatibility. While possible physical implementations exist, it is often unclear to how combine them with other technologies to create applications with an actual benefit. In fact, it is more than possible to create sensors and actuators at a very small scale – but only in isolation.

In addition to that, many of the other components of nanodevices like motors, antennas, and even clocks are technically either sensors, actuators, or a combination of both at the lowest conceptual level. It is thus sometimes difficult to separate the different influence factors into distinct categories. Yet, the simple presence of both sensors and actuators in a variety of areas highlights the extreme importance of the technologies. Without the ability to recognize and influence the environment, nanonetworks would be useless.

9

Energy

The applications and technologies presented thus far usually presuppose some kind of energy supply E to function. Energy exists in various different forms, ranging from potential energy, and kinetic energy, to chemical energy sources. At the nanoscale, there are even more potential mechanisms available that could be exploited to generate and harvest energy for nanodevices.

This chapter discusses the various different potential energy sources at the nanoscale and explains how they might be utilized in nanonetworks. We first list all the necessary definitions and words to efficiently talk about the topic. Then, we aim to get an intuition for all the potential energy sources that might be used for energy harvesting.

Next, we analyze the important topics of saving, storing, and harvesting energy. We start with different mechanisms to use less energy in, for example acoustic or electromagnetic nanonetworks. Afterward, we highlight several examples of nanoscale battery technologies or principles for harvesting energy. Even at the macroscale, batteries often have to be comparably big in comparison to the task that has to be performed. As an example, the battery of an electric vehicle constitutes up to a third or more of the combined weight of the vehicle.

At the nanoscale, things are even worse. Many circuits on conventional computer chips are already less than 100 nm small but require enormous amounts of energy in comparison. Extrapolating from that, batteries would have to be hundreds or even thousands of times bigger than the devices they supply with energy.

Finally, we analyze the energy requirements of nanonetworks in terms of communication for both molecular and electromagnetic technologies. In classical wireless sensor networks, 95% of the available energy is already used for communication purposes. The percentage is likely much worse at the nanoscale, where communication ranges are exceptionally short.

9.1 Energy Sources

At the macroscale, there are several sources from which, for example electrical energy can be generated. Most of them work based on turbines that convert rotational energy into electric energy. These technologies include wind turbines, nuclear power plants, and nearly all other large-scale power plants. The underlying source of energy is usually a (fossil) fuel or an already available source of mechanical movement.

Apart from that, another viable technology works based on fuel cells that can convert, for example, sunlight or chemical energy into electrical current.

Nanonetworks: The Future of Communication and Computation, First Edition. Florian-Lennert A. Lau.

At the nanoscale, the potential energy sources are more heterogeneous. While some of the macroscale sources also exist at the nanoscale, there are several new possibilities (Jornet, 2012). Here are several examples:

- Traditional batteries or nanoscale variations of batteries/capacitors can be used to power nanodevices, especially if the device requires a moderate amount of energy.
- Fuel cells like solar cells can be used to convert light or other types of chemical energy into electrical energy. Next to solar cells, other types of fuel cells exist. Among them are microbial fuel cells (MFCs), hydrogen fuel cells (HFCs), glucose fuel cells (GFCs), or enzymatic fuel cells (EFCs) (Ivanov et al., 2010). All of these function based on a similar principle.
- Thermoelectric generators create electricity from the temperature difference between two surfaces. They may be used to provide energy to nanodevices that operate at a high temperature.
- Some materials can generate electricity when subjected to mechanical stress or pressure. They may be used to harvest energy from naturally occurring movements inside the human body. This includes the pumping motion of the human blood or other kinetic energies in the human body (Li and Wang, 2011).
- Radiofrequency (RF) harvesting can be used to generate electrical energy from ambient electromagnetic fields.
- Magnetic fields can be used to induce electrical currents in conductive materials. While the devices that generate the magnetic fields are usually not nanosized, they may still provide the necessary energy to a nanonetwork.

As with macroscopic energy sources, nanoscale sources vary in many of their parameters. They include conversion efficiency, harvesting rate, energy density, or the possibility to store energy (Wang and Wu, 2012). Sometimes it is even possible to use an available source of energy directly without needing any conversion, which is the best-case scenario.

Definition 9.1 The *conversion efficiency* describes the efficiency of a *generator* to convert the available energy into an alternative form of energy. It may be calculated by

$$\eta = \frac{\text{work done}}{\text{heat absorbed}} = \frac{Q_1 - Q_2}{Q_1},$$

where Q_1 is the generated heat and Q_2 is the work done. The "work done" includes all types of losses due to friction and so on.

If not stated otherwise, the conversion from an arbitrary type of energy into electricity is assumed. Very efficient modern diesel engines may operate at up to 54.4% energy conversion efficiency (Takaishi et al., 2008). At the nanoscale, or with new technologies overall, it is very likely that such rates cannot be achieved initially. Nonetheless, new technologies sometimes offer the potential for highly efficient conversion rates.

Next to the efficiency of the conversion, the speed of harvesting energy or charging a battery/capacitor is of high importance.

Definition 9.2 The *harvesting rate* or *charging rate* of a generator or battery describes the speed at which the ambient energy may be converted into electricity.

While it is beneficial to have an efficient conversion of energy, it is not sufficient for most applications. If a nanodevice requires a set amount of energy to periodically send, for example, a message to other devices, it requires a sufficiently high energy input to do so. In many cases, a less efficient but faster source is preferable to the most efficient one. The chief goal is to provide enough energy for a nanonetwork to function, and efficiency comes second.

The last energy-related parameter of nanonetworks is the energy density of a battery of capacitors.

Definition 9.3 The *energy density* of a battery or capacitor describes the amount of energy stored per volume.

As an example, lithium-based batteries are typically very energy dense as lithium happens to be the lightest element in the periodic table currently used for battery technologies. As a result, most batteries of electric vehicles or devices that are often moved around are based on that element. At the nanoscale, the energy density is especially important as a nanodevice that has been applied to its target destination may often not be charged ever again as the devices are too small to even find, let alone manipulate again.

It is further important to keep in mind that storing energy in a battery usually means that some kind of conversion from, for example, electrical into chemical energy and back is necessary. It is not rare that up to 30% of the energy may be lost in that process. As a result, quality batteries are absolutely crucial at both the macroscale and the nanoscale.

9.2 Storing Energy

While there are various different types of storage technologies available at the macroscale, there are fewer potential candidates at the nanoscale. As mentioned before, a high energy density is essential and not many materials may provide what is needed.

The most frequently mentioned battery technology at the nanoscale is still based on lithium. *Lithium-ion batteries* (LiBs) still have the highest energy densities of all available battery technologies and thus make them a primary candidate for supplying energy at the nanoscale (Mustaffa et al., 2022). The primary difference between nanoscale LiBs and macroscale LiBs is their size and the materials used in their construction. At the nanoscale, LiBs are composed of nanoscale components, such as nanowires, nanoparticles, or thin films, which are designed to provide a high surface area and short diffusion distances for ions and electrons. These nanoscale components are usually made from advanced materials, such as silicon, CNTs, or graphene, which can improve the energy density, capacity, and cycling performance of the battery.

On the other hand, macroscale LiBs are composed of micrometer-scale components, such as electrodes, electrolytes, and separators, which are assembled into a battery cell. These components are typically made from materials like graphite for the anode, lithium cobalt oxide for the cathode, and liquid or polymer electrolytes.

While the basic principles of operation are the same for both nanoscale and macroscale LiBs, the use of nanoscale components may result in improved performance and new capabilities, such as faster charging, higher energy density, and longer cycle life. However, the fabrication and integration of nanoscale components into a battery cell also pose a variety of challenges. Among them are costs, scalability, safety, and of course the lack of accurate and precise tools for manufacturing.

9.2.1 Batteries Based on Zinc Oxide Nanowires

Zinc oxide nanowires (ZnOs) have been studied as an anode material for LiBs in recent years (Zhang et al., 2016). ZnOs have a very high surface area which gives them excellent electrochemical properties, and a potentially low cost.

During the charging of a LiB that uses ZnOs as the anode, lithium ions from the cathode move through the electrolyte and intercalate into the ZnOs at the anode. *Intercalation* describes the process of lithium ions entering into the crystal structure of the ZnOs, which causes the nanowires to expand slightly in size. This process is reversible and can occur over multiple charging and discharging cycles. When releasing the energy again, the process happens in reverse order, releasing the stored energy in the form of an electrical current.

9.2.2 (Super-)Capacitors

Supercapacitors are highly efficient capacitors and are sometimes able to store more than 100 times more energy than a regular capacitor. It is possible to fully charge a capacitor in mere seconds, while the same process may take a long time for conventional batteries. That said, a LiB of similar size is usually capable of storing 10 times more energy than that.

Apart from ZnOs as anodes for LiBs, they may also be used in other ways to store energy at the nanoscale as supercapacitors. Supercapacitors are energy storage devices that differ from batteries in that they store energy by accumulating and releasing charges at the surface of the electrodes. While batteries usually store energy chemically, supercapacitors store energy via pseudocapacitance as an electric potential between two very big surfaces.

In the case of ZnOs, their high surface area and good electrical conductivity make them promising materials for achieving relatively high energy density and short loading intervals at the nanoscale. The large surface area of ZnOs provides more sites for charge accumulation at the surface, while their good electrical conductivity allows for efficient charge transfer between the electrode and the electrolyte. Due to the short distance between both electrodes, extremely high amounts of energy may be stored in such a supercapacitor.

There are other potential materials with similarly high surface area compared to ZnOs like "bundles" of CNTs.

9.3 Energy Harvesting and Generators

As the potential energy stored in nanoscale batteries is small and the energy capacity of supercapacitors is even smaller, nanonetwork researchers started looking into alternative technologies to still provide nanodevices with sufficient energy. While it might not be possible to store enough energy to power most applications for longer periods, it might be possible to equip those devices with the necessary means to produce or harvest their own energy. This principle is well studied when it comes to technologies installed in remote places where changing batteries is not a feasible option. Other role models are space technologies that have to function more or less self-sufficient after launching them. As an example, many space probes are equipped with solar panels to create energy and small nuclear components that generate sufficient heat for the electronic components to function properly in the cold of space.

At the nanoscale, these harvesting principles have to function in very different environments. Neither nuclear energy nor solar energy sources are usually available, and, thus, alternative sources are necessary. As mentioned at the beginning of the chapter, there are luckily numerous potential sources available especially when it comes to environments like the human body that might be tapped into.

In the following paragraphs, we analyze the most promising energy sources in some detail. Before we start with that, we have a look at the fundamental principle behind most generators and try to understand how various types of energy may be turned into electricity.

9.3.1 The Generator

Generators are machines that convert mechanical energy into electrical energy. The aforementioned turbines that convert, for example, rising steam in nuclear power plants into electrical energy are special cases of generators. Generators were invented in 1831 by Michael Faraday and work based on the principle of electromagnetic induction.

The main part of a generator is a coil of wire that rotates inside a magnetic field. The rotating coil induces an electrical current in the wire. The entire mechanism closely resembles an electronic motor, but inverse.

The mechanical energy needed to rotate the coil can come from various sources including rising steam, moving gases, flowing water, or wind. At the nanoscale, several additional sources of mechanical movement might supply kinetic energy to a generator.

The amount of energy a generator can produce depends on the speed of rotation, the strength of the magnetic field, and the number of turns in the wire coil. At the macroscale, generators are capable of outputting GW of energy. Nanoscale generators are likely only capable of producing a few pW.

9.3.2 Harvesting Mechanical Energy and the Piezoelectrical Effect

As we have learned, nanoscale generators may harvest energy by converting mechanical energy from small-scale movements into electrical energy. As rotating coils in magnetic fields are difficult to implement at the nanoscale, nanoscale generators typically use *piezoelectric materials*, which generate electricity when subjected to mechanical stress, deformation, or pressure. This effect arises from the asymmetric crystal structure of most piezoelectric materials, which causes the displacement of electric charges when the material is deformed or strained. If done rightly, this effect may be exploited for energy harvesting (Erturk and Inman, 2011).

The piezoelectric effect can be direct and inverse. The direct piezoelectric effect describes the generation of electricity when mechanical stress or pressure is applied. The inverse piezoelectric effect describes the generation of a mechanical strain or deformation when an electrical field is applied to the material. These effects are, for example, used in special *scanning tunnel microscopes* that operate at the nanoscale (Binnig et al., 1986). These microscopes scan the surface of a material and require an even distance from the material for high-quality images. This may be achieved by piezoelectric scanner tubes that allow for sub-nanometer levels of accuracy.

Especially interesting piezoelectric materials are the aforementioned ZnOs (Wang and Song, 2006). Due to their small diameter and relative length, ZnOs are especially susceptible to various types of ambient movements, including bending, stretching, and vibrating. This also includes the pumping motion created by the heart.

As many of the most popular applications for nanonetworks take place inside the human circulatory system, the available energy sources in the blood are of special interest. Depending on the distance to the heart and the activity levels of the person, the pressure differences of the heartbeat provide a relatively regular input for ZnO nanowire generators. Given some simplifying assumptions, a nanogenerator should be able to produce around 0.01 pW/μ^3m. A microdevice of 128 μm^3 should thus be able to generate around 1.28 pW from just the pumping motion of the heart (Canovas-Carrasco et al., 2018).

The interested reader is encouraged to look at the work of (Pawar et al., 2023) for more detailed information.

9.3.3 Ultrasonic Energy

Ultrasonic energy harvesting is a special case of harvesting mechanical energy. In this case, the mechanical energy in the acoustic waves is turned into energy. Unlike electromagnetic waves, acoustic waves are actual particle movements that may be harvested using piezoelectric materials.

Ultrasonic waves are sound waves that have a frequency higher than the upper limit of human hearing, which is typically considered to be around 20 kHz. As a result, the waves cannot be heard by the naked ear, but they may be generated in various already-available ways. As its use in medical imaging suggests, ultrasound can penetrate human tissue relatively unhindered and may thus provide another source of energy to areas that are otherwise difficult to reach.

While low amounts of ultrasounds are relatively harmless, the amount of contained energy is strictly regulated in medical applications to ensure the patient's safety. As a result, fetal imaging is limited to less than 100 $\mu W/^2cm$. For adults, up to 720 $\mu W/^2cm$ is allowed. The duration of the exposure is limited to 60 minutes. That said, it might be possible to power a nanonetwork more or less constantly when less energy-dense sound waves are used.

According to the recent literature, it is even possible to power some implantable devices using ultrasound (Jiang et al., 2019).

9.3.4 Radiofrequency Harvesting

RF harvesting harvests energy from electromagnetic waves and converts them into electricity via an antenna. The antenna is typically designed to resonate at a specific frequency, which allows it to capture and convert energy from waves that have a similar frequency. The electrical energy may then be rectified and stored in a small battery or capacitor.

One of the challenges of RF harvesting in nanonetworks is that the amount of available energy is typically very low, on the order of microwatts, nanowatts, or even picowatts. As a result, the energy harvesting mechanism must be highly efficient to capture and store significant amounts of energy. Furthermore, the ambient electromagnetic waves can vary in frequency, amplitude, and direction, depending on factors such as the location and environment of the device. This means that the antenna and energy-harvesting component must be designed to be flexible and adaptable to ensure that they can capture and convert energy from a wide range of sources.

As of right now, RF is widely used as a technology to power wireless sensor networks and sometimes RFID networks (Parks et al., 2013). However, further technological advancements are necessary to enable the technology for the nanoscale. Currently, only between 0.02 and 40 $\mu W/\mu^3m$ may be harvested depending on various parameters (Mohrehkesh et al., 2017). However, it is unlikely that this number may be downscaled linearly.

9.3.5 Ambient Heat

The next type of available energy at the nanoscale is ambient heat/warmth. Heat is the "lowest" form of energy and a byproduct of many of the conversion processes we have seen before.

At the nanoscale, heat is a form of energy that arises from the motion of particles within a substance. The connection between heat and movement is immediately visible at the nanoscale as particles move faster if the ambient temperature rises. This can act as a catalyst for many (chemical) reactions that rely on direct contact with substances.

Thermal energy is sometimes called a "low" form of energy. This originates from the fact that the second law of thermodynamics states that any energy conversion process produces some heat as a byproduct. However, this does not imply that heat is a "bad" form of energy, but rather that it is a form of energy that tends to be difficult to recover or convert into other useful forms.

There are several ways in which thermal energy may be used as fuel in nanonetworks. The classical approach involves energy harvesting using the already-presented piezoelectric effect that works based on the close relationship between mechanical movements and temperature at the nanoscale.

Another effect that may be exploited is the *principles of thermoelectricity*, which is the conversion of temperature gradients into electrical energy. Thermoelectric materials can generate electricity when exposed to a temperature gradient. The phenomenon is based on the Seebeck effect, discovered by Thomas Johann Seebeck in 1821, which states that when two dissimilar metals are joined at two different temperatures, a potential difference or voltage is created between them.

The underlying principle of thermoelectricity is the use of thermoelectric materials, which are materials with a high Seebeck coefficient. The Seebeck coefficient measures the generated voltage per unit temperature difference across the material. Thermoelectric materials are typically made from at least two materials that have different thermoelectric properties. When a temperature gradient is applied across the material, the electrons in the material diffuse from the hot side to the cold side, creating a voltage difference.

The last and most direct way to use thermal energy skips any type of conversion and uses the ambient heat directly. A prime example is the binding processes of DNA sequences introduced in Chapter 4. Unless the ambient temperature is sufficiently high, no binding reactions can take place. DNA sequences rely on the Brownian motion introduced in Chapter 7. They have to physically hit each other and may stably bind together when appropriate nucleotides are present.

The higher the temperature, the faster the movements of tiny particles and the higher the number of collisions. However, the bindings of the DNA bases immediately break again if the ambient temperature is too high. DNA-based self-assembly systems thus have to operate at a sweet-spot temperature where the heat is sufficiently high for a fast reaction but low enough to allow for stable bindings (Winfree et al., 1998). In this way, ambient warmth may directly be used to power computation, construction, and even communication processes (Lau et al., 2019). As no conversion of energy from one form to another is involved, this type of energy harvesting is highly efficient.

9.3.6 Adenosine Triphosphate

ATP is a molecule that plays a fundamental role in the energy metabolism of creatures. ATP is the "energy currency" of cells as it is the main molecule cells use to store and transfer energy. ATP consists of an adenosine molecule, composed of adenine and ribose, linked to three phosphate groups. The high-energy bonds between the phosphate groups contain a significant amount of potential energy, which is released when the bonds are broken.

Cells use ATP for a variety of processes, such as muscle contraction, cellular respiration, and protein synthesis. When ATP is hydrolyzed, the released energy can be used to power chemical reactions or perform mechanical work within the cell. Nanodevices may be able to harvest energy from ATP through the use of, e.g. molecular motors. These motors are proteins that directly use the energy from ATP hydrolysis to generate motion. For example, the protein myosin uses ATP to move along actin filaments in muscle cells, generating muscle contraction. Similarly, the protein kinesin uses ATP to transport cargo along microtubules in cells.

As motor proteins naturally occur inside human cells and the current state of the art only suggests the exploitation of existing structures, the use of ATP as an energy source is limited to such environments. Yet, this form of energy is freely available and may be harvested from the human body within reason.

9.3.7 Fuel Cells

Fuel cells have been invented as early as 1839 by William Grove and are devices that convert the energy stored in a fuel directly into electrical energy through an electrochemical reaction (Basu, 2007). They work by combining a fuel such as hydrogen, methanol, or natural gases with an oxidizing agent like oxygen to produce electricity and heat/water as byproducts.

A generic fuel cell that consists of an anode, a cathode, and an electrolyte layer can be seen in Figure 9.1. The anode is the negative electrode where, e.g. hydrogen is oxidized, releasing electrons. The cathode is the positive electrode where, e.g. oxygen reacts with the electrons and protons from the anode to form water. The electrolyte separates the anode and cathode and enables ions to move between them while preventing the mixing of hydrogen and oxygen.

Some types of fuel cells are able to convert up to 60% of the energy in the fuel into electricity, compared to about 30% for conventional combustion-based power generation. Additionally, they produce significantly less pollution than traditional power generation methods as they only emit water and heat.

There are many different types of fuel cells that use different sources for creating energy. Among them are

- Solar cells/*photoelectrochemical fuel cells* (PECs),
- *hydrogen fuel cells* (HFCs),
- *methanol/ethanol fuel cells* (MEFCs),
- *microbial fuel cells* (MFCs) (Logan et al., 2006),
- *enzymatic fuel cells* (EFC) (Ivanov et al., 2010), and
- (enzymatic) *glucose fuel cells* (GFCs) (Kerzenmacher et al., 2008, Kulkarni and Slaughter, 2015)

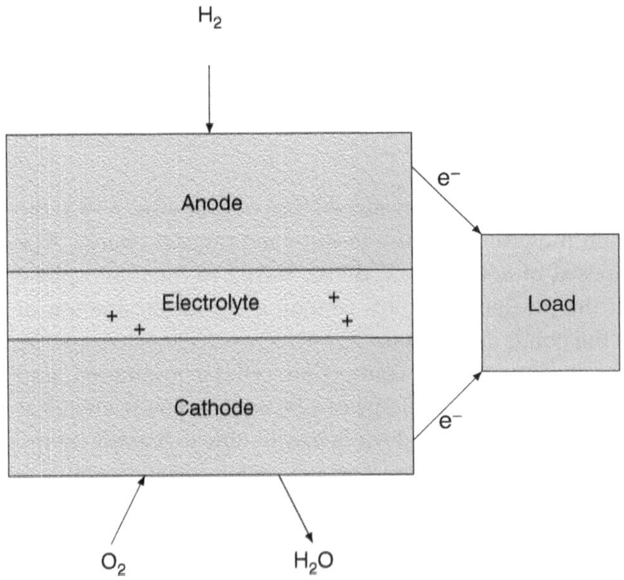

Figure 9.1 Diagram of a generic fuel cell based on hydrogen. The anode and the cathode are separated by an electrolyte layer. A chemical reaction that converts H_2 and O_2 into electrical energy that produces water as a byproduct. *Source*: Florian-Lennert A. Lau.

In the following, we analyze GFCs as potential EFCs for creating energy at the nanoscale. Since all fuel cells work based on more or less the same principle, understanding a single one suffices for a good intuitive understanding of the topic.

While the most well-known types of fuel cells convert alcohol or hydrogen into energy, a GFC uses glucose (Kulkarni and Slaughter, 2015). Glucose is a simple type of sugar and the most important source of energy for living organisms like humans. It was first isolated from raisins by the German chemist Andreas Marggraf as early as 1747. Due to its appearance in human blood and tissue, it is a potential fuel for nanodevices and nanonetworks as long as the host organism is not deprived of too much energy (Cosnier et al., 2014). While conventional battery technologies may also be able to power implants like pacemakers with energy, the frequent requirement for surgeries to recharge batteries renders most classic technologies inadequate.

The GFC was invented by Potter as early as 1911 and offers an alternative approach based on energy harvesting (Potter, 1911). The principle behind GFC is more or less the same as the HFC in Figure 9.1. As with most fuel cells, an anode and a cathode are separated by a layer of electrolyte. A chemical reaction oxidizes glucose to form gluconolactone while simultaneously releasing electrons. As long as sufficient oxygen and glucose are provided, the reaction continues to provide electrical energy and could thus supply implants with power without requiring any battery change. Recent publications tested the technology, for example, on rats and were able to provide up to 0.5 V/40 μW (Cosnier et al., 2014).

Outside of living animals, it is also possible to use the sugar contained in fruit to power an LED (Yoshino et al., 2013). The higher the sugar levels in the grape, the faster the LED blinks. Based on the correlation between the blinking speed and the sugar levels, it might be possible to create accurate blood sugar monitors for, e.g. diabetics.

9.4 Saving Energy

When it comes to the different energy sources, a potentially even more important topic should be discussed. No matter how efficient the hardware we use is, poorly programmed algorithms can always spend more energy than available. As a result, it is necessary to save energy wherever possible.

9.4.1 Communication

At the macroscale, between 40% and 90% of the required energy is spent in the communication process between nodes in wireless sensor networks (Heinzelman et al., 2000). The overall numbers depend on various factors such as the communication protocol, data transmission rate, distance between nodes, and the type of sensor nodes used. However, in general, communication in WSNs is a significant contributor to overall energy consumption.

At the nanoscale, this ratio is likely even worse as little computational power is available and antennas have to be as big as possible to extend the communication distance between devices. Even if all possible preprocessing steps can be performed on the devices themselves before transmitting any data, it is still likely that more than 95% of the energy is consumed for sending data. This is especially likely as the general communication distance is expected to be around 2 mm, as explained in Chapter 6. Thus, every bit of extra energy might just be invested in stronger signals to increase the signal range.

Especially in multi-hop networks, the problem is the broadcast storm that can lead to vast amounts of consumed energy. The problem is depicted in Figure 9.2. If more than a single node is

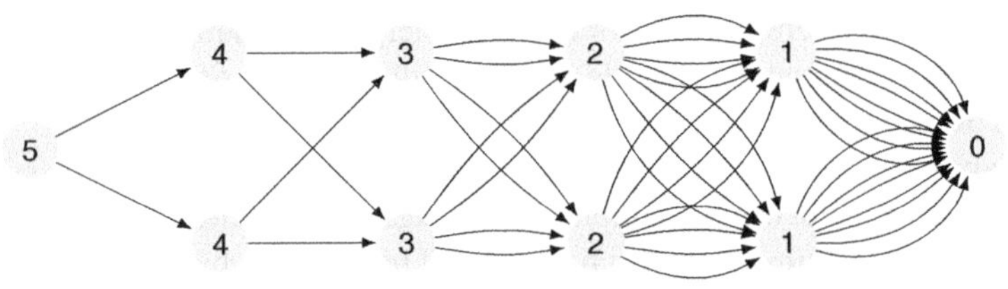

Figure 9.2 Broadcast storm problem in identity-free nanonetworks. As the neighborhood is unknown, devices potentially send messages to adjacent nodes multiple times, leading to an exponential number of messages. *Source*: Florian-Lennert A. Lau.

in range of a message, and forwards messages, an exponential number of messages is sent overall. As a result, the consumed energy is very high, regardless of the underlying hardware efficiency.

As a result, it is necessary to save energy by all means possible. We analyze the potential for saving energy regarding sleeping devices, preprocessing, efficient protocols, and many more aspects.

9.4.2 Electromagnetic vs. Molecular vs. Acoustic

First, it is necessary to distinguish between the different types of communication at the nanoscale in terms of energy consumption. Depending on the underlying principle, the potentially consumed energy can vary drastically.

While acoustic communication and electromagnetic communication are overall very similar, acoustic networks tend to consume less energy. Molecular communication however is even more difficult to assess in terms of energy requirements. While the propagation of the molecules usually happens via diffusion or flow processes that do not require additional energy, producing or collecting molecules does. If that energy is accounted for, up to 90% of the total energy requirement for such a system is used for communication alone (Farsad et al., 2016).

While not all molecular communication systems work based on that principle, the overarching problem should become clear. There is always an associated cost when it comes to communication. However, it is sometimes possible to "pay that price" at the macroscale where energy is not scarce and just use the molecules at the nanoscale to circumvent the energy problems. While this might not be possible in all cases, there are not many ways to reduce the number of required molecules to achieve a stable communication channel.

In the following, we mainly focus on protocols, techniques, and algorithms that reduce the energy requirements for both acoustic and electromagnetic communication.

9.4.3 Preprocessing, Encoding, and Aggregation

The first technique that comes to mind is simply based on reducing the number and size of messages sent over a network. As a big proportion of energy is required for sending, it makes sense to perform as many computational steps before transmitting any data. In general, there are several ways of reducing the amount of data and we discuss a number of them in this section. Among them are preprocessing, data aggregation/overhead reduction, compression, and encoding schemes.

The first and simplest step to reduce the amount of data is semantic. Instead of sending every piece of information over a network, the data can be scanned for relevant parts. Thus, it might be possible to only transmit data points that exceed a certain threshold or meet other predefined relevance criteria. Furthermore, it might be possible to clean data of outliers and other anomalies before considering transmission.

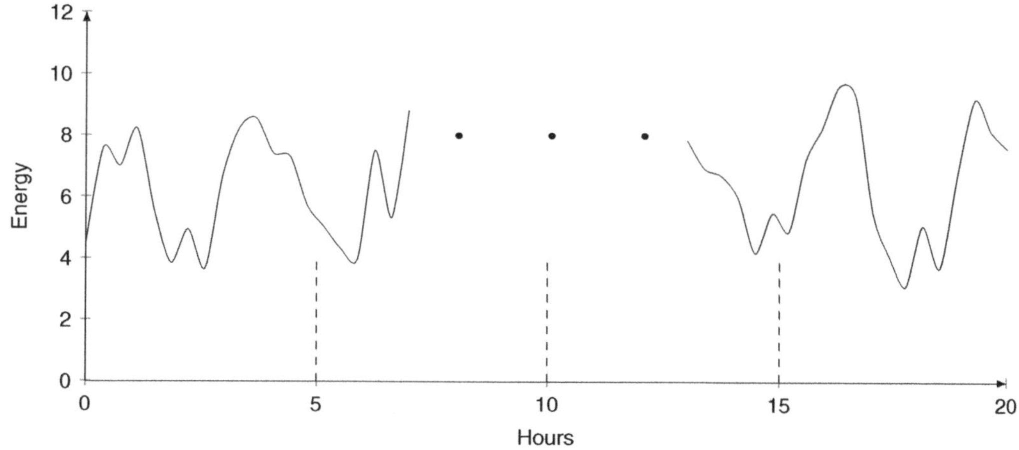

Figure 9.3 Comparison between harvesting and sending. While the harvesting itself might take several hours, a single sending event might only last several picoseconds. *Source*: Florian-Lennert A. Lau.

The next approach involves the aggregation of data before sending anything. While this might not be suitable for all applications, it is often possible to delay the transmission of measurements within reason to reduce the sending overhead. Usually, each package has a considerable amount of header/metadata that contains routing information. Aggregating a number of packages before sending further allows for the removal of redundancies.

Once all these steps are finished, the data can further be compressed by lightweight, lossless compression algorithms. Thus, the number of bits to transmit is further reduced.

Finally, the encoding of each message also plays a crucial role in the energy requirements. In general, messages are sent in binary form. Each of them is essentially a *word/string* containing nothing but "1" and "0." Each "1" usually requires energy as they are usually represented by a high voltage, frequency, or concentration. A "0" on the other hand is just a pause in an otherwise continuous transmission. Thus, each message should contain as many "0" as possible to minimize the sending cost.

While this is already an improvement, there are still standing waves involved in the communication process. They consume energy for the entire duration of the transmission. When energy is really scarce, it is usually better to send small pulses in TS-OOK procedures we learned in Chapter 6. Thus, only femtosecond long pulses are necessary and the time when no information is traveling may be minimized. As a result, devices are able to be in standby mode most of the time, requiring next to no energy. They only "wake up" through interrupts when tasks have to be performed.

An example of this can be seen in Figure 9.3. The nanodevices are idling and harvesting energy for the duration of seconds or even minutes. The dashed lines represent the femtosecond long pulses where data is actually transmitted.

9.4.4 Saving via Protocols

The aforementioned energy-saving measures are all limited to individual devices. However, most energy is consumed by repeatedly forwarding messages in multi-hop wireless networks. In general, this cannot be avoided due to the limited range of electromagnetic or acoustic signals. The routing of messages has the potential to increase energy efficiency tremendously. In fact, most effort in networking communities is invested in engineering new communication protocols that reduce the overall number of messages sent through a network.

Before we dive into detailed explanations of how to save energy, we have to specify a machine model and a general reference architecture we assume at the nanoscale. Nanodevices are generally believed to be extremely resource constrained (Lau et al., 2017). They possess little computational power and are likely not Turing-complete, have a communication range of a maximum of 2 mm, and may store next to no information. They are very likely not even capable of storing a unique identifier, and we thus assume an identity-free nanonetwork. In short, we assume the general reference architecture of hop-count nanonetworks presented in Chapter 6.

In general, we assume nanonetworks to form ad hoc when devices move/are moved within range of each other. As a result, the neighborhood of each device is ultimately unknown and only local information is available to each and every node. Thus, we must generally assume that messages exponentially multiply when sent over a network, which is displayed in Figure 9.2.

9.4.4.1 Destructive Retrieval

Based on that, the first objective to save energy is to avoid broadcast storms. Even macroscopic devices are not able to keep up with the energy requirements of such a scenario.

The intuitive solution to avoid a broadcast storm is to limit the number of messages each device is allowed to send. In the simplest case, devices are only allowed to forward a message once and will reset their own hop count afterward. Realistically speaking, nanodevices will deplete their energy fully for transmitting a single message and are "dysfunctional" for a while. Thus, a device is considered to be "destroyed" upon receiving and forwarding a message.

The drawback of such a procedure is quite clear. Whenever some event has been detected in the network, the hop-count topology has to be created anew. While this might sound like an inconvenience at first, frequently refreshing the neighborhood information is necessary in any case. Nanodevices can travel considerable distances away from each other in a matter of seconds relative to their size.

As an example, the blood flow is approximately 0.3 m/s in the aorta. In comparison, electromagnetic waves travel at roughly the speed of light (299 792 km/s in the vacuum). Assuming that the human body consists mainly of water, the speed of light changes to approximately 70 661.97 km/s. Given a payload of two bytes sending over 500 hops, each device will travel a maximum of 0.26 μm in the process. This allows us to assume a network to be quasi-stationary for the duration of a transmission. However, the neighborhood information is already outdated within a single second.

Figure 9.4 compares the quality of different approaches (Büther et al., 2018). The naive (circles) retrieval phase leads to an exponential blow-up of messages. Randomly (squares) deciding if a message is forwarded reduces the complexity slightly, but it is still exponential. Both destructive retrieval and randomized destructive retrieval limit the number of messages sent to $O(n)$, as expected.

9.4.4.2 Stateless Linear Path Saving

While limiting the number of messages is useful, the approach does not take local information into account. It would be much better if only a subset of nanodevices would forward a message. In the optimal case, only nanodevices situated on a direct path between the sender and the final receiver would forward any incoming message to save as much energy as possible.

However, a single hop count is not sufficient for that, but the aforementioned three hop counts in Chapter 7 are.

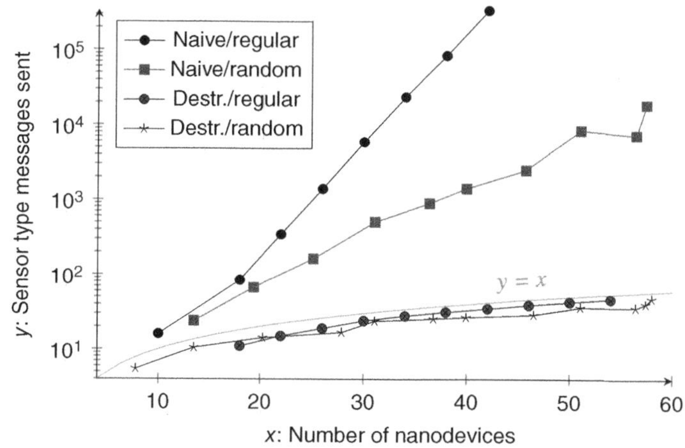

Figure 9.4 Number of messages sent during the retrieval phase in a hop-count nanonetwork. In the naive case, the number of messages sent grows exponentially and randomly dropping messages changes little about that. A destructive retrieval approach reduces the message complexity to linear growth.
Source: Büther et al. (2018)/with permission of Association for Computing Machinery.

Without any additional state information involved, it is possible to send data along a path in a curve-linear coordinate system (Tsioliaridou et al., 2017). For this, each nanodevice has to have at least three hop counts assigned. If the associated gateways are placed linearly independent, the hop counts span a curve-linear coordinate system that roughly resembles a Cartesian coordinate system over short distances.

It is now possible to add the source and destination information to each message. Any nanodevice that receives the message can now perform a check if it is on a direct path between the source and the destination and forward the message conditionally. Depending on the density of the network, it is possible to adapt this procedure to wider or thinner paths. In sparsely populated networks, it is usually better to forward messages even if a node is not directly on a path between two nodes, but two or three hops apart.

A primitive example can be seen in Figure 9.5. A source node at the bottom left sends a message to a node at the top right. All nodes that currently hold and propagate the message within a given time window are highlighted. Nanonodes forward a message if they are closer to the destination on at least one dimension. The condition can be easily adapted to having to be closer (within intervals) in all dimensions to implement the full stateless linear path routing (SLR).

9.4.4.3 Obstacles

While linear paths are an improvement in terms of energy efficiency, realistic environments are rarely like that. Once nanonodes have been deployed at their target destination, they are subject to the various kinds of passive movement introduced in Chapter 7. Thus, their ultimate position is at least to a degree randomized. An example of that can be seen in Figure 9.6.

The nanonodes are randomly distributed on a grid, which ensures that no areas in the network are entirely without nanonode coverage. Nonetheless, obstacles may still occur. This is either due to overuse of certain communication routes and thus no available energy or simply due to natural hindrances in, e.g. the human body like bones or tissue.

To still ensure the feasibility of routing protocols based on SLR, obstacles have to be dealt with (Arrabal et al., 2019). A proposed solution to the problem can be seen in Figure 9.7. Instead of

Figure 9.5 Signal propagation in a hop-count network that only forwards messages if they are closer to the destination in BitSim. *Source*: Arrabal etal. (2018a)/ With permission of IEEE.

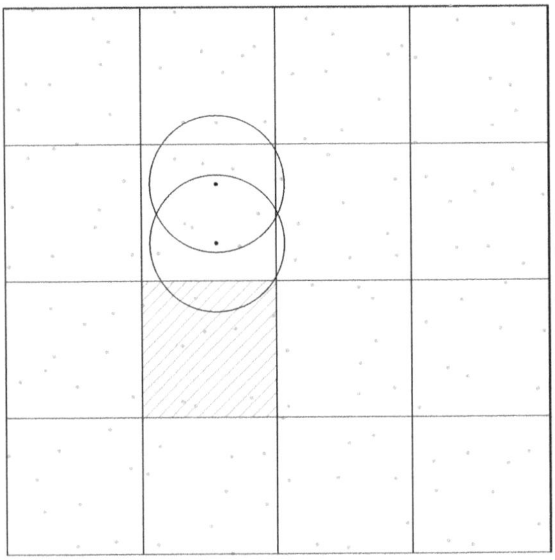

Figure 9.6 Obstacles on a potentially linear path. Various influences could lead to areas of a network not allowing for message forwarding. Among them are empty batteries, destroyed devices, the absence of nanodevices, and too much absorption.
Source: Florian-Lennert A. Lau.

getting lost on the way to the destination, the message is routed around the congested area in the middle.

The approach works based on *backoff flooding* (Arrabal et al., 2018a). This algorithm reduces the number of messages propagated through a network by employing a waiting interval before forwarding any messages. When a predefined number of additional messages have been received by the device, it only forwards it with a low probability.

For this to work, each nanodevice has to know the (approximate) local node density to derive a suitable message-forwarding probability. The simply *Density Estimator for Dense Networks* (DEDeNs) offer a suitable solution for low-cost density optimization (Arrabal et al., 2018b). Each node tries to estimate the local nanonode density by listening for small "test packages" that each device sends with a low but predefined probability. Based on the number of received packages, a good estimate of the local density can be made and a fitting probability for forwarding messages may be derived. This procedure has to be repeated frequently when used in mobile nanonetworks.

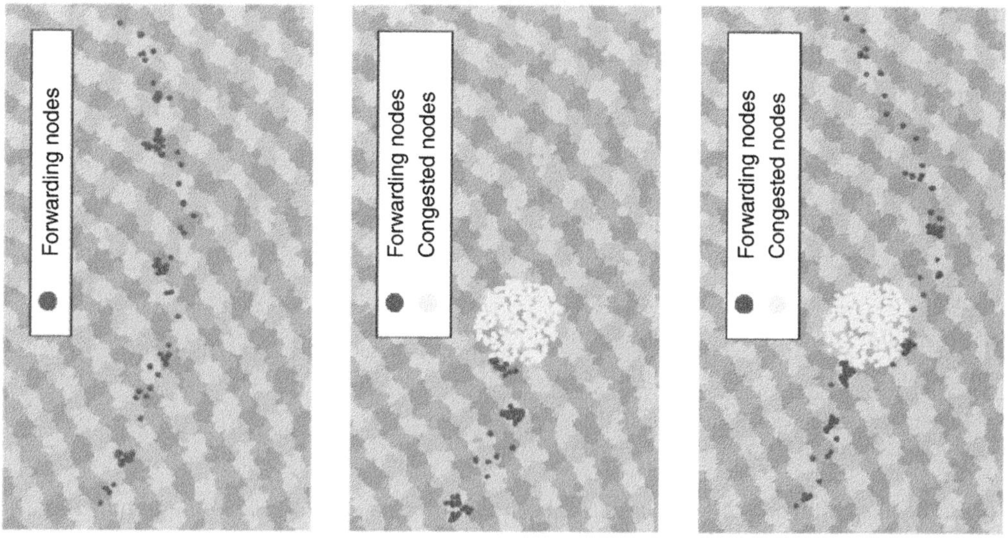

Figure 9.7 Example multi-hop propagation of a signal using SLR when encountering obstacles (Arrabal et al., 2019). *Source:* Dedu.

9.4.4.4 Ring Saving

The aforementioned procedure significantly increases the energy efficiency of nanonetworks. However, there is still one weak point. Each nanodevice still forwards a message to *all* other devices within range, even if an area is densely populated. To avoid this, a *ring-saving* approach may be employed.

The procedure is explained in Figure 9.8. In the middle of the circles, there is a transmitting nanonode that forwards messages. Given previously acquired node-density information, the center node is able to choose a `rangeSmall` value that minimizes energy expenditure while still maintaining connectivity in the network. Each message is only forwarded to devices within the resulting

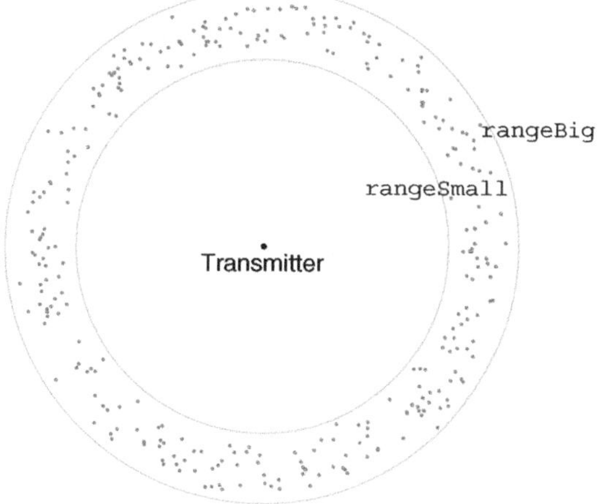

Figure 9.8 A transmitter sends messages only to devices within a certain range interval between `rangeSmall` and `rangeBig`. *Source:* Hoteit et al. (2022)/with permission of Association for Computing Machinery.

Table 9.1 Simulation parameters of the ring-saving algorithm.

Parameter	Value
Size of simulated area	6 mm × 6 mm
Number of nodes	10 000
Range	1 mm
RingBig	1 mm
RingSmall	Dynamic
Data package size	1003 bit
Control packet sizes	101/102 bits

Source: Hoteit et al. (2022)/with permission of Association for Computing Machinery.

interval. The algorithm only takes local neighborhood information into account and may thus be realistically applied to nanoscale electromagnetic network applications.

Table 9.1 shows the chosen simulation parameters to verify the procedure.

The simulation takes place in the Bit Simulator (Dhoutaut et al., 2018). Overall, 10 000 nanodevices participate within the nanonetwork of 6 mm × 6 mm. The maximum communication range is 1 mm, and packages are either 101/102 or 1003 bits long. While this package size might be too big for several applications, it might still save energy by activating much fewer nanodevices than before.

Figure 9.9 shows the simulation results of the naive approach compared to the ring-saving approach. In comparison, roughly 80% fewer nanonodes forward the original message. Nanonodes that forward the original message are colored in black while receiving nanonodes are colored in dark/bright. Only the first package of the first run is shown.

Figure 9.10 shows the results of combining SLR with ring saving to save even more energy. The three colors indicate the different hop counts. The black nanonodes are repeaters, and the

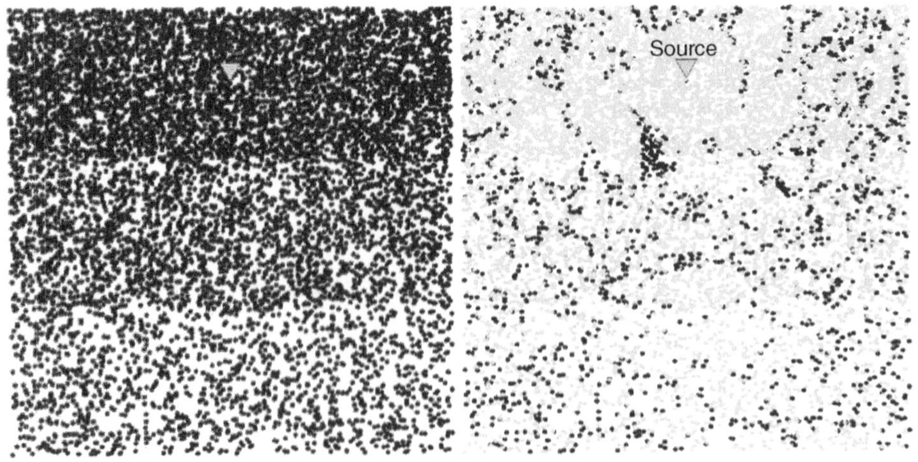

Figure 9.9 Naive broadcasting vs. ring saving. Black = repeater; dark/bright = receiver. 80% fewer messages are necessary (10 000 to 1949) (Hoteit et al., 2022). *Source*: Dedu (permitted).

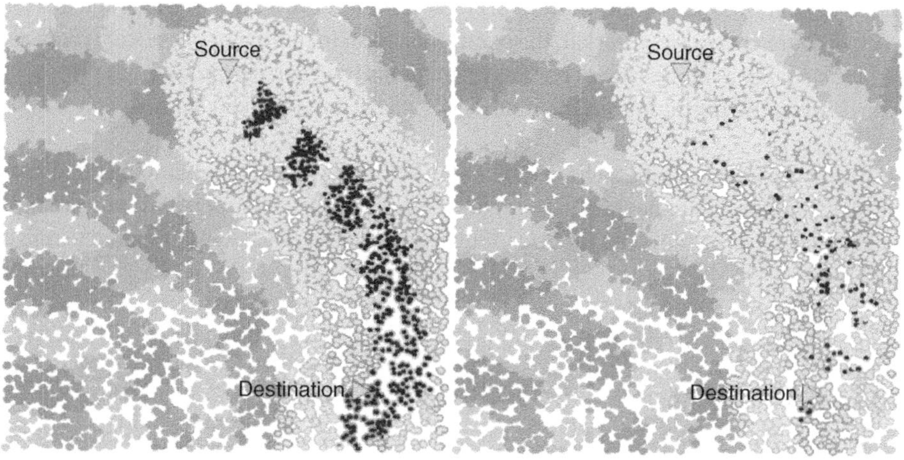

Figure 9.10 SLR vs. a combination of SLR and ring saving. Black = repeater; dark/bright = receiver. 85% fewer messages compared to ring saving only are necessary (901 to 129) (Hoteit et al., 2022). *Source*: Dedu (permitted).

dark/bright nanonodes are receivers. The source and destination as well as the path between both are marked by triangles.

The SLR forwards only 901 messages on its own. Combined with the ring-saving algorithm, only 129 messages in total are required to forward the message to its target destination. Compared to naive broadcasting, 98.71% fewer messages are necessary to get a message from point A to point B. This is a considerable improvement compared to before and might counterbalance the significantly bigger package size.

9.5 Summary

In this chapter, we have discussed the important topic of energy management in different kinds of nanonetworks. No matter how you approach the topic, sufficient energy is always necessary to provide nanonetwork activities with fuels of different kinds. Electromagnetic, acoustic, and even molecular communication all require energy, and there is no way to bypass this fact.

We have identified several different ways to store energy at the nanoscale. They include batteries and nanoscale ultracapacitors. While batteries take longer to charge, they can likely store at least 10 times as much energy as batteries of the same size. Yet, both of these might not be able to provide sufficient energy to power a nanonetwork for any significant period of time.

We have thus discussed the principle of energy harvesting as a remedy for this problem. The available storage technologies might not be able to power many operations, but it might be possible to repeatedly collect/harvest energy from the surroundings whenever necessary. As a result, nanodevices equipped with an energy harvesting module theoretically possess infinite amounts of energy and are only limited by the time-to-live of the module and its efficiency. We have discussed various different technologies based on the piezoelectrical effect, ultrasound, radio frequencies, and even ambient heat.

Given that energy is available in some form or another, saving energy is at least as important. Just as with macroscale computing, poorly designed algorithms can always lead to a waste of energy and computational resources. As a result, routing protocols that minimize the number and size of messages in a network are necessary. After all, the communication in nanonetworks likely requires more than 95% of the energy a nanonetwork consumes in total. However, designing such protocols is not a simple task given the limitations at the nanoscale, where devices likely do not even have a unique identifier.

We have discussed both SLR and ring saving as potential solutions for this problem in very dense networks. Instead of having all nanodevices forward messages through a nanonetwork, only a small subset of nanodevices is active at the same time, leading to astonishing efficiency compared to naive broadcasting.

10

Time and Randomness

In addition to memory and computing power, other important components in conventional computers are timers and random number generators. Many algorithms require simple time information to logically order, date, or sort events. Especially with sensory tasks, it is often important to know when a measurement was made.

In some situations, a relative sense of time is sufficient. In such cases, the only important thing is whether an event happens before or after another event. For example, when distributing medication, it is only important that this happens "after" the detection of symptoms or markers.

In addition, numerous methods require access to random numbers. In previous chapters, we have already seen numerous examples such as Backoff Flooding and DEDeN. Especially with algorithms that work with limited resources, randomness is often essential in order to "explore" the space of possibilities.

We first discuss the absolute basics of timing at both the macroscale and the nanoscale and present possible technologies to understand the topic sufficiently. Then, we analyze several mechanisms to synchronize computational models at the nanoscale via an external clock of, e.g. circuits.

Next, we analyze several methods for synchronizing the internal clocks of multiple nanodevices in a resource-constraint environment. Among others, we discuss well-known algorithms like Berkeley or Cristian but also special procedures like firefly algorithms. Afterward, we discuss the concept of logical time, where an absolute time in, e.g. milliseconds since January 1st, 1970, is not necessary. Based on that we discuss questions on data consistency at the nanoscale and explain how a distributed consensus among nanodevices may be established given limited resources.

Finally, we discuss the topic of randomness. There are at least two general types of random – pseudorandomness and true randomness. We try to understand how both types work and how they may be utilized at the nanoscale.

10.1 Time

For several applications, nanodevices need not only a logical understanding of time but an absolute one. As an example, medical sensor data is only relevant for a certain amount of time and irrelevant afterward. Furthermore, a lot of data is only relevant when temporally compared with data from other devices. All of that is only possible if nanodevices and nanonetworks in general have a timing component T.

Nanonetworks: The Future of Communication and Computation, First Edition. Florian-Lennert A. Lau.

The following paragraphs explain the basic physical principles that may be used to create an understanding of time for individual macroscale and nanoscale devices. We discuss oscillators, quartz, and atomic clocks and try to understand which principles might be applicable at the nanoscale.

Before we can start tracking time, we have to understand what time even is. From the most general perspective, time is an attempt at finding a discreet measure for "change." As various different objects and things change at different speeds, we have to measure time in *relation* to something else. Einstein suggested the speed of light in the vacuum as an ultimate point of reference for that comparison. A second, however, is defined as the fixed numerical value of the vibration frequency of cesium 133 atoms.

As a result, all devices in the world compare their own internal clocks with that frequency. The internal component responsible for the "counting" of time is an *oscillator*. The word is derived from the Latin word "swinging" and describes a preferably regular alteration between two or more states.

Most modern computer systems use quartz crystals to keep track of internal time because of their precise and stable oscillation frequency. The basic idea is that the crystal is cut to a specific size and shape so that it will vibrate at a precise frequency when an electrical charge is applied to it. This vibration is used as a reference frequency for timing electronic circuits in computers. Thus, the timing component T is technically a combination of special actuators and sensors, but its standalone importance justifies a special component.

In a computer, the crystal oscillator circuit typically consists of a quartz crystal, an amplifier, and a feedback loop. The signal of the crystal is amplified and fed back to itself.

These clock signals ensure that all operations occur at the correct time and in the correct order.

An example of quartz can be seen in Figure 10.1. An electrical signal is sent through the quartz to derive a time from it. To the right, there is a fully functional quartz oscillator.

While clocks based on, e.g. cesium are the most accurate, it is infeasible for every computer to have direct access. While it might be possible that nanodevices get their time directly from suitable atoms in their immediate vicinity, synchronization with external systems is still necessary. Thus, we discuss how to exchange time information between devices later in this chapter.

At the nanoscale, it remains unclear how the timing component may be physically realized. However, there are several physical effects that are potential candidates. Yet, the compatibility with other components remains an open question.

Figure 10.1 Several examples of quartz that are used in modern computers or other electronic devices. While they are too big for direct use at the nanoscale, the principles remain unchanged. *Source*: Stefan Riepl/Wikimedia Commons/Public Domain.

10.2 Synchronization

One of the key challenges in nanonetworks is synchronization. Synchronization ensures that all devices in the network operate on the same time scale and can coordinate their actions and measurements effectively. Nanonetworks often rely on precise timing and coordination among sometimes thousands of nanodevices in order to perform their desired functions. For example, if multiple nanodevices need to transmit data simultaneously, they must be able to do so in a coordinated manner in order to avoid interference.

In addition to the challenges posed by the small size of the devices, nanonetworks also face several other constraints, such as limited energy, computing power, and environmental heterogeneity. All of these factors make synchronization even more critical in order to ensure that the network operates effectively and efficiently. To avoid redundancy and wasted energy, individual devices need at least some indirect information or valid assumptions about the temporal behavior of other network participants.

There are several types of synchronization that we discuss in this section. The first and most obvious is the synchronization of the internal clocks of individual nanodevices. Second, the information processing of nanodevices also has to be synchronized. This problem is known as *clocking* in computers and ensures the deterministic behavior of, e.g. circuits. The last type of synchronization we discuss here is the synchronization of stored data or the system state among different nanodevices.

We start with the synchronization of the internal clocks. Figure 10.2 illustrates the problems. As atomic clocks are infeasible for individual devices, less accurate/precise clocks are used in general. Those oscillate fairly precisely, but there is a *drift* that creates a positive or negative distance from the ideal time. Depending on the quality of the, e.g. quartz, the drift is bigger or smaller and requires frequent synchronization with a verified clock as a "source of truth."

To adjust the internal time correctly, we can employ different synchronization algorithms that work, for example, via messages or other external signals. As package-based networks are most common nowadays, we focus on message-based synchronization schemes first. There is one last remark: You might be surprised by the number of comparably "old" algorithms. The resource

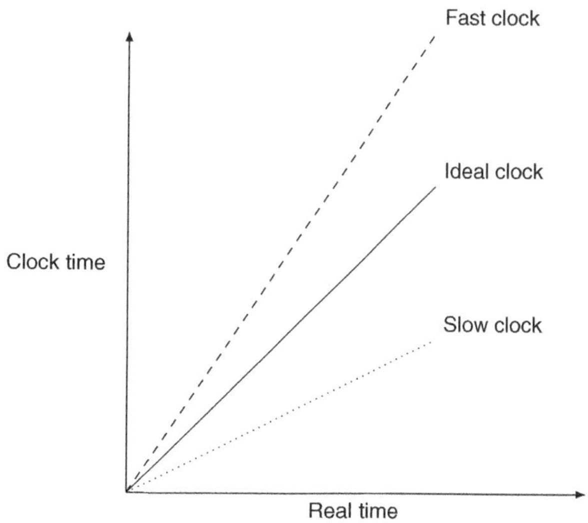

Figure 10.2 Concept of clock drift explained. *Source*: Florian-Lennert A. Lau.

constraints at the nanoscale are similar to the constraints in the early days of computer science, and given limited resources, only certain procedures are possible.

10.2.1 Cristian's Algorithm

The first and surprisingly simple algorithm that dates back to 1989 had has been developed by Flaviu Cristian Cristian (1989). It is a probabilistic method to establish a shared time in a computer network by periodically repeating a synchronization procedure. The algorithm works as follows:

1. One of the participants is the time server, and its clock is the point of reference.
2. The other participants send a request to the time server, asking for the current time.
3. The time server transmits its own time to each requesting participant.
4. Upon receiving the response, the requesting participant memorizes the time of arrival of the package.
5. The requesting participant estimates the *round trip times* (RTTs) that the packages required for travelling.
6. The requesting participant modifies its clock by half of the RTT.

The internal clocks of computers can be synchronized with a remaining error of only a few milliseconds if the latency remains somewhat stable. As the RTTs in the real world and especially in nanoscale applications might differ, the result of the synchronization can vary significantly in terms of quality. Furthermore, as the participating devices simply set their internal clocks to a specific value, there might be resulting inconsistencies, as some times might occur twice in a network or not at all. Modern approaches adjust the speed of their clocks to slowly even out clock drift.

While that might not be a reasonable assumption at the nanoscale, it is still a good starting point as the algorithm requires few resources and it might thus be feasible to implement it. Complex protocols like NTPv4 that are used nowadays require vast amounts of resources to function properly and will likely not be suitable for at least early applications.

10.2.2 Berkeley

Like Cristian's algorithm, the Berkely algorithm is designed for local networks and requires frequent repetition (Gusella and Zatti, 1989). The Berkely algorithm works as follows:

1. One of the participants is the time server, and its clock is the point of reference.
2. The time server polls the current time of other computers.
3. Each computer responds with its current time.
4. The time server collects the responses and computes an average, which represents the local "true" reference time. In some implementations, outliers are filtered away.
5. The time server sends the resulting time to all other computers on the network.
6. Each computer computes the difference between its own clock time and the computed time received from the time server.
7. Each computer adjusts its clock speed to even out the clock drift.

The Berkeley algorithm has the advantage of being more robust to network delays and clock drift than Cristian's algorithm. By taking an average of the responses from multiple computers, it reduces the impact of individual network delays and clock errors. Furthermore, it can account for outliers by simply cutting them away if they deviate from real time by too much. This may happen frequently as new devices often start with a "fresh" clock that is initialized on January 1st, 1970, 00:00:00 UTC.

One limitation of the Berkeley algorithm is that it assumes all computers on the network have a similar clock drift rate. If there are computers with significantly different clock drift rates, the algorithm may not be able to keep all clocks synchronized accurately. Yet, due to adjusting the clock speed instead of the absolute time, the algorithm is an improvement compared to Cristian's algorithm. After all, temporal inconsistencies could pose great risks in, e.g. medical nanoscale sensor networks, where lives are potentially at risk.

10.2.3 NTP

Both Cristian's algorithm and the Berkeley algorithm are used in the most widespread *Network Time Protocol version 4* (NTPv4). NTP is a protocol used to synchronize the clocks of computers over a network. NTPv4 is the current version of the protocol introduced in 2010. Among other things, this version improved the accuracy of the protocol, and new security features have been integrated.

The NTPv4 protocol uses a hierarchical system of time servers to synchronize the clocks of all participants. Each time server in the hierarchy is assigned a stratum number, with stratum 0 being the most accurate source of time and stratum 15 being the least accurate. Stratum 0-time servers are typically high-precision atomic clocks or GPS receivers.

The architecture can be seen in Figure 10.3. At the very top, there are potentially several atomic clocks that communicate directly with the stratum 1 layer. The stratum number represents the distance to an atomic clock.

When a client wants to synchronize its clock with a time server, it sends an NTP request packet containing its current time. The time server responds with an NTP response packet containing the server's time and other information, such as the round-trip delay and clock offset.

NTP uses several algorithms for time synchronization, including the *Simple Network Time Protocol* (SNTP) algorithm, the Marzullo algorithm, and the intersection algorithm. The SNTP algorithm is a simplified version of NTP that is designed for situations where only a basic level of time synchronization is required. It is often used in environments where precision time synchronization is not critical, such as in desktop computers and mobile devices. The SNTP algorithm uses a series of polls and responses between a client and server to synchronize their clocks.

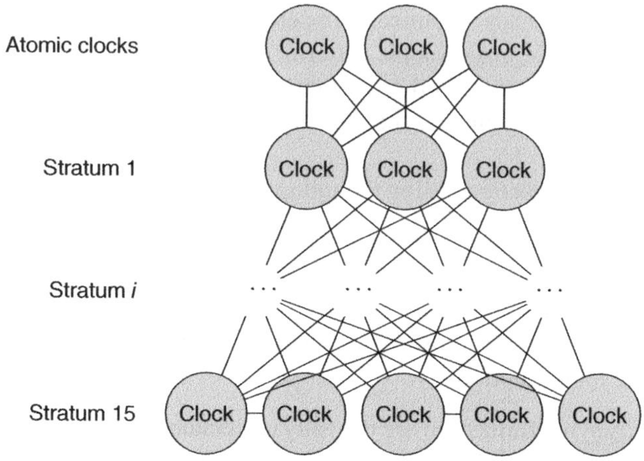

Figure 10.3 Layers' architecture of the NTPv4 protocol. All devices are divided into "stratum" that signifies their distance to the original time servers at layer 0 (atomic clocks). *Source*: Florian-Lennert A. Lau.

The Marzullo algorithm is a statistical algorithm that combines several time sources and computes the intersection of their readings to obtain a more accurate estimate of the current time. The Marzullo algorithm is particularly useful in environments where a high level of time accuracy is required, such as in financial trading systems and scientific experiments.

The intersection algorithm is similar to the Marzullo algorithm, but it is designed to work with only two time sources. By measuring the difference between the two sources, the algorithm adjusts the local clocks to compensate for any discrepancies. The intersection algorithm is often used in environments where only two sources are available, such as in peer-to-peer time synchronization.

The accuracy of NTP depends on the quality of the time sources and the network conditions between the client and server. In ideal conditions, NTP can achieve sub-millisecond accuracy. However, in practice, the accuracy of NTP can vary depending on the specific environment and implementation. To improve the accuracy of NTP, one could increase the number of time sources, use more accurate time sources, or optimize the network conditions between the client and server.

NTP uses the UPD transport-layer protocol and thus requires less overhead than comparable methods using TCP. NTP packets are usually between 48 and 68 bytes long, and the involved timestamps require 64 bits. These packages have to be exchanged periodically, depending on the internal drifts of the clocks of the participants.

While NTP is widely used in macroscale networks, the resource requirements might still be too high for first-generation nanonetworks. As a result, even more, minimal approaches might be necessary.

10.2.4 Fireflies

One of the least resource-intense procedures for synchronizing network participants is the nature-inspired *firefly algorithm*. While the procedure is named after fireflies, several organisms that create light use a similar procedure to synchronize themselves. In this procedure, each firefly produces a flash that lasts for a few milliseconds, and the flash period is roughly constant for each firefly.

The mechanism behind firefly synchronization can be explained using the concept of self-organization. Self-organization is a process where a system of individual agents, e.g. nanodevices, without any central control, spontaneously organize themselves into a coherent and structured pattern. It is crucial that the observable macroscopic pattern/order *emerges* from local rules alone without having access to the detailed states of other participants. In the case of firefly synchronization, each firefly acts as an individual agent, and the flashing pattern is the structured pattern that emerges through self-organization.

The process of firefly synchronization can be described using the following steps:

1. Each firefly starts flashing independently at a random time. Each flash is visible to its neighbors.
2. Each firefly adjusts its flashing period to match that of the majority of its neighbors whenever flashing is observed.
3. This procedure is repeated until all fireflies flash in unison. The algorithm continues running in the background and can compensate for disruptions.

The exact underlying effects of the firefly synchronization are not fully understood, but it can be modeled as a recursive feedback system. The light pulses emitted by each firefly act as messages that influence neighboring fireflies' behaviors. This process of mutual interaction and feedback eventually leads to the emergence of a synchronized flashing pattern.

The benefits of the firefly algorithm are immediately evident. In terms of resources, the individual devices require only the ability to generate random numbers, a timer that allows for periodicity, and the ability to store an integer. Furthermore, the algorithm is surprisingly tolerant of errors and failing devices.

In summary, it is difficult to imagine an algorithm that requires fewer resources while still taking all the nanoscale constraints into account.

10.2.5 Clocking

A related type and special case of synchronization is *clocking*. Unlike self-organized synchronization among devices in a network, clocking usually relies on a single central source of time and is used to synchronize different components of computational models or the internal components of single devices.

In classical computers, the clock is a component that provides time signals for synchronizing the components of a computer. The clock speed is the number of pulses per second and is usually measured in gigahertz. A clock speed of 1 GHz means that the clock generates one billion pulses per second.

The clocking mechanism in a PC is usually based on the aforementioned quartz crystals.

The generated clock signal resembles a square wave signal, which means that it alternates between two voltage levels. Computations and other operations may only be performed when the clock signal is also *high*, which is akin to applying a logical AND to the entire involved circuit.

The clock signal is distributed throughout the computer system via a clock distribution network, which consists of a series of clock lines and buffers. The clock lines are conductive traces on the computer's printed circuit board that carries the clock signal from the oscillator to various components in the computer. The buffers are components that amplify and synchronize the clock signal to ensure that it arrives at each component at the same time and with the correct timing.

The clock speed significantly impacts a computer's performance. A high clock speed allows for the execution of a higher number of operations per second. However, the clock speed is not the only factor that determines performance, but other factors such as the efficiency of the CPU architecture and memory subsystem also play a significant role. Yet, there are physical limitations when it comes to the speed of a clock as higher frequencies lead to more generated heat as a byproduct.

10.2.5.1 QCA Synchronization

Clocking is not only necessary in classical circuit-based computers but with nearly all computational models. At least the physical implementations of such models require some degree of external coordination to compensate for errors that are usually omitted in purely mathematical representations.

The Quantum-Dot Cellular Automata from Chapter 5 also requires synchronization as such circuits tend to behave rather chaotically given different wire lengths. As QCAs are based on the Coulomb interaction between electrons, a simple clock mechanism could be a magnetic field applied to all quantum cells. A function that describes the clock behavior can be seen in Figure 10.4.

The figure displays the behavior of a single quantum cell. Each cell repeatedly cycles through four phases in which the barriers between the quantum dots vary in strength. The strength of the barrier that is usually supplied by an external magnetic field is displayed on the y-axis. The x-axis displays the time and is separated into four different phases release, relax, switch, and hold.

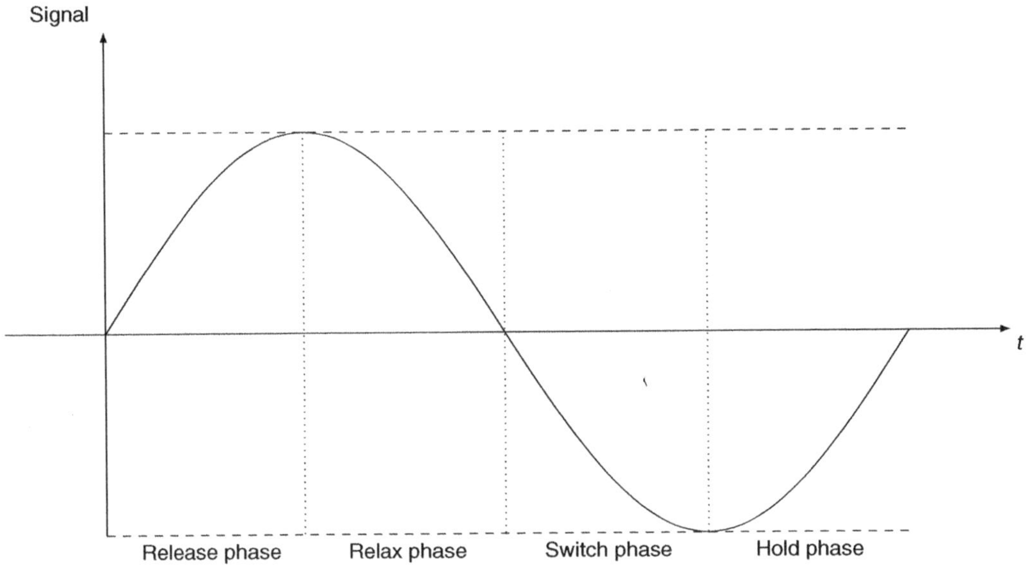

Figure 10.4 Clock behavior of quantum-dot cellular automata. *Source*: Florian-Lennert A. Lau.

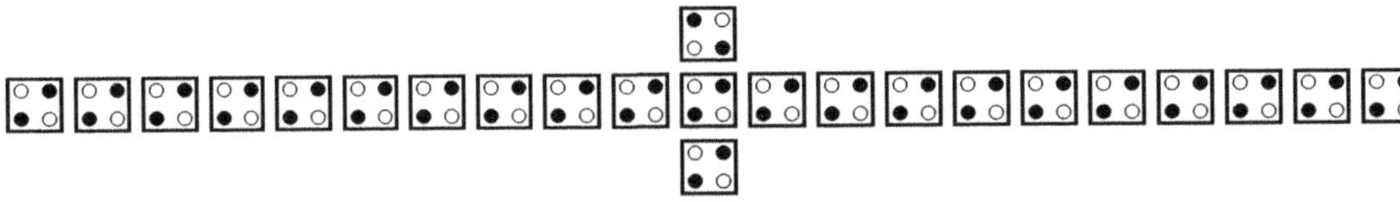

Figure 10.5 Dysfunctional quantum cell majority gate. Due to the long distances, a stable ground state may only be achieved after some time (Buss, 2017). *Source*: Florian-Lennert A. Lau.

In the hold phase, the signal is low and the quantum dot barriers are up. In the release phase, the signal is rising and the barriers are falling. Electrons start to move. In the relax phase, the signal is high and the barriers are down. Electrons move the most during this phase. In the switch phase, the signal is falling and the barriers are raised. This causes a new state to be stably assumed.

This clocking function is running globally in the background and therefore synchronizing the entire QCA. However, further measures are necessary to avoid certain kinds of errors. Figure 10.5 illustrates the problem using a special kind of majority gate.

The gate has three inputs to the left, top, and right. However, the distances to the output gate vary. This leads to an interesting behavior that does not match the expected behavior of propositional logic. In this case, the gate would simply forward the state of the top input for several units of time. Only once the signal of the left and right gates reaches the middle does the output change. As a result, this kind of majority gate outputs, e.g. "1" *until* it outputs "0." To avoid this, a signal should only be "valid" if a global clock signals that a system ground state has been reached.

In addition to that, wires that are too long can lead to further problems as each signal propagation leads to a loss of signal strength. As a result, after roughly 15 quantum cells, the state of the next one is completely undetermined and no clear assignment of "1" or "0" can be made.

10.2.5.2 Self-assembly Synchronization

The last type of computational synchronization we discuss in this section is the synchronization of self-assembly systems. This type of synchronization is fundamentally different from the others as there is no real external parameter that could be used to adjust the behavior of binding reactions.

While the global temperature dictates the overall speed of reactions, it does not provide a condition *in addition* to the logic computed in the assembly. As a result, computational synchronization has to be implemented in the computation/assembly process itself.

While self-assembly systems overall work non-deterministically, there are ways to control the assembly process in more detail. In fact, it is possible to incorporate "points of synchronization" into a tileset that forces an assembly to "wait" until other structures are finished before the remaining assembly may resume. This is technically only possible from a temperature 2 system onward as it is not possible to represent conditions otherwise.

That said, self-assembly can only represent a logical "happened before" time that can only loosely be mapped to real-time units. Given a concentration of tiles and a fixed environmental temperature, there is an expected number of binding reactions per second. Given simulations, one may derive a rough estimate for the expected assembly duration. That said, in the general case, such an estimate is not possible as the growth of an assembly is generally a probabilistic, non-deterministic process that cannot be described symbolically.

Yet, even deriving a real time for an assembly is no simple task, as the number of attempted binding reactions per assembly increases with the size of its growth front. The bigger an assembly is, the more collisions with other DNA tiles may occur which influences the overall duration.

In general, most computational models require at least a logical understanding of time that has to be implemented using physical timing components. How such a mechanism may be implemented may vary drastically from computational model to computational model. We have only briefly introduced a small number of such global synchronization schemes to give an inexperienced reader a solid intuition of the topic.

10.3 Logical Time

It is generally unclear if and how absolute timestamps may be implemented at the nanoscale and general resource constraints also affect any type of timing component of a nanodevice.

To offer a solution for those likely constraints, it might be possible to use a less strict form of *logical time*. Instead of requiring to know the time in relation to the definition of a second/date, we only require certain systems to know which event happened before which other event. While this is by no means sufficient for all proposed applications, some do not require a more strict form of timing.

Besides, the problem of establishing a global understanding of time in a distributed system is ultimately infeasible (Lamport, 1997). At the most fundamental level, this is simply due to the relativity of time. At a more specific level, it is due to the (re-)ordering of events, network latency, node failures, and message delays.

To address this challenge, distributed systems often rely on algorithms that can provide a partial ordering of events based on certain assumptions and heuristics. For example, the *Lamport timestamp algorithm* introduced in 1978, which assigns a timestamp to each event based on the maximum of its own clock and the timestamps of the messages it receives from other nodes (Lamport, 2019).

The idea behind the Lamport timestamp algorithm is that each node has a variable/"clock" that keeps track of the ordering of events. This clock is not based on physical pulses, but rather on rules that dictate how the clock changes in response to events/messages. The algorithm works as follows:

1. Each event in the system is assigned a unique identifier.
2. Each node in the system maintains a logical clock. The clock starts at zero and is incremented by one for each event that the node generates.

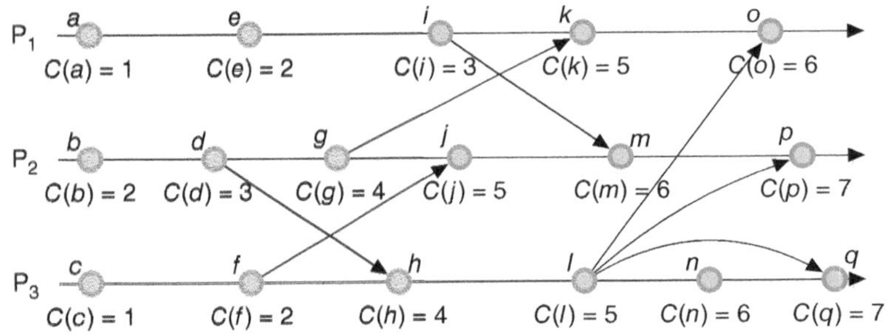

Figure 10.6 Example run of the Lamport algorithm in a system of three distributed processes P_1, P_2, and P_3. *Source*: Florian-Lennert A. Lau.

3. When a node sends a message, it includes its current timestamp in the message.
4. When a node receives a message, it updates its own timestamp as follows:
   - The node sets its own clock to the maximum of its current clock and the timestamp of the received message.
   - The node increments its clock by one to account for the receipt of the message.
5. All events with a timestamp can be logically ordered. If event A has a lower timestamp than event B, then we know that *A happened before B* in the system.

An example execution of the algorithm based on three processes can be seen in Figure 10.6. The example starts in the middle of the execution of the algorithm, and all processes already have a time greater than zero assigned. The three processes P_1, P_2, and P_3 either compute something on their own processor, send a message, or receive a message.

The required hardware for the Lamport algorithm is minimal, which makes it a perfect candidate for application at the nanoscale. All that is necessary is a single integer of fixed length, a component to communicate (which we presuppose), and the ability to compare two numbers with each other. It is likely not possible to find a less resource-intense algorithm that still guarantees the happened-before relation. As an example, technologies like *vector clocks* require information about the cocks of all other processes, which is likely infeasible in a network of potentially thousands of nodes.

10.4 Consistency

No matter if macroscale or nanoscale networks, many applications require the synchronization of state between devices. *State* in a network refers to both the content of the main memory of a device and all messages that are still being delivered. As messages are usually included in the global state, it can be quite tricky and oftentimes impossible to identify how the network state was at any given time.

In macroscale networks with few participants, it is often possible to record all incoming and outgoing messages and record the internal state at least to a degree. Thus, it is possible to compute the global state of such a system whenever requested.

In nanonetworks, the same procedure is likely not feasible as nanodevices possess little internal memory and messages are expected to be short. While this also reduces the size of the overall state, the number of expected participants in nanonetworks outweighs the other two factors. As a result, it is absolutely impossible to have the global state available at each nanodevice. Nanodevices have to rely on local information alone for decision-making.

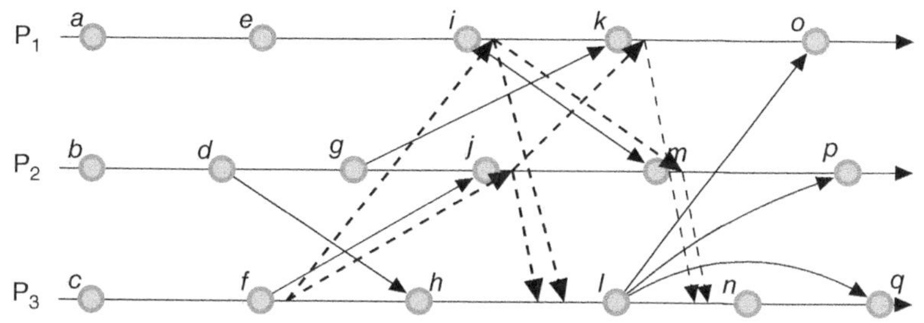

Figure 10.7 Example execution of the Chandy–Lamport algorithm to determine a consistent snapshot of the distributed system state among three processes. The dashed lines represent necessary messages to compute the distributed snapshot, and the regular lines represent the normal communication events. *Source*: Florian-Lennert A. Lau.

However, some simple applications might still work with minimal memory requirements and only very short messages. For such applications, it might be interesting to assess the global state of the network nonetheless. To solve this problem (Chandy and Lamport, 1985), introduced an algorithm to determine a cut or snapshot of the system state at any time.

An example run of the Chandy–Lamport algorithm can be seen in Figure 10.7. The objective of the algorithm is to determine a *consistent* global state that *might* or might not have taken place but is at least in line with temporal laws. The system in our example consists of three processes on three different devices that each possess an internal, logical clock.

The algorithm works as follows:

1. All processes in the distributed system work normally.
2. At some point, a process initiates the snapshot process by sending a marker message to all other participants.
3. Upon receiving a marker message *for the first time*, a process records its local state. This local state includes all local events, such as changes in variables, data structures, and any other necessary information.
4. As the marker travels through the system, each process appends information about the state of its outgoing channels. The channel state typically includes messages that have been sent but not yet received by the recipient process.
5. Once a process has recorded the state of all its outgoing channels, it records its local state again. This step captures any changes that occurred during the recording of channel states.
6. Upon receiving a marker, a process continues to propagate the marker along its outgoing channels, ensuring that the marker eventually reaches all processes in the system.
7. When a process has recorded its local state for the second time and propagated the marker to all its outgoing channels, it considers the snapshot process complete and informs all other participants.

In our example, process P_3 initiates the algorithm at event f and sends messages to all other participating processes. The messages reach P_1 after event i and P_2 after event j. Upon receiving the marker, the processes start tracking their local state and the messages that are still on the way. They further inform all other processes that they have received the marker. Once each process finishes, they inform all other processes and the snapshot is returned to the initiating process P_3.

While the algorithm does compute a logically consistent global system state, it is relatively expensive to do so. The message complexity of the algorithm is in On^2 as each process has to inform each other process on several occasions. In a nanonetwork of thousands of participants, this might be infeasible. Yet, if only a subset of certain "classes" of devices participate, it is still possible to determine a *relevant* "pseudo-global state."

10.4.1 Types of Consistencies

The problem of data synchronization between devices is closely related to the topic of *consistency*. Consistency describes a state where multiple copies or *replica* of data exist in a distributed system and all of them have the same state.

We generally distinguish between three forms of consistencies:

1. strict consistency,
2. sequential consistency, and
3. causal consistency.

Strict consistency is the strongest form of consistency in distributed systems. It guarantees that all operations or events in the system appear as if they occurred in a single, serialized order, regardless of the execution order on different nodes. In other words, all replicas observe/read the same order of operations. To achieve strict consistency, a distributed system must enforce a total order of all operations, which typically requires a centralized coordination mechanism such as a master node or a consensus algorithm. This level of consistency ensures that all replicas have a consistent view of the system state at all times. However, achieving strict consistency often comes at the cost of highly increased latency and reduced availability, as it requires synchronization and coordination among nodes.

Sequential consistency is a weaker form of consistency compared to strict consistency. It guarantees that operations are observed in the same order by all other processes in the system. However, the order of operations from different processes may not be strictly defined.

An example can be seen in Figure 10.8. The processes $P_1 \ldots P_3$ perform a series of read/write operations on a shared memory *e*. Process P_1 reads the respective values 1 and 3. Process P_2 reads 2 and 3. Process P_3 reads 1 and 2. While not all values occur in each process, they can still be sorted on a sequential timeline from 1 to 3.

In sequential consistency, the relative ordering of operations performed by different processes is preserved as long as those operations do not overlap. If two operations from different processes are concurrent (i.e. they do not have a causality relationship), their order of observation by other processes is not guaranteed.

To achieve sequential consistency, the system may employ locks, semaphores, or message-passing protocols. These mechanisms ensure that operations from a single process are globally visible to all processes in the same order.

Causal consistency is a more relaxed form of consistency that preserves causality between operations. It ensures that if one operation causally depends on another, the dependent operation is observed after the causal operation.

Causal consistency captures the cause-and-effect relationships between operations. If operation *a* causally affects operation *b*, all replicas of the system must observe *a* before *b*. However, for concurrent operations that are not causally related, their order of observation is not strictly defined.

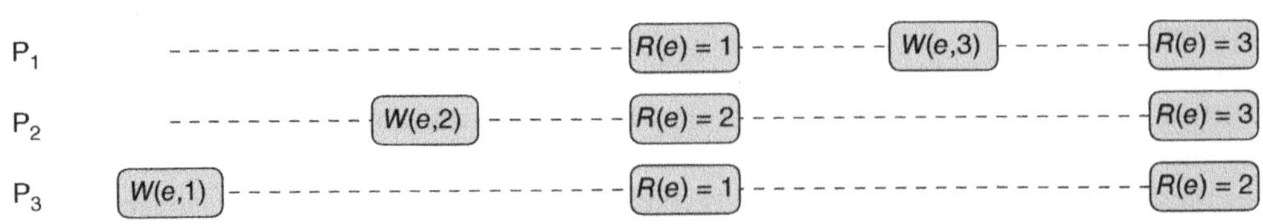

Figure 10.8 Example schedule of read/write operations that fulfills sequential consistency. It must be possible to sort all read events in a sequential order to fulfill the necessary criteria. *Source*: Florian-Lennert A. Lau.

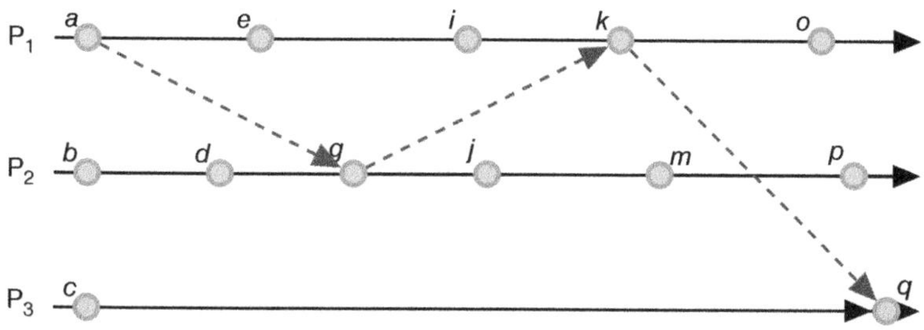

Figure 10.9 The dashed line has to follow the principle of transitivity. If *a* happened before *g* and *g* happened before *k*, then *a* also has to have happened before *k*. Yet, in such a system, it is completely unclear if *a* happened before *c*. Both events are called *concurrent*. *Source*: Florian-Lennert A. Lau.

P_1:		$R(e) = 1$	$W(e,3)$	$R(e) = 3$	$R(e) = 2$
P_2:		$W(e,2)$		$R(e) = 2$	$R(e) = 1$
P_3:	$W(e,1)$		$R(e) = 1$	$R(e) = 2$	$R(e) = 3$

Figure 10.10 Example schedule of read/write operations that fulfills causal consistency. Processes P_1 and P_3 are potentially influenced by each other, and the involved events must be read in the presented order by all processes to fulfill causal consistency. *Source*: Florian-Lennert A. Lau.

An example of concurrent operations can be seen in Figure 10.9. While the events *a*, *g*, *k*, and *q* have a strictly defined order, the same is not true for events *a* and *c*.

Achieving causal consistency typically involves capturing causal relationships between operations by using logical timestamps or vector clocks. These mechanisms allow replicas to determine the causal dependencies between operations and ensure that they are observed in the correct order.

Causal consistency is a tradeoff between strong consistency and the performance of a system. It allows for greater concurrency and reduces the need for synchronization and coordination between replicas, but it may introduce some inconsistency between unrelated concurrent operations.

An example of causal consistency can be seen in Figure 10.10. Process P_3 writes the number 1 onto resource *e*, and process P_1 reads that number a moment later and writes the number 3 onto the same resource. Therefore, process P_1 and process P_3 have a causal relationship with each other and all processes must read the same order of events to fulfill causal consistency.

It is worth noting that there are other consistency models as well, such as eventual consistency, where replicas may be temporarily inconsistent but eventually converge to a consistent state. These models offer a different balance between consistency, availability, and performance for special system requirements.

10.4.2 CAP Theorem

As we have learned, synchronizing data among replicas in distributed systems is not necessarily an easy task. There are many desired properties like consistency, availability, and partition tolerance. However, not all of them can be fulfilled at the same time.

The *CAP theorem* describes a law of distributed systems that addresses the trade offs between *consistency*, *availability*, and *partition tolerance*. According to the theorem, it is impossible to simultaneously achieve all three properties in a distributed system at once. Consistency means that all participants in a distributed system have the same state while operating. If a value is written to one system, any subsequent read from any node should return that value.

Availability means that any request to a system must produce a response, regardless of the state of the system. High availability ensures that the system remains operational and responsive even in the presence of failures.

Partition tolerance deals with the behavior of a distributed system in the event of network failures or network partitions. Another name for partition tolerance could be "network failure tolerance." Network partitions occur when nodes in a system are unable to communicate with each other, leading to a "split" of the network. Partition tolerance allows the system to continue functioning even when communication between nodes is disrupted or delayed.

The CAP theorem asserts that in the presence of a network partition, a distributed system must sacrifice either consistency or availability. In other words, it is impossible to have both consistency and availability during a network partition. This is because, during a partition, the system needs to make a decision on how to handle updates or reads in the affected nodes. There are two possible options:

1. *Consistency over availability (CP systems)*: In this case, the system prioritizes maintaining consistency over availability. When the state of a subsystem changes, the system halts and updates in order to maintain a consistent state across all nodes. This tradeoff ensures that all nodes have the same state, but it also means that the system may become unavailable.
2. *Availability over consistency (AP systems)*: In this case, the system prioritizes availability over consistency. When the state of a subsystem changes, the system continues to operate and process requests. Each node may have a slightly different state due to replication delays or conflicts, resulting in eventual consistency. This approach ensures that the system remains available at all times, but it may lead to temporary inconsistencies.

The CAP theorem assumes that partitions are a possibility in a distributed system as network failures are inevitable in real-world scenarios. However, it is worth mentioning that not all network partitions are the same. The CAP theorem focuses on the scenario where a network partition occurs and communication is completely severed between nodes.

Realistically, there is no one solution that fits all requirements of corporations. Many systems prioritize availability and allow for eventual consistency as it provides better fault tolerance and scalability in large-scale systems. However, some systems, particularly those dealing with critical data or financial transactions, may prioritize consistency and sacrifice availability.

It is important to consider the CAP theorem when designing distributed systems as it helps in making informed decisions about the trade offs and understanding the guarantees that can be achieved in different scenarios. The CAP theorem is just as important in nanonetworks as it is in real-world macroscale systems. One could argue that the unique environments at the nanoscale might change the usual priorities as network failures are likely much more common.

10.5 Randomness

The next major topic is randomness at the nanoscale. Randomness describes the lack of patterns or predictability in data or events. In disciplines like mathematics, statistics, physics, computer science, and as a result, nanonetworks, randomness is of great importance.

At its core, randomness implies that outcomes of certain processes are not influenced by any identifiable factors and cannot be predicted with certainty. Instead, random events occur based on chance or probability. Randomness can be categorized into two types: deterministic randomness and non-deterministic randomness.

Deterministic randomness or *pseudorandomness* refers to events that appear random but are generated by a predetermined process. For example, a *pseudorandom number generator* (PRNG) produces a sequence of numbers that appears random but is generated algorithmically based on a specific seed value and predefined rules. A PRNG will always produce the same sequence of numbers if initialized with the identical seed. A visual example would be the world generation in video games like Minecraft, where it is possible to create 100% identical maps given the right seed. While deterministic randomness is useful in various applications, it is not truly random in the strictest sense.

Non-deterministic randomness, or *true randomness*, can be derived from inherently unpredictable processes, such as radioactive decay or atmospheric noise. These processes cannot be predicted even if all initial conditions are known. This is either due to infeasible complexity or the inability to precisely determine the state of a system. True randomness is crucial in, e.g. cryptography or statistical sampling.

It is important to note that the question of determinism vs. non-determinism has no definitive answer as of yet. In fact, it is not clear if true randomness even exists or if the physical processes in our universe are fully determined and thus theoretically "computable." The question itself is likely as old as humankind and was already a topic of debate in ancient India where the Buddha declared a middle way between determinism and non-determinism (Gotama The Buddha, 1987b). While the Buddha discussed 62 kinds of wrong views in that specific lecture, they all boil down to two overarching principles that fit the question of determinism vs. non-determinism. He even described how the vast majority of people almost "instinctively" radiate toward one of two extremes: either "full determinism" or "full non-determinism," unable to see any middle ground. Despite all that, it might be interesting to note that many people were already addicted to dice games and other forms of gambling at that time.

10.5.1 Pseudorandom

In computer science, we mainly work with pseudorandomness as computers are in themselves not capable of producing true randomness. As most types of nanonetworks can be described as computational processes, they have the same limitations as macroscale computers. Pseudorandomness refers to the generation of a sequence of numbers or events that closely resembles a truly random sequence, but is actually produced by a deterministic process or algorithm. The term "pseudo" indicates that the generated sequence is not genuinely random, although it may exhibit statistical properties similar to randomness.

Pseudorandomness is widely used in computer science because true randomness is challenging to achieve in deterministic machines. Instead, algorithms known as PRNGs are employed to produce sequences of numbers that appear random within certain statistical measures.

PRNGs typically start with an initial value called a seed, which acts as the input for the algorithm. When the PRNG is executed, it performs a series of deterministic operations on the seed to generate a new number, which becomes the next element in the sequence. The generated number is then used as the seed for the next iteration.

Common examples of PRNGs are (linear) *congruence generators* (LCGs). An LCG consists of the following:

1. $n \in \mathbb{N}^+$ possible states,
2. a module $m \in \mathbb{N}^+ \setminus \{0, 1\}$,

3. factors $a_i \dots a_n \in \{0 \dots m-1\}$,
4. increments $b \in \{0 \dots m-1\}$, and
5. a seed $\sigma_i \dots \sigma_n \in \{0 \dots m-1\}$.

A possible example would be the following formula:

$$y_i = \left(\left(\sum_{k=1}^{n} a_k y_{i-k} \right) + b \right) \bmod m.$$

The LCG would be "linear" for $n = 1$.

The key property of a good PRNG is that the generated sequence should exhibit the following characteristics:

- It should appear random to an observer without knowledge of the seed or the algorithm. The generated numbers should not exhibit any recognizable pattern or correlation.
- The PRNG should have a long period of unique values that can be generated before the sequence repeats.
- The numbers generated by the PRNG should be uniformly distributed over the range of possible values. Each value should have an equal probability of being generated.

It is important to note that pseudorandom numbers might be suitable for many algorithms that rely on "restarting" a search in an otherwise infeasibly large space of options. However, they are not suitable candidates for cryptographic applications. The parameters of an LCG can be learned by tracking several values and creating and solving a linear equation system. A small number of samples are usually all it takes to find out the underlying parameters.

10.5.2 True Random

True randomness refers to the generation of numbers or events that are unpredictable and devoid of any discernible pattern or order. Unlike pseudorandom numbers, which are generated by algorithms and are deterministic in nature, true random numbers are inherently unpredictable and uncontrollable.

To obtain true randomness, various methods can be employed. Here are a few commonly used approaches:

- The first candidate is physical processes. True random numbers can be generated by leveraging physical phenomena that are inherently random. Several examples are as follows:
  - atmospheric noise,
  - thermal noise, or
  - quantum processes.
- Another method is based on random event timing that is considered to be unpredictable, like
  - the time intervals between user keystrokes or mouse movements or
  - network packet timings.
- Randomness can also be derived from physical sensors that detect natural processes with unpredictable characteristics, like
  - atmospheric noise sensors or
  - Geiger counters that detect random radioactive decay events.
- Chaotic systems, characterized by extreme sensitivity to initial conditions, can be used to generate random numbers. Those systems show complex and apparently random behavior, such as a double pendulum.

- Some unconventional methods employ physical devices that produce visually random patterns, like lava lamps or spinning disks with random patterns.

It is important to note that even with these methods, it is challenging to prove absolute randomness. Instead, we generally evaluate the randomness of a number generator by statistical tests and analysis, checking if the generated sequence exhibits properties expected of random data.

In practice, many applications utilize a combination of PRNGs with an initial seed derived from a true random source. This approach ensures a balance between efficiency and the unpredictability of random numbers. Such numbers can be obtained via web services that specialize in such services.

It is further important to note that true randomness is based on the assumption of non-determinism. If the entire universe could indeed be modeled by a huge "computation," true random numbers would be impossible to obtain. However, it would still be possible to obtain "good enough" random numbers nonetheless if it is *infeasible* to obtain the state of the system, for example due to the Heisenberg uncertainty principle.

Another example would be *computationally complex* systems, where the behavior is technically deterministic, but there is currently no known way to obtain the next state of such a system but via simulation. A simple example would be Langton's ant (Gajardo et al., 2002). An example can be seen in Figure 10.11.

The rules of the simulation are extremely simple. There is an infinite chessboard with an ant on top that is colored all in white. In each simulation step, the ant moves one step forward. If it steps on a white square, it changes directions clockwise and flips the color of the square. If it steps on a black square, it changes directions counter-clockwise and also flips the color of the square.

The figure shows the system after the simulation of 10 500 steps. It starts the process in a random-seeming pattern until, seemingly random, an order emerges and the ant starts creating a

Figure 10.11 Initial state of Langton's ant simulation after 10 500 steps. *Source*: Florian-Lennert A. Lau.

"street" to the left that continues infinitely. In the beginning, most people would not have expected such behavior, yet, many systems like that exist.

No matter if true random numbers exist or not, good enough random numbers may be obtained at the nanoscale nonetheless. The physical processes that are usually used to obtain such numbers are just as available at the nanoscale as they are at the macroscale, and even if it would be infeasible to obtain them frequently, it might still be possible to use a single of such numbers as a seed to create something akin to an individual chessboard starting position for Langton's ant.

10.6 Summary

In summary, time and randomness are extremely important topics at both the macroscale and the nanoscale. Especially in nanonetworks, synchronization of both internal clocks and data among nanodevices is crucial to cooperate toward a shared goal. For sensory tasks like monitoring the health parameters of a patient, accurate absolute time information is even more important as such data is only relevant for a certain amount of time and it is only useful when the exact time of measurement is available. Without good and resource-efficient algorithms, it is impossible to realize many proposed applications for nanonetworks.

The same is true for random numbers. Many resource-constrained algorithms heavily rely on random numbers to generate approximate results for a given problem. Oftentimes, it is simply infeasible to explore all possible solutions and frequent restarts of algorithms at random times are necessary. In a similar manner, it is often necessary to randomize the seeds of LCGs or require true random numbers for safety-critical applications.

11

Safety and Security

This chapter analyzes the topic of security in nanoscale systems. Unlike macroscale systems, this problem is either not well researched or not solvable with classical algorithms and methods given the different environments at the nanoscale. While it is in some cases possible to use identical algorithms as in the early days of computer science and the internet, many of them are insufficient by now. As an example, many procedures that use comparably short keys can now be broken given sufficient interest in it. Furthermore, these methods are only applicable if the underlying technology/computational model remains unchanged, which is not entirely clear. There are countless possible new technologies that might be used in nanonetworks and each of them offers a possible new attack route that must be taken into consideration.

As a result, we start with an analysis of the reference architecture of different nanonetwork types and try to list all the different network participants, the channels, and the corresponding attack types that may affect each of them. We further demonstrate a few peculiarities of the nanoscale and how an attacker might get access to the data on the hardware level. Then, we list the desirable properties that would satisfy modern security requirements. Among them are confidentiality, integrity, authentication, non-repudiation, availability, resilience, and authorization.

Next, we analyze both acoustic and electromagnetic nanonetworks in terms of these properties and list possible attack types and answers to them. We further list several possible algorithms that fulfill some key requirements to demonstrate possible but maybe insufficient approaches.

Finally, we discuss the entirely new molecular communication networks based on either regular molecules or DNA-based message molecules. In that context, the classical algorithms are not applicable as the data rate usually does not allow for encryption. Transmitting a sufficiently strong key alone would sometimes take minutes, let alone the payload of the message.

As a result, other methods are necessary and we first demonstrate the infeasibility of the standard approaches. This also includes technologies like *One-Time Pads* (OTPs) that are provably secure but do not work with either molecules or DNA. We then explore potential technologies that might still work at the nanoscale, especially given molecules or DNA messages as a communication medium. Among them are simple avoidance protocols that expose fewer weaknesses and *steganography*. Unlike classical systems, steganography attempts to *hide* the data among noise or artificial junk data instead of rendering it incomprehensible. Thus, the attacker might suspect that communication is taking place but is unable to tell when or what part of the "molecule cocktail" actually contains the useful information.

Nanonetworks: The Future of Communication and Computation, First Edition. Florian-Lennert A. Lau.

11.1 Nanonetwork Safety Analysis

We start this chapter with an analysis of two popular medical scenarios and the suggested applications. They are the immediate treatment of diseases, while it is still emerging and the continuous monitoring of health parameters is connected with the reporting of that data in case anomalies occur. Based on these, we can suggest several possible nanonetwork architectures and the respective weaknesses they might face.

Both of these scenarios may be solved in various ways, and each of them offers its own unique strengths and weaknesses. As we cannot know all possible future solutions, we focus on the common ideas in the state of the art that might be subject to future change. Both scenarios have been presented in Chapter 4, and the interested reader is encouraged to look back if the knowledge is no longer present. The most relevant components are shown in Figure 11.1.

No matter what explicit application, some general components are almost always required. Among them is an external entity that might store the data and make judgment calls based on it, an interface/gateway that connects the macroscale network and the (in-body) nanonetwork (Alabdulatif et al., 2023). There are likely many gateways and even more nanodevices that participate in some proposed applications. That said, as all the nanodevices have limited memory, an attacker might gain next to no knowledge in attacking them – the interface is a much better candidate for acquiring data.

Next to the devices themselves, we also have at least two conceptual communication channels that are necessary. First, we have the connection between the macroscale and the gateway, and second, we have the connection between the gateway and the devices as well as between the nanodevices themselves. Conceptually speaking, these channels are fundamentally different in their parameters. While the macro–micro link is likely always based on electromagnetic technologies like Bluetooth or WLAN, the micro–nano or nano–nano links are very heterogenic. They might be implemented by any of the previously presented communication technologies or even combinations of them. The biocyber interface must be able to understand and accumulate data from all the different sources.

11.1.1 Classical Attack Types

Based on the components, we can now analyze them in terms of their classical weaknesses. The most obvious attack types target the classical components of the network. These attacks that are special to the nanoscale are discussed later.

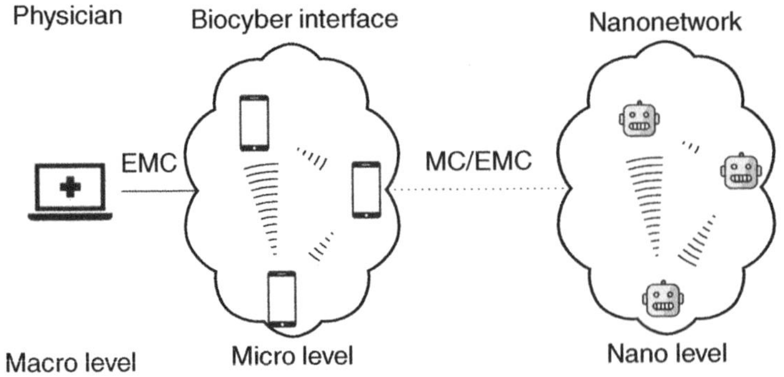

Figure 11.1 Different components of a nanonetwork that may be subject to attacks (Stelzner et al., 2016a). *Source*: Stelzner et al. (2016a)/with permission of Association for Computing Machinery.

First, it is possible to attack databases, the communication devices of physicians, various macroscale communication links, or the biocyber interface itself. Since most of these use established technologies, the corresponding attack types and solutions apply. On the most general levels, an attacker might be able to

- read parts of or the entire encrypted/decrypted messages,
- manipulate parts of or the entire encrypted/decrypted messages,
- gain control over parts or the entire system (deactivating, fake data generation, ransomware, harmful "malfunctioning"...), or
- impersonate one or more of the participants.

This list is exhaustive in the sense that there are no more additional objectives an attacker might have. Specific attack types vary in terms of the detailed implementation of such attacks. As an example, an attacker might gain full access to the system by means of zero-day exploits, planned backdoors, social engineering, or other approaches. Nonetheless, the result is always the same: The attacker can control the system or parts thereof as if he were the owner. The same is true for reading/manipulating data messages.

Furthermore, it is important to keep in mind that there is only a limited number of goals an attacker might have. They are either to control the behavior of an individual or to gain some kind of monetary advantage. However, the details of how this is achieved may vary too.

11.1.2 Classical Secure System Properties

From the given attack types, we can derive a number of desirable properties of a secure system (Zafar et al., 2021). Classically, they are sorted into the three categories of confidentiality, integrity, and authenticity (Menezes et al., 1996). However, we go by a more advanced scheme that also includes several additional properties. We only briefly explain them here and only go into depth when necessary. The interested reader is encouraged to select one out of many great books on the topic.

- The first property is confidentiality. Confidentiality ensures that sensitive information is only accessible to authorized individuals, institutions, or devices. Ensuring confidentiality involves mechanisms such as encryption, access controls, and data classification to prevent unauthorized disclosure or access. Confidentiality measures protect against unauthorized reading or interception of data, both while in transmission and on routers or devices.
- The next property is integrity. Integrity ensures that data or messages are accurate, complete, and unaltered. Maintaining integrity ensures the consistency and trustworthiness of data throughout its entire lifecycle. To achieve that, checksums, digital signatures, and access controls help prevent unauthorized modifications or tampering. Integrity measures verify the correctness of data and protect against unauthorized or accidental modifications.
- Next comes availability. Availability ensures that a system or resource is accessible and usable when needed. It involves preventing or mitigating disruptions that could render a system or resource unavailable. Possible attacks are message floods/*denial of service* (DoS) attacks on critical infrastructure. Additional redundancy, fault tolerance, load balancing, and a disaster recovery plan assist in maintaining continuous availability when faced with difficulties.
- Another important property is authentication. Authentication verifies the identity of users or systems that attempt to access a system. To achieve that, it is necessary to validate the credentials provided by the entity, such as usernames, passwords, or cryptographic tokens. However, in the simplest case, a *medium access control* (MAC) address already provides some authenticity.

- The next property is authorization. Authorization determines the actions or operations that authenticated entities are permitted to perform. It involves defining and enforcing access controls based on roles, permissions, or other criteria. Authorization mechanisms ensure that users or systems have the necessary privileges to perform specific actions and prevent unauthorized activities. As an example, while a doctor should be able to initiate certain medical procedures, an average person should not be able to do that.
- Next comes non-repudiation. Non-repudiation ensures that involvement in a process cannot be denied. Techniques such as digital signatures and secure audit trails help establish non-repudiation, making it difficult for entities to disclaim their responsibilities or actions.
- Finally, we have resilience. Resilience is the ability to withstand and recover from various attacks or system failures. It involves implementing mechanisms to prevent, detect, and respond to security incidents. Resilience measures include backup and recovery strategies, redundancy, fault tolerance, and incident response plans.

A system that fulfills all these properties is by conventional means sufficiently secure. Such systems are at least somewhat secure, even if there are yet undiscovered attack types and problems with social engineering. That said, it is unclear if and how all those properties may be implemented at the nanoscale. While some are comparably simple, others may be impossible to ensure given the limited available resources.

11.2 Attack Types

We now come to a list of specific attacks that have emerged over the years. All of them can be sorted into the four previously presented categories but vary greatly in their implementations. Some of them affect the messages, some of them affect the nanodevices, and yet others are *side channel* attacks that bypass the security models by, e.g. social engineering.

We start this section with an introduction on how to gain access to the physical systems themselves as that might not be an easy task with nanonetworks. We then discuss several universal attack types that all systems may be affected by. Next, we focus on electromagnetic and acoustic systems specifically, followed by molecular and DNA-based systems.

11.2.1 Gaining Physical Access

In some cases, nanonetworks behave similarly to regular systems and possible attacks are more or less identical. As an example, it might be possible to target the gateway/biocyber interface and thus proceed as usual. However, this type of attack only allows for indirect information from/above the system as the nanonetwork itself is still "hidden" behind such interfaces. Additional steps are necessary if an attacker wants to gain full control or at least insight into such systems.

The first challenge in attacking any nanoscale system is gaining physical access to the system itself. As mentioned before, many new resource constraints apply to such systems which make them naturally more difficult to attack. In many cases, there is either no remote access to a system at all once it has been deployed or whatever access there is might be of a very limited nature.

As a result, gaining insight into, e.g. a medical nanonetwork requires additional steps. In the case of electromagnetic or acoustic nanonetworks, it might be possible to extract data in non-invasive ways. However, gaining access to molecular communications or DNA-based communication usually requires obtaining a physical sample of the communication medium. From such a sample, parts of or the entire communication may be inferred.

Reading either clear or cipher texts from the system might require the attacker to obtain a blood or other tissue sample. Since many nanonetwork applications are medical, this means that the person might be attacked directly or indirectly by targeting any part of a blood test chain. In any case, getting access to a physical blood/tissue sample can be much harder than simply intercepting digital messages and it is much more difficult to remain unnoticed.

Manipulating messages or systems might be a bit simpler but is still more difficult than with classical systems. In this case, a substance, nanodevices, or signals must be artificially introduced into the nanonetwork/host body. Like with extracting information, it might be necessary to gain access directly to the bloodstream or tissue of the target. This would mimic the way other nanoscale systems are administered in the first place.

Other ways would involve hiding nanodevices or messages in, e.g. food items that the target ingests unknowingly. However, this involves the problem of the first path filtering through the liver. Whatever enters the human body is first broken down in the liver to avoid intoxication. As a result, many molecules are broken down into smaller pieces and might lose their intended functionality. This may be avoided by hiding the devices/messages in, e.g. a liposome to avoid degradation. Other approaches could be motivated by certain drugs that have "extras" attached to them that get broken down in the liver, resulting in the desired structure and maybe some waste products.

Gaining a (limited) degree of control over a nanonetwork requires similar steps. However, depending on the degree of desired control, such an attack might be simpler or more difficult to perform. If the goal is just to stop the nanonetwork from functioning properly, it might be much simpler than gaining full remote access. As an example, molecular communications might be disabled by introducing channel noise. DNA-based networks might be disrupted by manipulating the ambient temperature and electromagnetic/acoustic communication could be interrupted by introducing interference signals. Full control, however, requires bypassing any security measures and the same might be true for attempts at impersonating a system or person.

11.2.2 Universal Attacks

While some types of attacks are unique to the type of nanonetwork, some are universally applicable. Among them are eavesdropping, message manipulation, and DoS attacks. In the following, we briefly introduce different, sometimes specific attack types and suggest countermeasures. However, the involved algorithms are explained in the section thereafter.

11.2.2.1 Message/Cipher Eavesdropping

The most well-known type of attack on both nodes in a network and on communication channels themselves is *eavesdropping*, where an attacker gains access to either the cipher or the clear information.

In unencrypted networks, attackers can intercept all data, including login data and other critical information. *Encryption* prevents this by altering messages in such a way that they no longer openly display any information about the original message. Any attacker needs additional information to access the original content of the messages.

11.2.2.2 Injection Attack

Injection attacks can be manifold and aim at manipulating encrypted or plain text network traffic. As an example, it might be possible to identify even encrypted packages as commands to another system. That intercepted message may then be duplicated to interfere with the target system.

It is also possible to change parts of the messages themselves or introduce entirely new messages, thereby undermining the authenticity of the original message. This may be prevented by signing message procedures like MD5 or other sufficiently complex cryptographic hash functions that generate a unique string of text that serves as an identifier. The now-fingerprinted message is much harder to interfere with unknowingly.

11.2.2.3 Denial of Service

The next type of attack that universally targets all kinds of networks is a *denial of service* (DoS) attack, where an attacker aims at interfering with the general availability of an application. In this case, he is not so much interested in gaining information or access, but in causing either economic or real harm – this is especially true for medical nanonetworks.

There are at least three types of attacks that would lead to a network's reduced availability:

- message flooding,
- resource depletion attacks, and
- request overloading.

If an attacker has gained sufficient control over a network, he may manipulate network traffic in a way that interferes with the normal function. As nanonetworks usually require multi-hop communication, each new message has to be forwarded through the network, thereby occupying the communication channel. During that time, it might be difficult or impossible to send other messages, thereby severely restricting access to such a system.

In the same or similar ways, it is possible to deplete the resources of a nanonetwork by forcing it to perform "useless" tasks until the internal batteries are depleted or the energy harvesting mechanism is no longer able to keep up. Next to energy itself, it might also be possible to deplete other resources that are crucial for a network to function properly and that are of a limited nature.

The last type of attack is the DoS attack as we know it, where individual devices like the gateway are flooded with requests so that the entire system is inaccessible for the time being. This may be achieved by either filling up the request buffers of the devices or by completely occupying the communication link that leads up to the nanonetwork. Those links usually have a more limited bandwidth than the rest of the network it is connected to.

11.2.3 Attacks on Wireless Nanonetworks

Sensor networks and small nanonetworks exhibit several common traits (Zafar et al., 2021). They both operate as ad hoc networks comprised of interconnected (sensor) nodes. One notable challenge they face is the lack of distinct device identification, making it difficult to uniquely identify individual devices within the network. Furthermore, these networks tend to possess a substantial amount of redundancy, which can lead to a high failure rate.

The application of sensor networks extends beyond civilian domains and finds relevance in military operations as well. Given the sensitive nature of the information transmitted within these networks, robust communication security is crucial. Consequently, considerable research efforts have already been dedicated to addressing security concerns in these networks.

We now discuss a number of attack types that are more or less unique to wireless networks of any kind that may affect nanonetworks in the same way as WSNs. The "uniqueness" is due to the fact that wired connections are likely infeasible at the nanoscale. Yet, the mentioned attack types do also work in a wire-connected network if it operates at high speeds.

11.2.3.1 Black Hole Attacks

If a node offers a shorter route to the destination than others, more messages will be sent through it. An adversarial node can exploit this by falsely offering a very short route and not forwarding any messages that pass through it. This effectively cuts off all nodes that intended to send messages through this route from the network. This type of attack is known as a black hole.

11.2.3.2 Wormhole Attack

A wormhole attack aims to extract or modify network data. An attacker must control multiple nodes that offer a shorter route to the destination than other nodes. As a result, many messages flow through this route where they can be altered or intercepted. Such attacks only work on routing protocols that take route lengths into account.

11.2.3.3 Replay Message Attack

In a message replay attack, the attacker replays valid messages. This is possible when an attacker eavesdrops on and stores messages. Such attacks can be prevented using unique timestamps. If a message is too old, it is discarded in such systems. Implementing this method in nanonetworks is not easy since universal time measurement for individual devices poses challenges.

11.2.3.4 Man-in-the-Middle Attack

While *man-in-the-middle* (MITM) attacks are technically applicable to all kinds of multi-hop networks, they mainly apply to wireless nanonetworks as molecular communication networks might be able to communicate directly in closed systems like the blood flow.

A MITM attack is a cyber-attack where an attacker intercepts and alters the communication between two parties without their knowledge. By inserting themselves between the sender and recipient, the attacker can eavesdrop, modify, or manipulate the messages being exchanged. The MITM attack acts as the communication partner and may thus gain access to different kinds of information. Encryption, secure protocols, and caution with untrusted networks help mitigate the risk of MITM attacks.

11.2.3.5 Malware Attack

Finally, we come to several attacks that work on the hardware of the devices themselves. In this case, malicious code is applied to the devices themselves. In the context of nanonetworks, this is usually only possible in a very limited manner as nanodevices themselves are often only *configurable* and not fully programmable. Thus, this type of attack may mainly affect the gateway devices.

Malware is typically used to gain some degree of control over a system. Among them are botnets, a special kind of Trojan horse/rootkits where the attacker unknowingly gains access and control over a system, and ransomware, where the attacker encrypts parts of or the entire system in exchange for money. Another option would be the use of keyloggers to gain additional confidential information.

11.2.3.6 Device Tampering

Finally, it is also possible to directly interfere with the hardware of a nanonetwork. When it comes to the nanodevices themselves, it is only really possible to interfere with the production process. Once deployed, it should be almost impossible to make additional changes as devices are mainly configurable and not universally programmable. The only realistic implementation of this attack type at the nanoscale would be the introduction of a sufficient number of additional devices that interfere with the existing ones.

The gateway, however, offers more possible vulnerabilities due to its size. In this case, it might indeed be possible to make targeted changes to the hardware and thus alter the behavior of a nanonetwork adversely.

11.2.4 Attacks on DNA/Molecular Nanonetworks

Attacks on molecular and DNA-based communication systems are fundamentally different in nature as the underlying communication medium uses other physical effects. As a result, some established attack types are impossible, but new attack types also exist. As an example, molecular communication inside the human bloodstream likely does not require multi-hop communication and a MITM attack is likely much more difficult due to the high communication latency and low bandwidth. Similarly, wormhole or black hole attacks will be more difficult to perform. Yet, the most general attack types of eavesdropping, message manipulation, and denial/delay of service are still possible but likely more difficult to perform.

11.2.4.1 Attractant/Repellant Attacks

Some types of molecular communication rely on attraction or repelling effects to increase the chance of detection by other nanodevices. These mechanisms usually monitor the concentration of molecular markers that indicate the proximity to other devices. As molecular communication of this type only infers the actual positions of other devices, it is possible to modify the respective substance's concentration to redirect the communication. As a result, it might be possible to completely interrupt or at least delay the communication.

11.2.4.2 Molecular DoS/Congestion Attack

Similarly, it is possible to influence the chemical balance in other ways. In fact, once an attacker knows which molecule is used for communication, he could interfere with its concentration to modify the bits themselves or "congest" the entire channel with junk information. This is both true for simple molecular communication and DNA-based communication.

11.2.4.3 Chemical/Physical Disruption

In the same way, it is possible to disrupt the communication channel in various ways. Unlike electromagnetic communication, various environmental variables have to be "just right" in order for it to work. As an example, increasing the environmental temperature results in faster-moving molecules which might interfere with the device synchronization. In the case of DNA-based communication and computation, the effect of temperature changes is even more severe as bindings might break or form in unintended ways which might render the nanonetwork completely useless.

There are several other environmental variables like the pH, other molecules, electrical charges, and many more. All of these have the potential to interfere with communication that is already slow and prone to errors. Thus, molecular communication protocols have to be designed with extreme care to mitigate all those errors and still allow for sufficiently good network services.

11.3 Securing Nanonetworks

In this section, we discuss various methods to allow the implementation of the seven desired properties for secure systems. We start with simple procedures like low-power advanced encryption standard (AES), cyclic redundancy check (CRC), and other algorithms for wireless communication systems. After that, we also propose several ideas to secure molecular communication systems in non-classic ways.

11.3.1 Low-power AES

The *Advanced Encryption Standard* (AES) is the most widely used symmetric encryption algorithm since 2002 (FIPS, 2009). It operates with key lengths of 128, 192, or 256 bits, encrypting plaintext in 128-bit blocks. Round keys are derived from the input key, which is then used to encrypt the plaintext. The number of rounds varies based on the key length.

For a 128-bit key length, there are 10 rounds and thus 10 round keys. Before the first round, the first round key is bitwise XORed with the plaintext. Then, four operations are performed on the input text for nine rounds: SubBytes, ShiftRows, MixColumns, and AddRoundKey.

- SubBytes substitutes each byte of the text with another byte defined in a lookup table.
- ShiftRows cyclically rotates the rows of the divided text.
- MixColumns linearly combines and mixes the columns of the divided text.
- AddRoundKey is the aforementioned XOR operation with the round key.

In the last round, MixColumns is not performed. This process produces encrypted text that cannot be converted back to plaintext without the original key.

There are various methods for encrypting multiple consecutive blocks. One of them is "electronic code book," where each block is encrypted individually. Consequently, two identical plaintext blocks will result in the same encrypted text.

While such a symmetric encryption method is feasible at the macroscale, at the nanoscale, this is likely not the case. Panu Hamalainen et al. described a new implementation of the AES algorithm designed to consume minimal space and energy (Hamalainen et al., 2006). It utilizes an 8-bit core while maintaining the encryption process and key length of the standard AES. However, this implementation can only encrypt and does not support decryption. Different versions are possible, each optimized for size, energy consumption, and speed.

Since the encryption speed of all known versions of AES exceeds the transmission speed of nanodevices, we can prioritize lower gate count or reduced energy consumption. Lower energy consumption is preferable if sufficient gate space is available on the device since available energy is the limiting factor for transmission speed. Thus, the area-optimized circuit should only be chosen if it does not fit on the selected nanodevice. The algorithms assume very high throughput, which may not be necessary for some applications in nanonetworks. Therefore, reducing the clock frequency can save more energy.

Messages in nanonetworks are likely to be shorter than the AES block length. To utilize the full key length, multiple messages could be combined and encrypted together using a single key. Reducing the key length would require redesigning the algorithm and may make it more susceptible to brute-force attacks, among other potential drawbacks.

While there are some alternative low-cost versions of algorithms like AES, it is unclear if they are sufficiently well suited for ensuring security at the nanoscale.

11.3.2 One-Time Pads

OTPs are a symmetric encryption method that provides perfect security when used correctly. They involve the use of a random binary string that is as long as the plaintext. If such a key is reused, an attacker can gain more and more information about the key with each additional message.

To encrypt a message, each bit of the plaintext is combined with a bit from the key using an operation like addition or XOR. This process produces the ciphertext, which is the encrypted version of the original message.

The decryption requires the receiver to use an identical key an operation to reproduce the plaintext. The security of OTP depends on the true randomness of the key, its length, and the one-time use of the key. Randomness ensures that no patterns are discernible, while length ensures that the key covers the entire message.

When these principles are strictly adhered to, OTPs provide perfect secrecy. This means that without the key, the ciphertext reveals no information about the plaintext. It is crucial to securely share the key between the sender and the receiver using a trusted and confidential channel. Deviating from the key properties or reusing the key significantly weakens the security of OTPs. That said, as the performed computation is so simple, most nanodevices should be able to perform it.

11.3.3 Cyclic Redundancy Check

A *Cyclic Redundancy Check* (CRC) is a checksum computed from the bits of a message (Peterson and Brown, 1961). It was developed by William Wesley Peterson in 1961. The checksum is computed by dividing the message by a generator polynomial, and the remainder of this division becomes the CRC that is added as metadata to messages. The receiver can divide the received message, including the appended CRC, by the same generator polynomial. If the result has no remainder, it is highly likely that the message was transmitted correctly. The choice of the generator polynomial depends on the message length and the desired error detection probability. The polynomial also determines the length of the CRC.

CRC can be implemented on hardware using shift registers, making it suitable for nanodevices with limited resources. The errors that CRC can detect depend on the chosen generator polynomial. If the polynomial has an even number of terms, the algorithm can detect errors where an odd number of bits have been changed. If the polynomial has an odd number of terms, the algorithm can detect errors where an even number of bits have been changed.

In the case of communication using TS-OOK, errors typically occur only when a 0 and a 1 are sent simultaneously by two devices. This situation could be utilized to not only detect but also correct errors by examining which bit changes lead to the correct CRC.

11.3.4 Low-power Hashing

Hash functions are another method used for integrity checking (Yuksel et al., 2004). Cryptographic hash functions provide both integrity and authenticity. By combining the message and a key, a checksum is generated, which the receiver can use to verify the message's origin and integrity. The receiver also requires the corresponding key. Yüksel et al. presented three implementations of an algorithm designed for sensor networks.

In these algorithms, the message is divided into blocks. Each block is added to a corresponding block of the key, and these block pairs are added together. The result is a checksum with the same size as one of these blocks. The receiver can then compute the same checksum and verify if the sender possesses the key and if the message has remained unchanged.

The most efficient implementation of these algorithms consumes 11.6 μW of power at a frequency of 500 kHz. This power consumption value depends on the frequency, and one-third of the incoming energy is lost during the operations. The paper by Yüksel et al. does not discuss energy consumption at lower frequencies.

The majority of the energy consumption is due to leakage current, which decreases with smaller gate sizes. Optimizing the algorithm for nanodevices could significantly reduce energy consumption, making it applicable in nanonetworks.

11.3.5 Medium Access Control

A possible method for achieving authenticity and integrity might come for "free" as a byproduct of MAC methods. MAC ensures that only a certain number of participants may use the communication channel at any given time to avoid collisions and thus errors. As a result, each participating nanodevice requires a MAC address that uniquely identifies it in the network. Without such an address, a nanodevice simply cannot participate.

It is further possible to assign those addresses only to trusted devices and thus use a "whitelist" approach. As a result, newly introduced, possibly malicious devices can only listen to or "disturb" the communication but not participate in an organized manner. If the procedure is designed properly, even non-repudiation is ensured as messages can be traced back to devices, if they are capable of possessing one or more unique identifiers. If such identifiers are not possible and shared among, e.g. groups of devices, it is not possible to trace a message or action back to a single nanodevice. The resource constraints at the nanoscale sometimes render easy problems hard to solve.

11.3.6 Gateway Security

To achieve all the other kinds of security requirements, it is likely necessary to utilize the gateway or biocyber interface. The individual nanodevices themselves are simply not powerful enough to allow for any kind of sophisticated rights management or another system. As of right now, the devices in nanonetworks will likely only fulfill one or two functions at best.

As a result, the nanonetwork itself is best hidden behind a "gateway firewall" that restricts any kind of unauthorized access to the nanonetworks – that is, for networks where such a post-deployment access/control is even possible or feasible. If access is indeed possible, even if limited, it must be ensured that all external traffic is routed through the gateway to perform additional security checks. Luckily, by the very design of basically all reference architectures, the gateway(s) is the only point that connects the nanonetworks with the remaining IoNT.

Apart from that, we can assume that the gateway itself is more computationally powerful compared to the nanodevices themselves. As a result, it is likely that the gateway, and by indirection the nanodevices themselves, can be secured by classical security measures. Even if such a gateway is much less powerful compared to conventional computers, it is still sufficient for most purposes and the expected limited number of users. For now, it is a reasonable assumption to just use the already established technologies. If that does not work for whatever reasons, it might be a good idea to gain inspiration from the early days of computer systems where resources were also scarce.

11.4 Molecular and DNA-based Security

Most of the well-known security principles cannot be directly applied to molecular communication or DNA-based molecular communication specifically (Mahjabin et al., 2023). The latency in such systems can be extremely high, and the bandwidth is likely so small that adding even a few bits to a message could lead to additional minutes of waiting time before a message is received. Current experiments succeeded in achieving data rates of a few bits per second under optimal conditions and using macroscale devices. It is likely, not possible to directly transfer those results to nanoscale systems.

It would be suitable to compare the bandwidth with submarines that are currently only capable of receiving and transmitting a few bits per minute when submerged sufficiently deep underwater. The absorption of the water is simply too high and only signals with a very short wavelength are able to reach sufficient depth. Yet, this requires radio towers and antennas of incredible size and the low frequency limits the bandwidth tremendously.

11.4.1 Infeasibility of Classical Algorithms

But first, a few words on the likely infeasibility of classical security procedures. While molecular communication systems have a much lower bandwidth and a severely higher latency, they are still likely operating in a medical and thus time-critical context. Measured data is only relevant for some time, and the amount of time necessary to reach a doctor is of the essence.

In many cases, security protocols require additional metadata to be added to any existing message sent over a network. It is not rare that this leads to a potential blow-up of those messages by hundreds or even thousands of bits in some cases. As an example, Hoteit et al. (2022) assumed messages of roughly 1000-bit length that would be completely infeasible for molecular communication. In such scenarios, it would be best to just transmit a chosen and very small number of bit strings.

As a result, methods that generate overhead, as with classical protocol stacks, are likely not suitable for molecular communication in general and the same applies to DNA-based communication.

However, DNA-based communication faces some additional restrictions as such systems are also able to compute via self-assembly. However, each and every step would manifest itself physically for every person who would be able to get a sample of the communication channel. Computations in tile-based self-assembly systems always include the entire "state history," including the keys and inputs and everything else. Thus, classical encryption simply does not work.

The most immediate demonstration of this problem can be seen with OTPs. While OTPs appear like a lightweight, computationally feasible method for nanonetworks to achieve limited security, they do not work with DNA. The input number would always be part of the resulting message molecule, and thus, it would not be obfuscated in any way. Even if the input number was *not* part of the message, the glues must still contain the information of the key, otherwise, it would not be possible to decrypt the message again.

The same problem applies to any other kind of classical encryption scheme that relies on string manipulation based on a secret. They do not work unless it is possible to "decouple" the computational steps sufficiently from the resulting message.

11.4.2 Steganography

One possible solution to deal with the problems of molecular communication security would be *steganography*, which was first mentioned by Johannes Trithemius as early as 1499 (Morkel et al., 2005). Using steganography, secret information or messages are hidden within other non-secret data. The goal of steganography is to conceal the existence of hidden information, instead of making it unreadable.

Unlike cryptography, steganography aims to hide the message in plain sight. If an attacker is not aware that a hidden message exists, they are unlikely to attempt to decode or intercept it.

Steganography techniques vary depending on the medium used to conceal the information. An example would be image steganography. In this case, the hidden message could be embedded within the pixels of an image. This can be achieved by modifying the least significant bits of the pixel values, which are typically imperceptible to the human eye. The changes are subtle enough to not significantly alter the visual appearance of the image.

Steganography can be used for various purposes, ranging from covert communication and information concealment to digital watermarking and copyright protection. While steganography can successfully hide information, additional encryption techniques are often used together with steganographic methods to ensure the security of the hidden message.

Yet, steganography might be a prime candidate to conceal information in nanonetworks. While it might not be possible to fully hide the messages, it might be possible to create enough *junk information* to delay or completely deter any attacker. When a number of messages are present at the same time in the communication medium, it is not easy to decide which one is relevant and which one is not. As a result, the effort any attacker must make to gain information could be arbitrarily high.

In this case, the fact that some messages are sent is not hidden, but enough "noise" is added to the communication channel that it is unclear when *relevant* messages are sent. The same is also applicable for electromagnetic communication but requires a lot of energy that is usually not available at the nanoscale. In this case, the communication medium would be filled with additional *noise* that makes it hard to detect where messages begin and stop. Especially with TS-OOK, this would make it much harder to detect any beginning/end frames.

In summary, molecular communication might benefit most from steganography, while electromagnetic communication might benefit from classical measures.

11.5 Summary

In summary, nanoscale cryptography faces some unique challenges and it is unknown if classical security measures can be applied to nanonetworks. While gateways are likely powerful enough, the nanodevices themselves are so limited that most security measures are either impossible to implement or infeasible to perform.

Another unique property of nanonetwork security is the limited remote access to such systems. This is an inherent advantage that shields such systems from unwanted interference. In many cases, an attacker must gain access to the living body to extract a sample from the communication channel or to introduce their own nanodevices to interfere. This is very different from classical security, and the respective countermeasures are not in the digital domain.

That said, the attack types themselves are pretty much identical as the objectives of potential attackers do not change.

To achieve some degree of security nonetheless, it might be possible to use encryption for electromagnetic and acoustic communication and steganography for different types of molecular communication. That said, since the final technologies are yet to be determined, it is unclear which measures need to be taken. Nonetheless, we can be certain that some security measures are necessary as especially medical nanonetworks operate on critical data that are highly confidential.

12

Nanonetwork Concepts and Architectures

This chapter takes all the information we generated thus far and combines it into several functional architectures for nanonetworks. It is literally all prior chapters introduced are crucial for many of the proposed networks and cannot function without them. We start with an analysis of ad-hoc networks and mobile networks to derive valuable information from macroscale networks for their nanoscale counterparts. Afterward, we recapitulate the most important peculiarities of nanoscale networks as opposed to comparable macroscopic networks. Then, we explain *Internet of Nano-Things* (IoNT) and *body area networks* (BANs) in detail, as they serve as hybrid types of networks that attempt to holistically bridge the gap between macroscopic and nanonetworks.

Next, we explain different architectures for acoustic and electromagnetic nanonetworks. After that, we have a look at nature-inspired network types that emerge from local rules only like swarms. Then, we analyze bacteria in terms of their potential to form functional nanonetworks.

Finally, we analyze molecular nanonetworks and especially DNA-based nanonetworks in depth and explain several tools for evaluating and simulating networks. As of right now, most nanonetwork concepts cannot be realized physically, with the exception of certain nature-inspired networks based on bacteria or DNA-based nanonetworks.

12.1 From Macro to Nano

When approaching the topic of very small networks, it is natural to first try and use the already available technologies and apply them to the new setting. In the context of nanonetworks, the closest match would be mobile networks that form on the fly without any or very little static infrastructure. Nanodevices are expected to move around a lot and have to adapt to sometimes rapid changes in their environments.

Ad hoc networks and *mobile ad hoc networks* (MANETs) are decentralized wireless networks that can communicate with each other without the need for a centralized infrastructure (Ebers, 2016). These networks are often formed spontaneously, on-the-fly and several aspects of such networks resemble nanonetworks. In an ad hoc network, each mobile device provides all network functionalities at once, including sensing/acting or routing/forwarding messages. These devices form a temporary network whenever they come within range. As the devices move, the network topology changes regularly and devices may join or leave the network dynamically.

Nanonetworks: The Future of Communication and Computation, First Edition. Florian-Lennert A. Lau.

MANETs have a variety of characteristics that they share with nanonetworks. Among them are decentralization, dynamic network topologies, peer-to-peer, limited range, delay tolerance, resource constraints, and specialized applications:

- MANETs operate without a central authority or infrastructure. Nodes participate in the routing and forwarding of data packets, allowing communication between devices.
- MANETs have a dynamic topology due to the mobility of devices. The network's structure changes as nodes move, join, or leave, requiring adaptive routing protocols to maintain connectivity.
- Each device in a MANET is both a sender and a receiver. Intermediate nodes may also participate in forwarding packets to reach their destination.
- The range of communication between nodes in a MANET is typically limited, often within a few hundred meters or less. Beyond this range, devices cannot directly communicate without the help of intermediate nodes.
- MANETs rely on dynamic routing protocols that discover and maintain routes between nodes (Hail, 2018). These protocols determine the most efficient paths for data transmission, considering factors like hop count, signal strength, and energy efficiency.
- Devices in MANETs are often resource constrained in terms of battery power, processing capabilities, and memory. This limitation affects the design of routing algorithms and protocols to optimize energy usage and minimize overhead.
- MANETs can solve a variety of unique problems, such as disaster recovery operations, military operations, sensor networks, and collaborative environments where infrastructure-based networks are unavailable, impractical, or inefficient.

All of the above-mentioned characteristics carry over into nanonetwork research, but there is still a number of unique characteristics that only occur at a very small scale. Nanonetworks operate in unique and challenging environments that set them apart from traditional networks. These special environments present a range of distinctive characteristics and constraints, shaping the design and operation of nanonetworks.

First, nanonetworks often deal with extreme participant numbers. Unlike conventional networks with a limited number of nodes, nanonetworks can comprise a massive number of nanoscale devices, leading to complex interactions and communication patterns. The sheer scale of participants introduces scalability challenges and necessitates efficient protocols for information exchange.

Another defining characteristic of nanonetworks is the high topological dynamism. The network topology is constantly changing as nanodevices move within the environment, making it essential to adapt to dynamic connections. This dynamism poses challenges in terms of routing, resource allocation, and maintaining connectivity, requiring robust algorithms and protocols that exceed the aforementioned requirements of MANETs by far.

Moreover, nanonetworks are subject to stringent time and space limitations. Nanodevices have limited computational capabilities and memory, imposing constraints on data processing and storage. Additionally, the physical space available for communication and interaction is often restricted, requiring efficient utilization and optimization of resources. The same is also true for the availability of precise/accurate timing mechanisms, random number generators, and partially as a result, the ability to move or sense/manipulate environmental parameters.

Heterogeneity is another critical aspect of nanonetwork environments. Nanodevices can vary in terms of size, capabilities, and communication modalities. The heterogeneous nature of these devices presents challenges in achieving interoperability and seamless communication, demanding adaptive protocols and mechanisms to accommodate diverse characteristics.

Unlike conventional networks with centralized control, nanonetworks have minimal central control instances. Decision-making and coordination are distributed among the nanoscale devices themselves, making it crucial to develop self-organizing and self-regulating mechanisms to enable efficient operation and collaboration. Nanonetworks have to make decisions based on local information only as the often extreme number of participants makes accurate knowledge about the global state infeasible.

Biocompatibility is another critical consideration in nanonetworks. Nanodevices must often operate without causing harm or adverse effects to the surrounding biological system. This necessitates the design of biocompatible materials and protocols, ensuring the safe integration and operation.

Energetic availability poses yet another major challenge in nanonetwork environments. Nanodevices typically have limited energy resources, making energy efficiency a crucial concern. Energy harvesting, power management, and energy-aware protocols are essential for sustaining the operation of nanonetworks over extended periods.

Finally, waste disposal is a significant aspect of nanonetwork environments. Nanodevices generate waste and byproducts during their operation, which need to be properly managed and disposed of to prevent environmental contamination or adverse effects on the host biological system. This is especially true for the nanodevices themselves that have to be disposed of once they are no longer operational or no longer desired.

In conclusion, the special environments of nanonetworks encompass extreme participant numbers, high topological dynamism, time and space limitations, heterogeneity, minimal central control, limited control capabilities, biocompatibility, energy availability, and waste disposal concerns. Addressing these unique challenges requires the development of tailored solutions, protocols, and mechanisms to enable efficient and reliable communication and operation in nanonetworks. The following nanonetwork types have to take all of them or at least a subset of those considerations into account based on the desired application.

12.2 Nanonetwork Role Models

We now summarize the components and aspects from prior chapters into fully functional concepts for nanonetworks. We start with network types like IoNT and BANs that are somewhat similar to macroscopic computer networks. Afterward, we analyze acoustic and electromagnetic nanonetworks as well as "networks-on-chips" as special cases of EMC networks. Finally, we have an in-depth look at nanonetworks based on molecular communication and DNA-based nanonetworks in particular.

12.2.1 IoNT

The *IoNT* is an emerging paradigm in nanonetworking and the *Internet of Things* (IoT) alike. It focuses on integrating nanoscale devices into the broader IoT ecosystem. The "classic" IoT is a network of physical devices, vehicles, appliances, and other objects embedded with sensors, software, and connectivity capabilities. They usually collect and exchange data over the internet. The IoNT further incorporates nanodevices into the IoT.

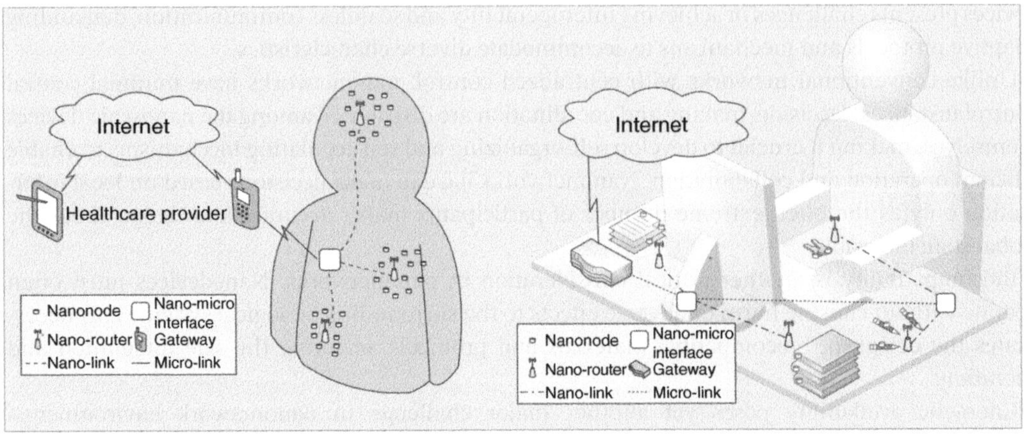

Figure 12.1 IoNT reference architecture, as proposed by Akyildiz in 2010 Akyildiz and Jornet (2010). *Source*: Akyildiz and Jornet (2010)/With permission of IEEE.

The IoNT solves problems to which classical IoT devices have no good answer as of right now. By integrating nanodevices into the IoT, it is possible to collect data at an unprecedented level of granularity, which enables new sensing capabilities and creates novel applications across various domains.

The IoNT can be described by a set of usually mandatory components and their interaction as a *reference architecture*. Figure 12.1 shows an overview of the architecture. There is a set of established IoT devices. In addition to that, there are new nano-routers, nanonodes, and nano–micro interfaces. As we learned in Chapter 6, they are necessary to bridge the gap between the nanoscale and the macroscale. While many of the components resemble macroscale devices, the nano components are usually envisioned to be inside of the human body. There, they might gather medical data and supply them to healthcare providers.

While there is no standardized reference architecture for IoNT yet, it is possible to outline the typically involved components and participants. They are nanodevices, gateways, channels, and cloud infrastructure:

- At the core of IoNT are the nanoscale devices themselves. These devices can be nanosensors, nanoactuators, or computational units that operate at the nanoscale. They are responsible for collecting data, performing computations, and executing tasks within the IoNT system. Nanoscale devices have limited resources and unique properties, requiring specialized communication protocols and interfaces.
- Nano gateways serve as an interface between nanoscale devices and the larger IoT infrastructure. It acts as a bridge, aggregating data from multiple nanoscale devices and translating their communication protocols into standard IoT protocols that can be understood by other IoT devices or the cloud. The nano gateway may also handle security, authentication, and data preprocessing tasks.
- Since nanoscale devices are extremely small, traditional communication protocols may not be directly applicable. Nanoscale communication techniques, such as molecular communication or terahertz communication, are employed to establish connectivity between nanoscale devices and larger IoT devices. These techniques often leverage wireless, optical, or even molecular signaling methods to transmit and receive data.

- The cloud infrastructure forms the backend of the IoNT system. It comprises servers, storage, and computing resources that allow for complex analytics and long-term storage. The cloud further provides scalability, persistent storage capabilities, advanced analytics, machine learning, and artificial intelligence algorithms to gain insights into the massive amount of data.

The IoNT supports various applications that are unique for nanodevices. These applications can span different domains, such as healthcare, environmental monitoring, agriculture, or industrial automation. Examples include real-time monitoring of physiological parameters, pollution detection, precision agriculture, or smart manufacturing. Yet, most proposed applications nowadays are of a medical nature.

Without a holistic infrastructure, the data collected by in-body nanonetworks is likely not accessible. As a result, some integration of nanonetworks into larger systems is necessary.

12.2.2 Body Area Networks

As we learned in Chapters 3 and 6, a typical envisioned nanonetwork architecture is BANs. BANs are a type of WSN that consists of wearable or implantable devices that monitor and collect data from the human body. These devices are typically equipped with various sensors, such as accelerometers, heart rate monitors, temperature sensors, and electrocardiograms.

The primary purpose of BANs is to enable continuous and real-time monitoring of physiological parameters and other health-related information. The collected data can be used for medical purposes, such as remote patient monitoring, early detection of health issues, and providing personalized healthcare services. BANs are a *component* of the IoT and are sometimes envisioned to be part of the IoNT. The IoNT enables BANs the seamless integration of body sensors with other healthcare systems, medical professionals, and data analysis platforms.

An example of a BAN with nanoscale components can be seen in Figure 12.2. The displayed network consists of a variety of possible devices on all scales. The small dots represent the nanoscale components in a BAN. As it is unclear how they communicate with a larger infrastructure, they are envisioned to send their data to a suitable gateway (device at the armpit and hip) that gathers, filters, and preprocesses the data and communicates it to the microscale or directly with devices of a regular BAN.

The devices on the ankles, wrist, hip, and right biceps are regular macroscale devices of a BAN. They include sensors for measuring pulse, blood pressure, or simply walking patterns. There are also implantable devices included in a BAN, in this case, situated at the heart region where a pacemaker could be situated. All kinds of other implantable or wearable devices could be included in the BAN. Interestingly, some of the implantable devices already use energy harvesting techniques based on fuel cells that we have discussed in Chapter 9 to avoid frequent invasive battery changes.

Finally, some of the larger devices may transmit their collected data to the internet, as indicated by the laptop to the top left. This connection allows BANs to participate in the larger IoNT infrastructure.

12.2.3 Swarm-Based Networks and Self-organization

One of the main idols for resource-constrained networks is nature inspired. For example, certain insects like ants or bees, birds, fish, and many mammals have developed simple strategies that improve their chance of survival tremendously. Those strategies are often the result of millions of years of evolution and sometimes a primitive "culture" of passing down behaviors from one generation to another.

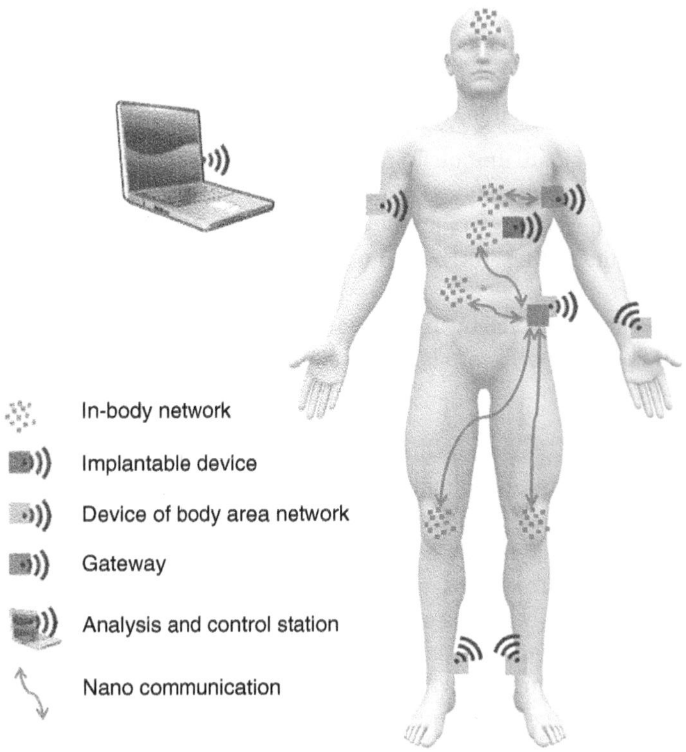

Figure 12.2 Multi-layered body area network that includes macroscale, microscale, and nanoscale devices that are used for monitoring and potentially improving a patient's health. *Source*: Dressler and Fischer (2015)/with permission of Elsevier.

Interestingly, none of these species follow complicated behavioral protocols or algorithms to form networks. They do so based on often surprisingly simple rules or algorithms. For example, three rules are sufficient for these animals to form swarms, flocks, schools, or herds – they are as follows:

1. move toward the center of those you see around you (cohesion),
2. move away when someone gets too close to you (separation), and
3. move roughly in the same direction as your neighbors (alignment).

These three rules are sufficient to ensure that a group of animals stays together while not interfering too much with each other. Two examples of how to implement this behavior can be seen in Figure 12.3 on the left. We focus on one individual in the center of the figure. On the left, there are concentric zones that represent the rules. The innermost zone implements repulsion and may not be entered by others. The next zone implements "alignment," and the center animal tries to roughly imitate their movements. The outermost zone implements "attraction," and the center animal tries to stay within a distance. The same behavior can be achieved by dividing the other animals of a group into sets of nearest neighbors and so on (see the right figure). The three rules are sufficient for a stable network to form.

Such swarm behaviors have been an inspiration for designing and understanding nanonetworks. Swarm behavior describes the collective behavior of a group of entities that collaborate, exhibiting emergent properties that exceed the capabilities of the individuals. Nanonetworks aim to achieve similar collective behaviors through distributed communication and coordination. There are

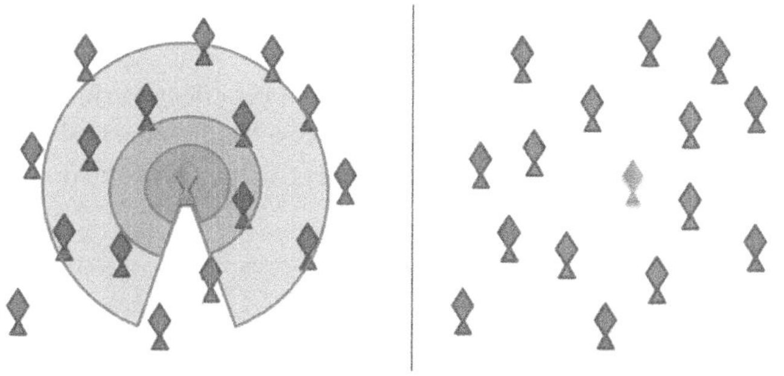

Figure 12.3 Example of a swarm-based network implementation using concentric zones. *Source*: Murphd84/Wikimedia Commons/Public Domain.

many interesting properties that might be achieved based on local knowledge and interaction alone without requiring any central controlling entity.

Swarms utilize self-organization, where individual agents follow simple rules or local interactions to collectively achieve complex behaviors. Similarly, nanonetworks strive for self-organization by allowing individual nodes or agents to communicate and make decisions based on just local information. By sharing information and coordinating actions without centralized control, nanonetworks can achieve scalability, adaptability, and fault tolerance.

Furthermore, Swarm behavior relies on decentralized decision-making. Nanonetworks also adopt a decentralized approach, where nodes make autonomous decisions based on their local observations and interactions. Decentralization allows nanonetworks to be robust and resilient as individual nodes can adapt and respond to changes in the network or environment without relying on a single point of failure.

Communication is vital in swarms as it enables individuals to share information, coordinate actions, and maintain cohesion within the group. Similarly, nanonetworks rely on communication protocols to exchange information among network nodes. By sharing data, status updates, or control signals, the nodes in nanonetworks can collectively perform complex tasks.

Swarms are also adaptive and error resistant when it comes to changes in the environment or group dynamics. They can reconfigure their behavior or formation. Similarly, nanonetworks strive for adaptability and flexibility, allowing nodes to dynamically adjust their behavior, reorganize, or optimize resource usage based on changing conditions or network requirements.

Furthermore, swarms can dynamically scale up or down in size without significantly impacting the overall behavior. Nanonetworks also aim for scalability, enabling the network to accommodate a large number of nodes while maintaining efficient communication and coordination.

Finally and most surprisingly, swarm behavior often leads to emergent properties, where the collective behavior of the group is more than the sum of its individual parts. An example would be the search algorithm that colonies of ants use to collect food. By dynamically marking and following paths and otherwise performing random walks, complex ant networks may emerge. Through local interactions and coordination, a nanonetwork can display behaviors beyond the capabilities of individual nodes.

The advantages of such procedures are quite clear. Organized group behavior always increases the chance of survival of animals. If a group of animals attacks or flees at once, individual prey or predator have a more difficult situation to deal with. The same is true for the human military as it evolved over the ages. Some European tribes developed "shield walls" where every soldier stood in a line protecting each other and advancing slowly. Another row of people behind them was equipped with weapons, and groups without such a strategy had almost no chance of survival, even if they outnumbered them.

In summary, the naturally emerged swarm strategies are often so simple yet effective that they can be implemented on nanonetworks or sometimes individual nanodevices. Nature nearly always has to function in a resource-constrained environment, and certain aspects are perfect role models for nanoscale systems.

12.3 Nanonetworks

With all that we have learned thus far in mind, we can now precisely define the environments and different necessary components as well as their specific implementations in detail. Each chapter has provided us with the necessary information that allows us to be as precise and accurate as possible in, for example, models for simulation or simply to communicate knowledge adequately. We start this overview with wireless network types and later analyze nanonetworks based on molecular communication.

12.3.1 Acoustic Nanonetworks

As we have learned in Chapter 6, acoustic nanonetworks focus on communication and networking among nanodevices using sound waves as the medium of information transfer. The idea behind acoustic nanonetworks is inspired by the advancements and problems in nanotechnology, where it has become possible to fabricate primitive nanostructures with dimensions on the order of nanometers. However, due to their small size, traditional wireless communication methods are possibly not suitable for reliable and feasible nanonetwork formation.

Acoustic waves offer a potential alternative for communication at the nanoscale. Acoustic signals can propagate through various media at different speeds and can interact with nanoscale devices effectively. They may be able to avoid some of the problems electromagnetic nanonetworks face.

Table 12.1 shows a list of relevant parameters and restrictions that acoustic nanonetworks have to accommodate for. While not every parameter might be relevant for each particular type of nanonetwork, it is easier to understand when each table of parameters lists the same entries.

As of right now, most publications envision acoustic nanonetworks to work similarly to the CMOS technologies we are used to. They assume that further miniaturization will enable nanoscale devices that possess more or less identical capabilities as macroscale computers but with limited resources. As a result, possible construction materials are CMOS circuits or CNTs that may be used to form circuits while providing nearly identical functionality.

As the devices themselves are tiny, the available memory is also limited, and as a result, little computational power is available. As of right now, transistors are usually 20 nm in diameter and it seems difficult to reduce their size further. There is also a hard limit for further miniaturization as a certain number of atoms is necessary to represent binary information in a stable form. A generous assumption would be that CMOS/CNT-based nanodevices might be able to compute simple circuit problems with constant or no memory requirement or problems that require logarithmic amounts of memory at best. That said, the number of participants might compensate to some degree for the limited resources as each device possesses at least a tiny amount of resources. Assuming that a nanonetwork with 10 mm range requires 64 000 participants to ensure connectivity, an acoustic nanonetwork might require 100 times as many (Stelzner and Traupe, 2019). This is due to the assumed communication range of around 100 μm (Freitas Jr., 1999).

Table 12.1 List of currently expected parameters and constraints current concepts of acoustic nanonetworks might face.

Parameter	Value	Parameter	Value
Construction material	CMOS/CNT	Energy source	Electricity/nanogenerators
Participants/body	6 400 000	Energy requirement	Low
Computational power	Low	Timer	Quartz/CMOS
Memory size	Constant/logarithmic	Random generator	CMOS/CNT
Communication range	100 μm	Safety/security	Sync/async
Communication speed	1500 m/s	Biocompatibility	Low
Bandwidth	Medium	Life expectancy	Unknown
Movement	CMOS/CNT	Waste management	Unknown
Movement type	Molecular motors + passive	Error vulnerability	High
Sensors/actuators	CMOS/CNT		

Source: Florian-Lennert A. Lau.

The speed of the communication itself is limited by the speed of sound in a watery medium and might vary but is expected to be around 1500 m/s (Hogg and Freitas Jr., 2012). Figure 12.4 illustrates the overall working of acoustic communication. Each robot represents a large group of nanodevices with similar but limited capabilities. Due to the small size, they are only able to observe and influence their direct surroundings in a limited manner and thus possess only local information about the network state. The bandwidth itself however might be around 200 kHz (Hogg and Freitas Jr., 2012), which is a considerable amount at the nanoscale. Due to error mitigation strategies, it might be necessary to reduce the effective data rate by introducing redundancy to messages.

Like all nanoscale devices, acoustic nanonetworks are subject to all kinds of passive movements including diffusion and flow. In addition to that, some scientists envision acoustic nanodevices to actively move via modified flagella or simply small electrical motors.

The necessary energy for all the desired actions and observations might pose a bigger problem. Most scientists assume an energy harvesting mechanism based on the piezoelectrical effect or chemical reactions. They might be able to provide around 10 pW (Hogg and Freitas Jr., 2012). Given generous assumptions about conversion loss and other sources of error, acoustic nanodevices might transmit data at 100 pW if they are only active 1/10 of the time and possess an appropriate means of storing energy.

As with many applications, acoustic nanonetworks are envisioned to perform primarily medical tasks like health parameter monitoring (Hogg and Freitas Jr., 2012). As a result, some means of timing the observations are necessary and those are likely based on small quartz.

It might further be necessary to conserve energy by employing SLR with a combination of ring-saving, as presented in Chapter 9. For this to work, acoustic nanodevices also require a means to generate random numbers. Many algorithms for synchronization and density approximation, as well as many other tasks, require random numbers. The same is also true for cryptographic algorithms that ensure security.

The biocompatibility of acoustic nanodevices is similar to electromagnetic devices. It is unknown as of right now how the human body might react to CMOS or CNT-based technologies. Furthermore, it is necessary to use sufficiently weak sound waves to avoid string heat generation.

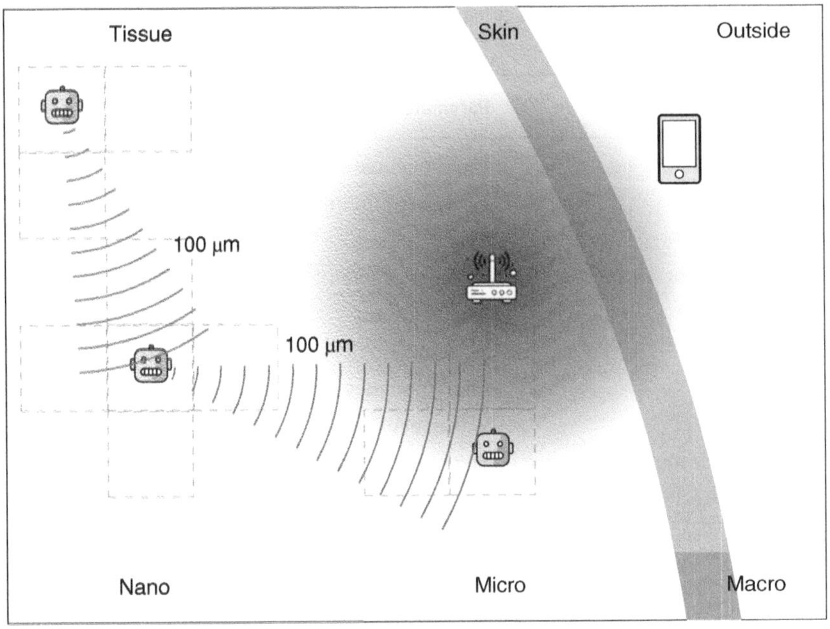

Figure 12.4 Example of an acoustic nanonetwork. The devices communicate over a range of approximately 100 µm using mechanical waves. *Source*: Florian-Lennert A. Lau.

Finally, life expectancy (e.g. due to device failures) and waste management are currently unknown as no real acoustic nanoscale devices exist as of right now.

12.3.2 EMC Nanonetworks

Electromagnetic nanonetworks are likely the most often envisioned type of nanoscale network. The reasons for this are at least twofold. First, using a well-established technology allows us to utilize a vast amount of already established research from macroscale networks as they function based on essentially the same technology. Second, it is much easier for scientists to enter a domain if they can use their existing expert knowledge. As it is often said, we are standing on the shoulders of giants and do not create science anew on our own.

Nonetheless, given radically different external circumstances, it is sometimes necessary to come up with radically new paradigms. The old and tested technologies might face irreconcilable limitations. That said, the final communication technology has yet to be determined and EMC nanonetworks are a viable candidate just like the others.

Table 12.2 specifies the capabilities and materials of electromagnetic nanonetworks in detail. As before with acoustic nanonetworks, We now try to define electromagnetic nanonetworks as accurately as possible.

Overall, electromagnetic nanonetworks likely function based on similar construction materials. Most literature suggests that they function based on either further miniaturized CMOS technology or similar CNT building blocks.

As the communication distance is a bit higher, fewer total participants are necessary to achieve full connectivity throughout an average human body. All of these likely have very little memory of either constant or logarithmic size. As a result, electromagnetic nanonetworks only have limited and local knowledge about parts of the total network.

Table 12.2 List of currently expected parameters and constraints current concepts of electromagnetic nanonetworks might face.

Parameter	Value	Parameter	Value
Construction material	CMOS/CNT	Energy source	Electricity/nanogenerators
Participants/body	320 000	Energy requirement	High
Computational power	Low	Timer	Quartz/CMOS
Memory size	Constant/logarithmic	Random generator	CMOS/CNT
Communication range	2 mm	Safety/security	Sync/async
Communication speed	0.75 * speed of light	Biocompatibility	Low
Bandwidth	Medium	Life expectancy	Unknown
Movement	CMOS/CNT	Waste management	Unknown
Movement type	Molecular motors + passive	Error vulnerability	High
Sensors/actuators	CMOS/CNT		

Source: Florian-Lennert A. Lau.

Unlike acoustic nanonetworks, electromagnetic networks have a much higher communication speed of roughly 0.75% of the speed of light. Similarly, the potential bandwidth is much higher as the terahertz band allows for the encoding of a tremendous amount of information in the frequency. The additional possible redundancy may allow algorithm designers to compensate for many possible error sources.

As with all nanodevices, electromagnetic devices are subject to all kinds of passive movements. Active movement is likely based on artificial molecular motors. One example would be artificial flagella that react to the external application of a magnetic field to generate rotational forces and thus propulsion.

As the underlying technology is more or less identical to acoustic nanonetworks, so are the actuary and sensory capabilities and energy sources. One difference would be that much more energy is necessary to power a static electromagnetic wave to transmit information. That said, many suggestions like OOK may avoid that high energy cost by only transmitting information for incredibly short pulses.

Timers, random generators, and safety algorithms are likely the same as with acoustic nanonetworks as the majority of the involved components are assumed to be identical. As a result, electromagnetic nanonetworks face the same challenges and unanswered questions about biocompatibility, waste management, and life expectancy.

Figure 12.5 shows an example reference architecture. Unlike acoustic waves, electromagnetic waves propagate equally in all directions and are not influenced as much by other mechanical factors like blood flow. Yet, there are still molecular absorptions due to signal inference with hydrogen happening which leads to weaker signals the further the signal travels. Due to the limited distance, it is necessary to include a special gateway with additional capabilities that gathers, preprocesses and transmits the data either to the microscale or directly to the macroscale.

12.3.2.1 Nanonetworks on Chips

Another use case for terahertz communication is networks on chips. Nowadays, most processors are based on multi-core architectures, where a single chip contains multiple independent processor cores (Lemic et al., 2021). These cores compute in parallel and use the on-chip memory to share data

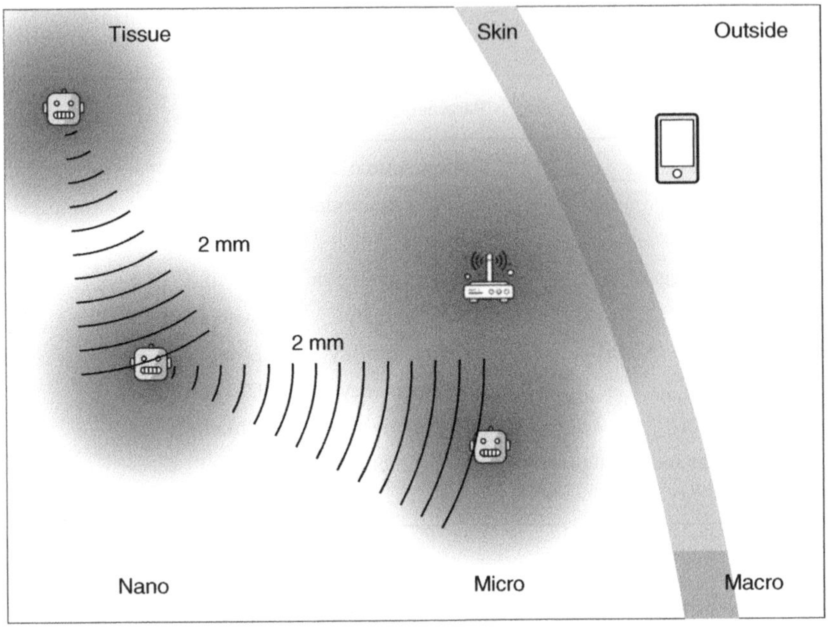

Figure 12.5 Example of an electromagnetic nanonetwork. The devices communicate over a range of approximately 2 mm using electromagnetic waves. *Source*: Florian-Lennert A. Lau.

and synchronize their executions. To enhance performance and error resilience, there is a trend to integrate more cores within the same chip. However, this makes coordination and synchronization of parallel/concurrent instructions on the on-chip difficult. If all cores require a direct connection to each other for the fastest possible communication, $O(n^2)$-wired connections would be necessary.

Extensive research has focused on improving on-chip interconnects, with bus-based interconnects being replaced by more efficient *Networks-on-Chip* (NoCs) (Marculescu et al., 2008). In the past, NoCs were mainly formed via wired connections, but they faced challenges in terms of delay, power requirements, and area overhead as more cores were integrated. To connect n different components fully, n^2 wires are necessary which might not be feasible due to difficulties in creating wire crossings. This led to the proposal of alternative interconnect technologies, including wireless on-chip communications.

Wireless communication offers a few advantages for intra-chip networks, including reconfigurability, improved scalability, throughput, and a lower energy consumption. However, current *Wireless NoCs* (WNoCs) are primarily utilized for long-range point-to-point links to decrease the average hop count of traditional NoC solutions. They are employed to enhance wired NoCs due to the relatively large sizes of metallic antennas required for wireless communication in the mmWave band.

An example of the architecture is shown in Figure 12.6. An array of communicating components is placed on a chip. Due to the nature of wireless communication, each device may communicate with other devices without requiring any physical connection. As long as only a limited number of devices have to communicate at any given time, such an architecture can reduce the cost of parallel processing.

Recent proposals have suggested the use of nanoscale WNoCs with graphene nanoantennas, which are significantly smaller than metallic antennas and can utilize THz frequencies for inter-core communication. These nanoscale WNoCs provide a potential solution to

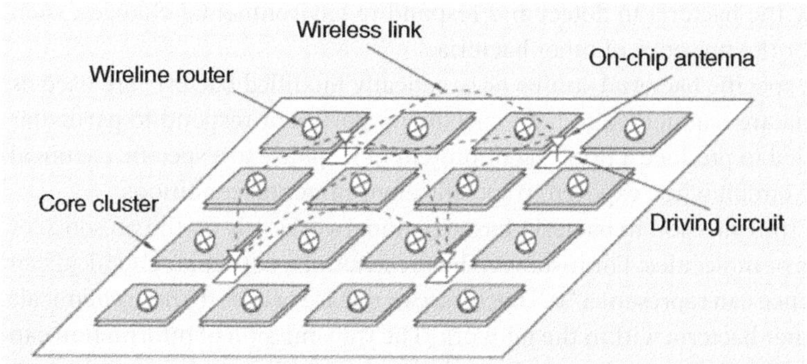

Figure 12.6 Example of an architecture for a nanonetwork employed on an integrated circuit to enable direct high-speed component communication (Abadal et al., 2013). *Source*: Abadal et al., (2013) / With Permission of IEEE.

the requirements of area-constrained, latency-bound, and throughput-intensive on-chip communication. The concept of nanoscale WNoCs using graphene nanoantennas has been further developed.

In on-chip communication, the traffic typically consists of short control messages for cache coherence, data consistency, and synchronization, along with larger data transfers. Communication can involve unicast, multicast, and broadcast transmissions depending on the application's memory access patterns. Latency is a primary concern as delays in packet delivery can significantly impact computation, but throughput is also important to avoid the throttling of computing cores. The throughput requirements for on-chip networks can reach up to 100 Gbps and even higher for communication-intensive architectures.

WNoCs face challenging requirements for end-to-end latency in the sub-microsecond range with high reliability comparable to on-chip wires. Security concerns related to on-chip communication are not as significant, and the mobility of network nodes is not anticipated. Energy efficiency is crucial to prevent overheating of the chip due to high energy dissipation.

Considering the requirements of different application domains, electromagnetic nanocommunication in the THz frequency band using graphene-based nanoantennas shows promise. The THz band can support high bit rates, can provide new communication mechanisms suited for nanodevices' limited capabilities, enables efficient medium sharing, and offers a large bandwidth. The THz frequency band has low interferences and is expected to remain so due to its low utilization, high attenuation, and large bandwidth. Yet, despite the potential of electromagnetic nanocommunication, there are numerous unresolved challenges.

12.3.3 Bacteria-based Nanonetworks

One of the more controversial types of nanonetworks are *bacteria-based nanonetworks* (Cobo and Akyildiz, 2010). Bacteria-based nanonetworks mainly function based on already available, naturally occurring life forms that only have to be slightly modified to behave like a nanonetwork. While it is likely simple to have certain kinds of bacteria transmit information via, e.g., plasmid rings, it is likely very difficult to modify such a network further. Bacteria can usually only be modified in a limited range before they either die or do not perform the desired function anymore.

Bacteria naturally communicate with each other using signaling mechanisms. Among them are chemical signals, plasmid rings, or even the exchange of entire DNA packages. For example, bacteria can release molecules into their environment, which act as signals to other bacteria.

These chemical signals allow the bacteria to detect and respond to environmental changes, such as the presence of nutrients or the presence of other bacteria.

Within the nanonetworks, specific bacterial strains or genetically modified bacteria are used as "nanodevices." These bacteria are engineered to have specific properties or respond to particular signals. They could be modified to produce a fluorescent protein in response to a specific chemical signal or to activate a genetic circuit when exposed to specific environmental conditions.

Information encoding and transmission in bacteria-based nanonetworks rely on the presence or absence of specific chemicals or molecules. For instance, the presence of a particular chemical can represent a "1," while its absence can represent a "0." Bacteria can release or absorb these chemicals to transmit information to other bacteria within the network. The transmission of information can occur through diffusion, where chemicals propagate through the surrounding medium, or through direct physical contact between bacteria.

Bacteria also possess inherent computational capabilities that can be harnessed within the nanonetworks. They can perform simple computations using their biochemical pathways and genetic regulatory systems. By manipulating the genetic makeup of bacteria or introducing synthetic gene circuits, researchers can create bacterial logic gates that enable very basic computational operations. These computational abilities allow bacteria to process information locally or collectively as a distributed computing system.

However, there are several challenges and limitations associated with bacteria-based nanonetworks. Ensuring reliable communication in noisy and dynamic environments is a significant challenge. Changing environmental conditions, such as variations in temperature or pH, can affect the bacteria's communication capabilities. Additionally, addressing security and privacy concerns is crucial to prevent unauthorized access or manipulation of the network. Furthermore, scaling up the complexity of computation in bacteria-based nanonetworks remains a significant research challenge.

Table 12.3 lists possible parameters and constraints that bacterial nanonetworks might be subject to. Unlike acoustic or electromagnetic nanonetworks, bacterial nanonetworks are made of possibly hundreds of thousands of bacteria and other (bio-)chemicals.

Table 12.3 List of currently expected parameters and constraints current concepts of bacterial nanonetworks might face.

Parameter	Value	Parameter	Value
Construction material	Bacteria/DNA	Energy source	Chemical/light
Participants/body	Hundreds of thousands	Energy requirement	Low
Computational power	Low	Timer	None
Memory size	Constant/logarithmic	Random generator	None
Communication range	Direct contact	Safety/security	Secret systems
Communication speed	10 μm/s	Biocompatibility	High
Bandwidth	Medium	Life expectancy	Hours–days
Movement	Flagella/cilia	Waste management	Free
Movement type	Flagella + passive	Error vulnerability	Medium
Sensors/actuators	Gap junctions		

Source: Florian-Lennert A. Lau.

Each bacteria only possesses very primitive means of computation and a limited amount of memory. While bacteria do carry a vast amount of DNA, most of that cannot be used for storing information and it is difficult to access any encoded information for communication or computation purposes. Thus, it is best to be conservative about the numbers while acknowledging the potential.

The communication parameters vary considerably from other types of wireless nanonetworks. Bacteria can communicate in a number of different ways, but most of them require direct physical contact to exchange, for example, plasmid rings made of DNA. An example of the process can be seen in Figure 12.7, which shows a reference architecture for bacterial nanonetworks.

The entire environment is filled with a variety of chemical substances that the involved bacteria might use for chemotaxis to find each other. The bacteria themselves can move at around 10 μm/s using flagella, but this value differs a lot according to the specific kind of bacteria and if the other types of passive movement act in favor of the communication process. As a result, that is also roughly the speed of communication. The bandwidth is simply the ratio of communication speed and the amount of transmitted information. If plasmid rings are used, the information per ring exchange might be between a handful to several hundred kilobases.

The remaining reference architecture is more or less identical to other types of nanonetworks. A gateway connects the nanoscale with the macroscale and other IoNT devices connect the nanonetwork to the remaining infrastructure.

Sensors, actuators, and means of producing energy are all naturally available. By changing little about the bacteria themselves, the mitochondria, cilia, and, e.g., gap junctions just function as usual and thus provide the necessary infrastructure.

However, bacteria do not have an understanding of time or randomness and synchronization is thus likely difficult to achieve. That said, some bacteria are able to synchronize their behavior

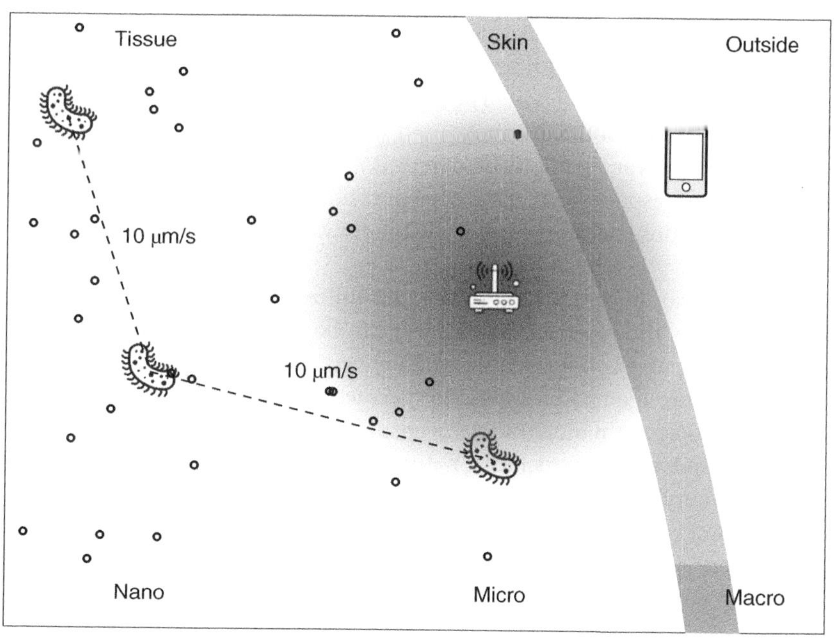

Figure 12.7 Example of an bacterial nanonetwork. The devices communicate over a range of approximately 10 μm/s using plasmid rings. *Source*: Florian-Lennert A. Lau.

via Quorum sensing. Other CMOS components are likely necessary to fulfill those tasks, but the compatibility is unclear. Similarly, security is likely difficult to achieve as five logical gates per cell are not enough to perform any meaningful cryptographic algorithm.

Yet, bacteria have the advantage of being biocompatible in many cases, they reproduce and thus have a potentially near-infinite life expectancy and the human body is skilled at getting rid of dead bacteria. However, bacteria are likely subject to erroneous behavior like most other nanoscale processes.

All in all, there is likely some potential for bacterial nanonetworks as they are relatively simple to implement and thus serve as a proof of concept for nanonetworks overall. They might act as a stepping stone, but it is likely only possible to modify the bacteria's behavior within strict limits. This restricts the possible applications considerably, and it remains to be evaluated if the benefits of immediate availability and compatibility outweigh the cons of limited programmability and configurability.

12.3.4 Molecular Nanonetworks

Molecular nanonetworks implement nanonetworks using molecules as the medium for information exchange. These networks compute and communicate at the molecular level, possibly providing solutions for molecular recognition, drug delivery, and many others.

The molecular communication model consists of three main components: transmitters, receivers, and communication channels. The transmitters and receivers are typically composed of specially designed molecules, or in early stage testbeds macroscale machines, that can emit and detect signals, while the communication medium serves as the pathway for signal propagation. The process begins with the transmission of molecules from a transmitter. The transmitter is engineered to encode information in a specific way, such as by modulating the concentration, pauses between signals, or other characteristics of a signal.

The communication channel is typically a fluidic environment, such as a solution containing molecules or a biological system like the blood flow. The signal moves around, influenced by flow and diffusion as well as other properties of the channel.

When the signal reaches the receiver, it might be detected. The receiver is sensitive to the encoded signal and can recognize/respond to it. This detection process may involve changes in the molecule's structure, conformation, or energy state, which allows it to discriminate the signal from background noise.

Once the signal is detected, it can be further processed and utilized for various applications. For instance, in molecular sensing, the received information can be used to detect the presence of specific molecules or environmental conditions. In drug delivery systems, the received signal can trigger the release of therapeutic agents in response to specific molecular cues.

One of the key challenges in molecular nanonetworks is ensuring reliable and accurate communication. Due to the small scale and inherent stochastic nature of molecular systems, various noise sources as well as "channel memory" can affect the transmission and reception of signals.

Table 12.4 shows an overview of various parameters and constraints of molecular nanonetworks. Unlike other nanonetwork types, most construction materials are possible candidates. All that is necessary is a controllable container that contains communication molecules and releases them on demand.

The number of participants is different from the previously presented networks and might vary tremendously according to the respective application. Molecular nanonetworks do not need to cover, e.g., the entire body over multiple hops. All they have to do is release a sufficiently

Table 12.4 List of currently expected parameters and constraints current concepts of molecular nanonetworks might face.

Parameter	Value	Parameter	Value
Construction material	CMOS/CNT/DNA/bacteria	Energy source	Electrical/chemical
Participants/body	Unknown	Energy requirement	Very low–high
Computational power	Low	Timer	Quartz/CMOS/other
Memory size	Constant/logarithmic	Random generator	CMOS/CNT/other
Communication range	Arbitrary within body	Safety/security	Sync/async/other
Communication speed	30–40 cm/s	Biocompatibility	High
Bandwidth	Very low	Life expectancy	Variable
Movement	Diffusion/flow	Waste management	Variable
Movement type	Passive	Error vulnerability	High
Sensors/actuators	CMOS/CNT/DNA/bacteria		

Source: Florian-Lennert A. Lau.

high number of molecules that will eventually reach the destination. In a closed circulatory system like that of a human body, two molecular nanodevices might directly communicate with each other over several meters of distance. This is a significant improvement over other network types.

However, the high distance comes at the price of a very low information density and communication speed. An example of a reference architecture of such a molecular nanonetwork can be seen in Figure 12.8. Molecules in the bloodstream move at roughly 30–40 cm/s, diffuse at speeds between 1 and 53 μm/s, and can often only transmit a single bit of information every few seconds. As a result, it is exceptionally difficult to retrieve a larger amount of time-critical information from such a nanonetwork.

In addition to that, it is unclear how a gateway may collect and preprocess molecular information. They likely require a sufficient number of suitable multi-use receptors that may "consume" binding molecules to avoid detecting the same molecule multiple times. Furthermore, varying distances from such a gateway would require a gateway to store various different concentration thresholds. However, those thresholds may be reached in several different ways if a number of nanodevices release molecules. It is not possible to identify the transmitting nanodevice with certainty.

The computational power and memory size are limited by similar factors compared to other types of nanonetworks as identical materials might be used. Biological components offer no real improvement in that area, and possible hybrid solutions are subject to the same limitations. It is likely that the individual memory size is constant and the computational power is likely limited to several logical gates or simple circuits (or equivalents) of simple circuit classes like AC^0.

Sensors, actuators, energy sources, and timers/random generators may be implemented from any type of compatible material, and there is no clear consensus among scientists on what might be the best material. Yet, molecular nanonetworks can additionally choose among biological components and are thus potentially more powerful.

The energy requirement itself depends on several factors. If the molecules themselves have to be created on the fly, the requirements could be very high. If containers are filled in advance and released into the communication channel, then very little additional energy is necessary. That way, the required energy is used at the macroscale where it is no limiting factor.

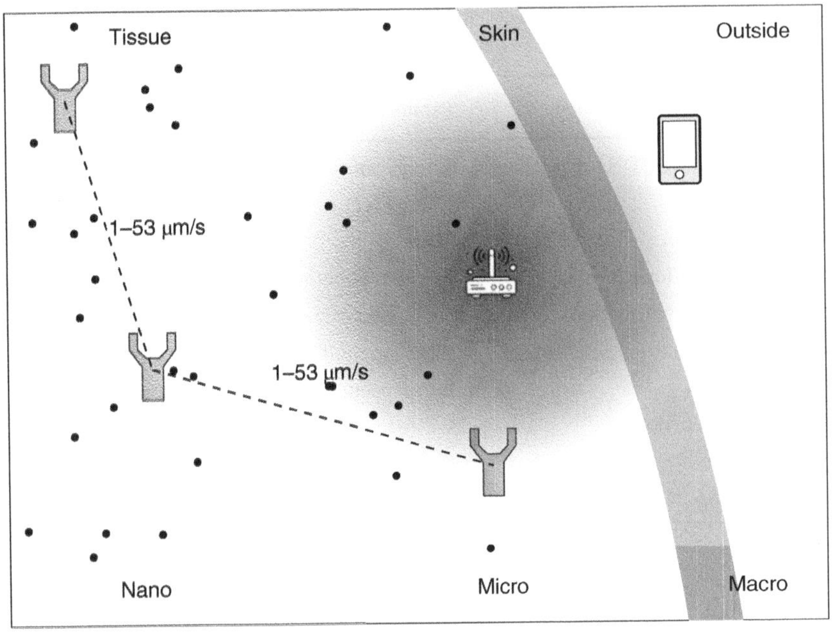

Figure 12.8 Example of a molecular nanonetwork using simple DNA molecules at 25 °C in water as information carriers. The devices communicate over a range of several meters using message molecules or ions. *Source*: Florian-Lennert A. Lau.

Biocompatibility, waste management, and life expectancy all depend on the specific materials used, and there is no simple answer.

Finally, molecular nanonetworks are likely more prone to errors than the other contenders. Due to the fact that molecules persist (channel memory) for some time in the communication channel, molecular inferences are almost guaranteed. As a result, it is already difficult to achieve standard molecular communication between two parties, let alone an entire network of possibly communicating molecular nanodevices. Furthermore, there are infinite possibilities to achieve identical molecule concentrations at the receiver if both the initial concentration and distance between the sender and the receiver are variable.

12.4 DNA-Based Nanonetworks

The most unknown type of nanonetwork is DNA-based nanonetworks. DNA-based nanonetworks are a special case of molecular nanonetwork that uses artificial DNA molecules and their self-assembly properties to implement various proposed functionalities of nanonetworks. Due to the limited general knowledge about such nanonetworks, we discuss them in a bit more detail in the following paragraphs. An example of such a nanonetwork can be seen in Figure 12.9. The figure shows roughly the same architecture as introduced in Chapter 5.

The DNA-based nanonetwork may function in a petri dish or any other artificial environment to perform, e.g., diagnostic tests or it may be applied in a human body (Lau et al., 2021a). A number of tiles, nanobots, and nanosensors are always present in the respective environment. The nanosensors are typically programmed to react with certain markers and release additional tiles upon detecting something (Li et al., 2018; Benenson et al., 2004). Those released tiles are the

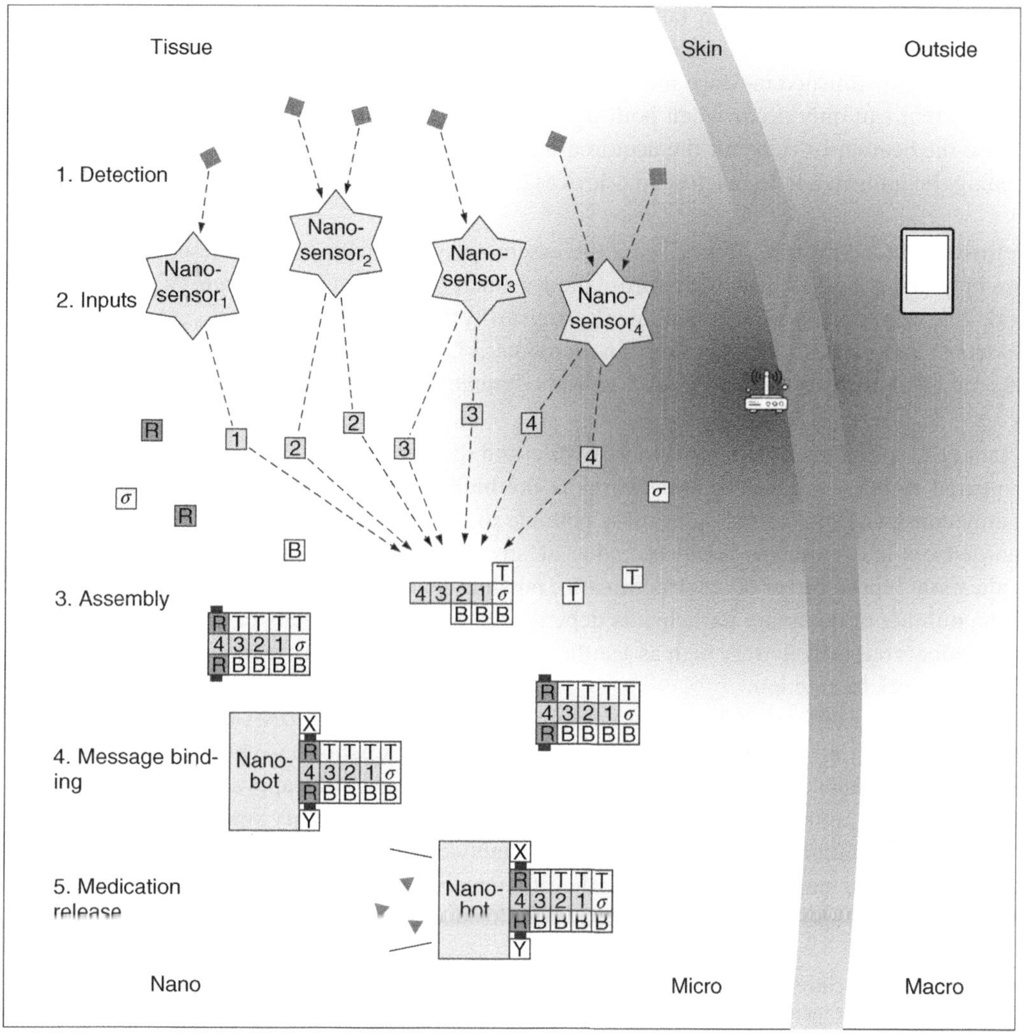

Figure 12.9 Reference architecture of the DNA-based nanonetworks introduced in Chapter 5. *Source:* Florian-Lennert A. Lau.

input for computation and also the missing building material to finish specific message molecules (Li et al., 2018; Andersen et al., 2009).

Definition 12.1 A *message molecule* $\mathcal{M}_{\Phi}$ is a tileset T that computes a Boolean formula Φ and forms a ligand upon successful completion of the computation.

The assembly process of such a message molecule also encodes a computation (Lee et al., 2004). In Figure 12.9, a simple 4-bit AND is depicted. The entire structure may only form when all building blocks especially the four input tiles have been released in sufficient quantity by the nanosensors. Once present, Brownian motion leads to collisions between tiles and assemblies, which allows some fitting tiles to permanently bind with assemblies. As the process is rather fast and subject to extreme levels of concurrency, small message molecules may form after a relatively short time, depending on the concentration of tiles and the ambient temperature.

In favorable conditions, it may take around 10 minutes for such a message molecule to successfully perform its computation, which is signaled by the addition of a ligand on the left (Cheng et al., 2005). The now-finished message molecule can interact with other nanobots via another binding reaction that can only occur when both parts of the ligand are present. The nanobot may then release medication or transmit the acquired information to a gateway that is capable of further sending the information to macroscopic devices that are part of a larger IoNT.

Definition 12.2 Let $\mathcal{T} = (T, \sigma, \tau)$ be a tile assembly system with a finite tileset T. σ is a seed assembly, and $\tau \in \mathbb{N}^+$ is the temperature of the system. A *DNA-based nanonetwork* $\mathcal{N}_\Phi$ is a nanonetwork. The components of the nanonetwork are given by the tuple $\mathcal{N}_\Phi = (\mathcal{N}_{Se}, \mathcal{N}_R, \mathcal{M}_\Phi)$, where $\mathcal{N}_{Se}$ is a set of nanosensors and $\mathcal{N}_R$ is a set of nanorobots. $\mathcal{M}_\Phi$ is a tileset for a set of message molecules. $\mathcal{M}_\Phi$ can be correctly assembled from $\mathcal{T}$. Φ is the function computed by the message molecule.

Table 12.5 specifies DNA-based nanonetworks and their constraints further. The first difference compared to other types of nanonetworks is the building material. DNA-based nanonetworks mainly consist of DNA or RNA, but it is possible to integrate other materials. For example, the nanosensors and nanobots could be made of any material that allows the storage of tiles or medication. Liposomes or other structures are possible candidates.

The number of necessary participants depends on the storage capacity of the nanosensors and the nanobots but is likely very high as a sufficient concentration of tiles must be achieved for such a nanonetwork to function.

Another major difference lies in the computational capabilities of DNA-based nanonetworks. The computation no longer takes place on a logical unit of a nanodevice but inside of the communication channel where less severe size and resource constraints apply. As a result, there is potentially more computational power and a bigger DNA-based memory available.

The communication aspects of DNA-based nanonetworks are basically identical compared with regular molecular nanonetworks. The only difference is the material used for communication. Furthermore, the speed of diffusion depends on the size of the structure. As message molecules

Table 12.5 List of currently expected parameters and constraints current concepts of DNA-based nanonetworks might face.

Parameter	Value	Parameter	Value
Construction material	DNA/RNA/other	Energy source	Electrical/chemical
Participants/body	Arbitrary but high	Energy requirement	Low
Computational power	Medium	Timer	None
Memory size	Constant/logarithmic	Random generator	CMOS/CNT
Communication range	Arbitrary within the body	Safety/security	Steganography
Communication speed	30–40 cm/s	Biocompatibility	Very high
Bandwidth	Medium	Life expectancy	Very high
Movement	Diffusion/flow	Waste management	Filter organs
Movement type	Passive	Error vulnerability	Arbitrarily low
Sensors/actuators	CMOS/CNT/DNA/bacteria		

Source: Florian-Lennert A. Lau.

can be of arbitrary size, it is not possible to give a definitive value here. Instead, we just list the speed of the blood flow in major vessels.

The bandwidth of DNA-based nanonetworks is bigger than that of other molecules as DNA may encode information in its bases and not just in the presence of molecules or their concentration. Furthermore, DNA also inherently has an addressing mechanism that allows only for specific bindings to preprogrammed nanodevices.

Sensors and actuators could be made from any material, and it is not possible to be more specific. Timers and random generators cannot really be realized using just DNA, but synchronization can be achieved by designing special message molecules that have to wait for specific input tiles.

Another major difference is the energy source of DNA-based nanonetworks. Unlike other types, they use the ambient temperature to function. A high temperature leads to fast-moving molecules and thus fewer stable bindings. A low temperature leads to slower movements and more stable bindings. Somewhere in the middle, there is a sweet spot where computations may be performed. That energy can and often must be supplied externally, and it might simply be possible to use the natural human body temperature as an energy source.

Unlike other types of nanonetworks, biocompatibility and waste management come "for free." The human body has very few mechanisms against DNA, and it is to be expected that such networks could work inside the bloodstream if coating problems by other substances can be avoided.

The life expectancy depends on the area of application. While DNA can be a stable information storage for thousands of years, the filter organs in a living creature may inhibit network functionality after a relatively short time. Thus, DNA-based nanonetworks might be potential candidates for, e.g., medical diagnostic procedures akin to a PCR test first.

Finally, DNA-based networks require other types of security as every step of a computation physically manifests and each step can always be traced back to its origins. As a result, it is likely better to just hide relevant information in a constant stream of artificial "noise," which could simply be junk information that is also present. Even simple yet effective procedures like one-time pads are not applicable in DNA-based nanonetworks.

In the following paragraphs, we analyze several potential DNA-based nanonetwork applications that further illustrate their general functionality. Unlike other network types, DNA-based nanonetworks are less intuitively "universal" and often require unique and creative solutions.

12.4.1 And – The Distributed Consensus

Figure 12.9 shows a possible implementation of a 4-bit And by a DNA-based nanonetwork. Nanosensors 1–4 represent four very large amounts of different nanosensor types, which emit tiles when a (disease) marker is detected, which are representative of the detection. These then assemble into message molecules in a stochastic process, with a distributed consensus being computed between the different types of nanosensors. The complete molecule can only be assembled under certain conditions. This includes a suitable temperature, a sufficiently high concentration of tiles, and the presence of all required building materials.

If all conditions are met, a large number of message molecules will be assembled. Once the message molecule is fully assembled, the 4-bit And is considered computed and the result can be received by a nanobot. The decision problem computed during the assembly process corresponds, among other things, to a distributed, unanimous agreement among the various nanosensor types. In addition, erroneous positive detections can be avoided in the same way since a single erroneous nanosensor is no longer able to induce an entire system to make an erroneous diagnosis.

A DNA-based AND nanonetwork can now be defined as a tuple analogous to message molecules and nanostructures.

Definition 12.3 A tile-based *AND* nanonetwork $\mathcal{N}_{\text{And}}$ is a tuple $(\mathcal{N}_{\text{Se}}, \mathcal{N}_R, \mathcal{M}_{\text{And}})$. $\mathcal{N}_{\text{Se}}$ is a large set of nanosensors, $\mathcal{N}_R$ is a large set of nanobots, and $\mathcal{M}_{\text{And}}$ is a tileset for message molecules that computes the operation AND.

The desired behavior of this nanonetwork can be confirmed in simulators such as the ISU TAS or the NetTAS. The simulation has shown that a message molecule of 15 tiles is expected to be fully assembled after about 3000–4000 binding reactions. This includes both the successful ties, and the ties that have broken off again. Furthermore, 0-strength glue is taken into account in the simulation, which further increases the total number of reactions required and corresponds to realistic conditions.

The result and (indirectly) tileset of the simulated message molecule $\mathcal{M}_{\text{And}}$ are shown in Figure 12.10. The tiles to the left show a receptor. The remaining tiles assemble into the message molecule.

Only when the message molecule and receptor are fully assembled, an interaction between the two can occur. Figure 12.10 shows the bound message molecule for the problem AND. The binding of a message molecule is only visualized for the message molecule AND since the process is identical for all message molecules.

Figure 12.11 summarizes the result of 100 simulation runs in a histogram. The number of binding reactions required appears to be log-normally distributed. A distortion toward higher values is observed. No message molecule was completed in less than 1500 reactions.

In reality, countless binding reactions take place almost simultaneously. This can be emulated by the 2HAM module in the ISU TAS/NetTAS or by the kTHAM module in NetTAS. It can be confirmed that erroneous interactions of unfinished message molecules or tiles are extremely unlikely under the given conditions.

As shown in Table 12.6, numerous substructures exist at the same time since 2HAMs do not need a seed assembly. This allows interactions between any two tiles or subassemblies that meet the temperature requirement τ. With cleverly chosen glues, there are no unwanted interactions between subassemblies. In order to create broken bindings, there must be multiple simultaneous failures.

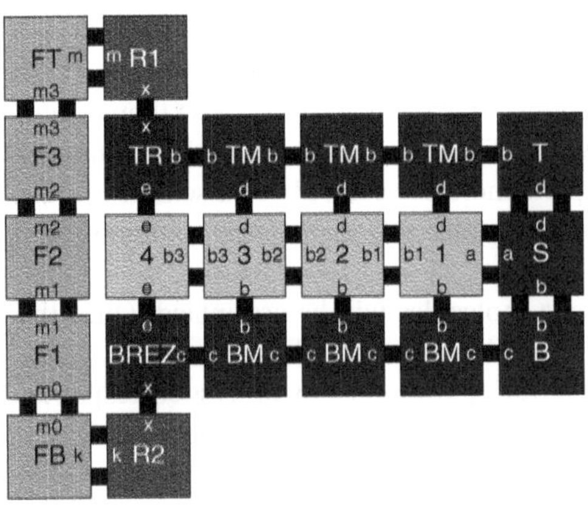

Figure 12.10 Binding of a 4-bit message molecule $\mathcal{M}_{\text{And}}$ in 2HAM to a receptor (the tiles to the left). *Source*: Lau (2020)/with permission of Elsevier.

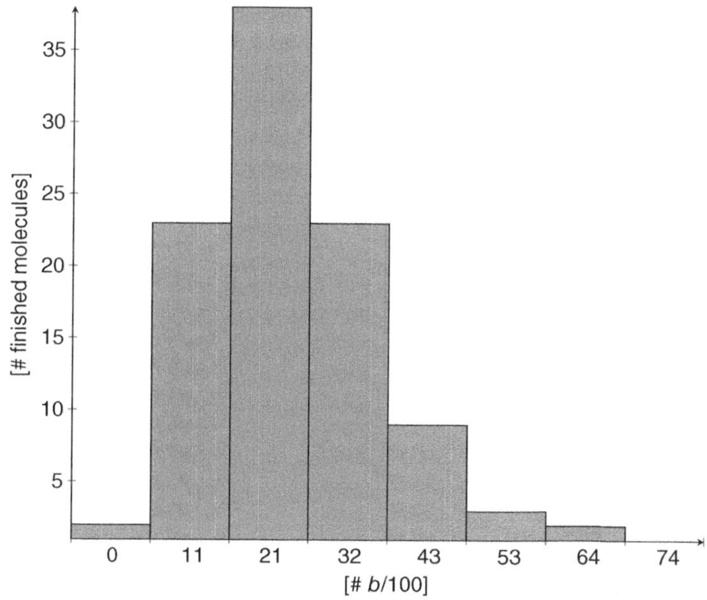

Figure 12.11 Result of 100 kTAM simulations of the message molecule $\mathcal{M}_{\text{And}}$. The histogram shows the number of complete molecules in relation to the required binding reactions *b*. *Source*: Lau (2020)/with permission of Elsevier.

Table 12.6 Number of subassemblies of size *k* of message molecule $\mathcal{M}_{\text{And}}$.

Size	1	2	3	4	5	6	7	8	9	10	11	12	13	14	15	22
Number	18	12	11	11	11	12	11	11	10	9	7	5	3	2	1	1

It can be seen that there are many ways to create smaller assemblies. Large assemblies become rare, and a binding at the receptor of the whole message exists only once (size 22).
Source: Florian-Lennert A. Lau.

No simulation run has revealed an erroneous interaction between the tiles of the receptor and the ligands of the message molecules. Even after 100 000 binding reactions, no errors are observed in the considered message molecules and they remained stable. The error probability for the composition of a single message molecule was 0%.

12.4.2 Thres – Exceeding a Critical Threshold

The threshold operation forms an important basis for a large number of nanonetwork applications. A threshold is required, for example, to carry out a vote among different devices, if it is sufficient that, for example, three out of five nanodevices agree. Furthermore, threshold values are required in order to recognize whether a defined threshold value has been exceeded, which, for example, indicates the presence of a disease.

In the following, we analyze a DNA-based nanonetwork that computes a simple threshold value. Figure 12.12 shows the tileset used for a threshold of size three out of four values in total.

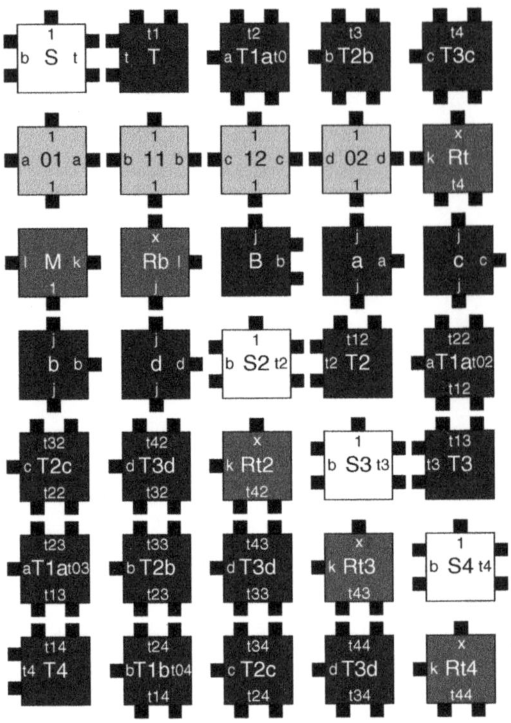

Figure 12.12 Tileset for a 4-bit message molecule $\mathcal{M}_{\text{Thres}}$. There are tiles for 4 different message molecules that represent the different ways to reach a threshold. *Source*: Lau (2020)/with permission of Elsevier.

The associated nanonetwork is shown in Figure 12.13. Unlike before, message molecules can take different forms, and for each of those, a unique seed tile exists, which is represented by the different ordering of the displayed message molecules in Figure 12.13. For each possible subset of three tiles, there are specific framework tiles in the medium. While there is a fifth option where all four tiles are present, such a DNA-based nanonetwork would evaluate to be true before such an assembly could form – hence, we omit that option. For each of the tile types σ_i, T_i, and B_i, four different versions exist. The rest of the scenario works analogously to the 4-bit AND.

Definition 12.4 A tile-based THRES nanonetwork $\mathcal{N}_{\text{Thres}}$ is a tuple $(\mathcal{N}_{\text{Se}}, \mathcal{N}_R, \mathcal{M}_{\text{Thres}})$. $\mathcal{N}_{\text{Se}}$ is a very large set of nanosensors. $\mathcal{N}_R$ is a very large set of nanobots, and $\mathcal{M}_{\text{Thres}}$ is a tileset for message molecules that compute the function THRES.

It is possible to verify each of the four molecules in the ISU TAS, yielding identical assembly times compared to the AND network. However, the concurrent simulation of all four molecules at once in the NetTAS is of greater interest.

Figure 12.14 summarizes the result of 50 simulation runs in a histogram. The number of binding reactions required also appears to be log-normally distributed. A bias toward higher numbers can be seen since a complete assembly in just a few steps is hardly possible. No message molecule was completed in under 6500 reactions.

In reality, countless binding reactions take place almost simultaneously. Since we analyzed four assemblies at the same time, we had to divide the resulting number of attempted bindings by four. That way, the resulting numbers are more or less identical to the simulation results of the AND message molecule. Overall, the number of binding reactions required correlates with the time required

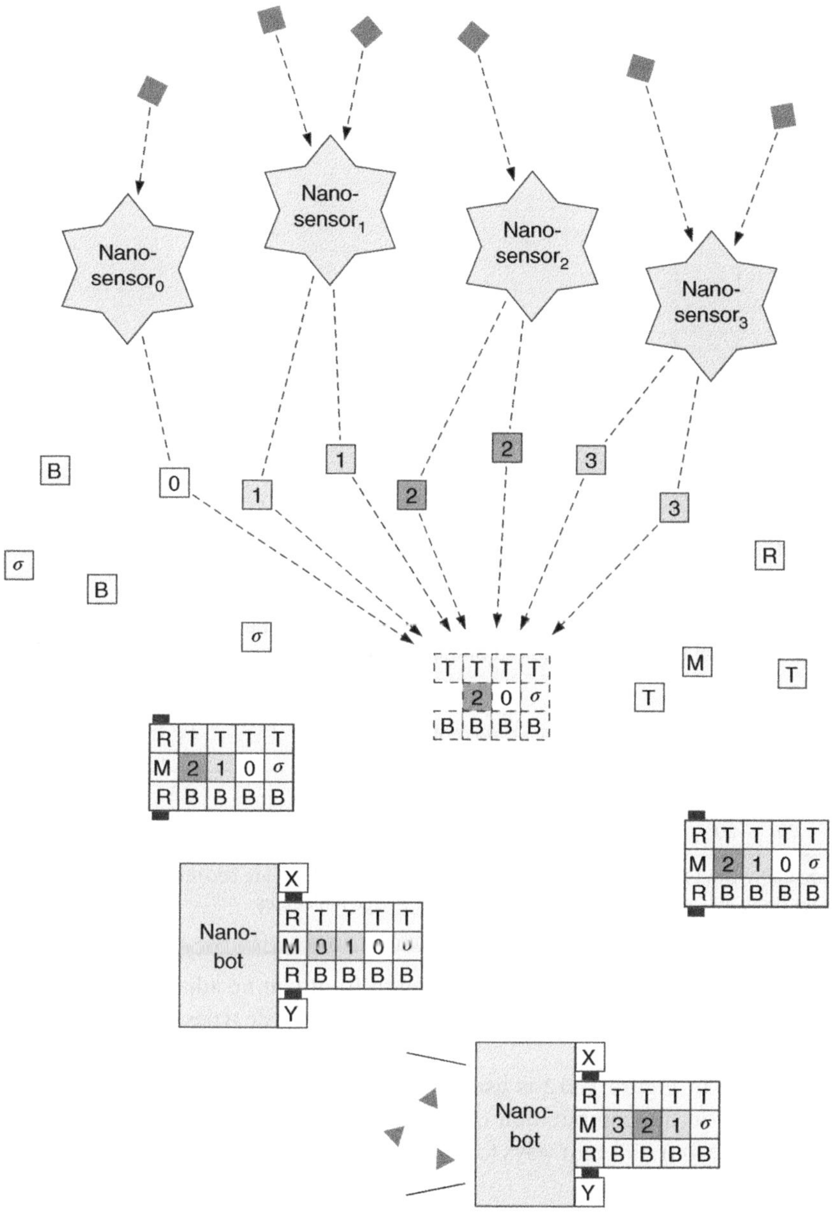

Figure 12.13 Model of the problem THRES as a DNA-network. Nanosensors$_i$ release input tiles 0–3 once an event is detected. This starts the assembly process of message molecules and a later reaction to the finalization. *Source*: Lau (2020)/with permission of Elsevier.

to complete an assembly. Yet, it is not a simple task to derive a real-time estimate for the assembly duration in seconds as the number of attempted binding reactions increases, the bigger an assembly gets.

The simulation runs have shown that the presented architecture is not very error-prone – 7 out of 100 assemblies showed permanent growth errors, which, however, could not spread. The probability of an incorrect assembly of a message molecule is therefore 3.5%. The individual rows assemble

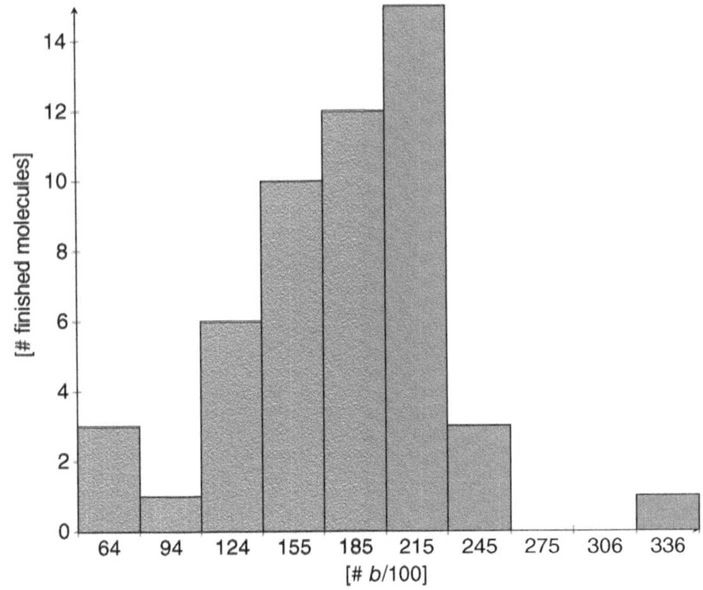

Figure 12.14 Result of 50 kTAM simulations of the message molecule $\mathcal{M}_{\text{Thres}}$. The histogram shows the number of complete molecules in relation to the required binding reactions *b*. *Source*: Lau (2020)/with permission of Elsevier.

Table 12.7 Number of subassemblies of size *k* of message molecule $\mathcal{M}_{\text{Thres}}$.

Size	1	2	3	4	5	6	7	8	9	10	11	12	13	14	15
Number	35	24	20	20	20	24	24	24	24	20	16	12	8	4	4

It can be seen that there are many ways to create smaller assemblies. Large assemblies are becoming rarer, and size 15 assemblies only exist four times. These correspond to the four full message molecules.
Source: Florian-Lennert A. Lau.

one after the other so that only about two parts of the growth front can be adjacent at any one time. The simulation time increases multiplicatively with the number of tile types and the possible binding positions.

The tileset of size 35 shown in Figure 12.12 is used for the 2HAM simulation. Only the tiles 01, 11, 12 and 02 are involved in the computation of the threshold value and are conditionally distributed by nanosensors. The four different ways to reach a threshold of height 3 are encoded by the framework tiles.

Table 12.7 shows that numerous substructures exist at the same time. The 2HAM simulation also shows that there are no probable interactions of the subassemblies that can lead to incorrect behavior. There are no assemblies that consist of more than 15 tiles – after 15 steps, the simulation process reaches a fixed point at which no new subassemblies are created. A Thres message molecule exists for each assembly of size 15.

12.4.3 Add – Basic Arithmetics

The last basic operation implemented by a DNA-based nanonetwork is the function problem Add. Adding numbers is absolutely crucial for a variety of protocols even at the nanoscale. For example, even the simplest hop-count routing procedure requires the participating nanodevices to increment

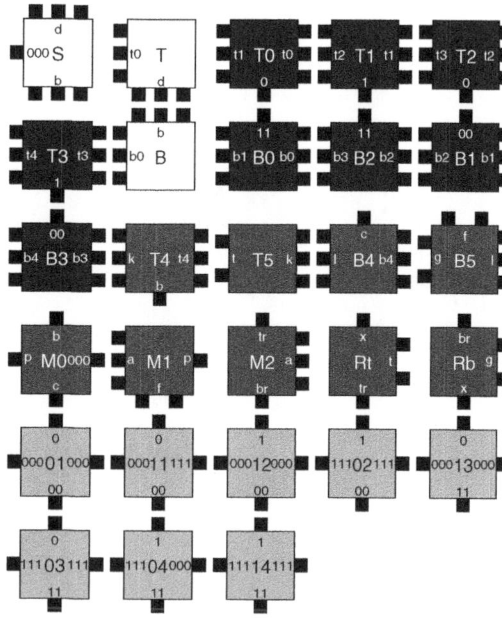

Figure 12.15 Tileset for the message molecule Add. The glues are chosen in a way that avoids possible overlaps between the top input and bottom input. We use a temperature 3 system as additions at temperature 2 require much more space and as a result more tile types. *Source*: Florian-Lennert A. Lau.

numbers, which is a special case of the Add function/binary counting. Additionally, nearly all complex applications require some set of arithmetic operations. The DNA-based nanonetwork here serves as an example to illustrate the arithmetic capabilities of tile-based self-assembly systems.

Figure 12.15 shows the tileset used for the Add nanonetwork from Figure 12.16. In this scenario, nanosensors emit complete binary numbers as assemblies or their components in equal quantities. These then form a framework of tiles in which the addition can take place.

The addition itself works in the same way as a binary adder. As shown in Figure 12.16 in the middle, the assembly process starts with a seed tile on the right and the binary inputs on the top and bottom. These now provide three different types of glue, all of which must fit so that the individual bits of the result can bind stably in the middle. The glues at the top and bottom encode the inputs, and a potential carry-over is taken into account from the right. The process is repeated until the full number is formed. It is important to note that the result of an addition is potentially one bit longer than both inputs.

Only when the last tile is part of the message molecule, a ligand can form and the result can be communicated. Since this is a functional problem, the result cannot be directly transmitted to a nanobot. Only the information of a successfully performed computation can be immediately processed via a binding reaction. In the case of three-dimensional tiles, the back of the input tiles can be used to apply additional glues on which additional binding processes or processing can take place.

An Add nanonetwork is defined as a tuple analogous to message molecules and nanostructures.

Definition 12.5 A tile-based *Add* nanonetwork $\mathcal{N}_{\text{Add}}$ is a tuple $(\mathcal{N}_{\text{Se}}, \mathcal{N}_R, \mathcal{M}_{\text{Add}})$. $\mathcal{N}_{\text{Se}}$ is a very large set of nanosensors. $\mathcal{N}_R$ is a very large set of nanobots, and $\mathcal{M}_{\text{Add}}$ is a tileset for message molecules that compute the function Add.

To verify the functionality, kTAM and 2HAM from, e.g., the ISU TAS or the NetTAS can be used. For the kTAM simulation at temperature 3, the parameters $G_{\text{se}} = 5.95$ and $G_{\text{mc}} = 16.0$ were chosen. The simulation showed that a message molecule of size 24 was fully assembled at the expected value after about 14 000 binding reactions. This includes both adding and detaching tiles. The probability of an incorrect assembly of a message molecule is 14%.

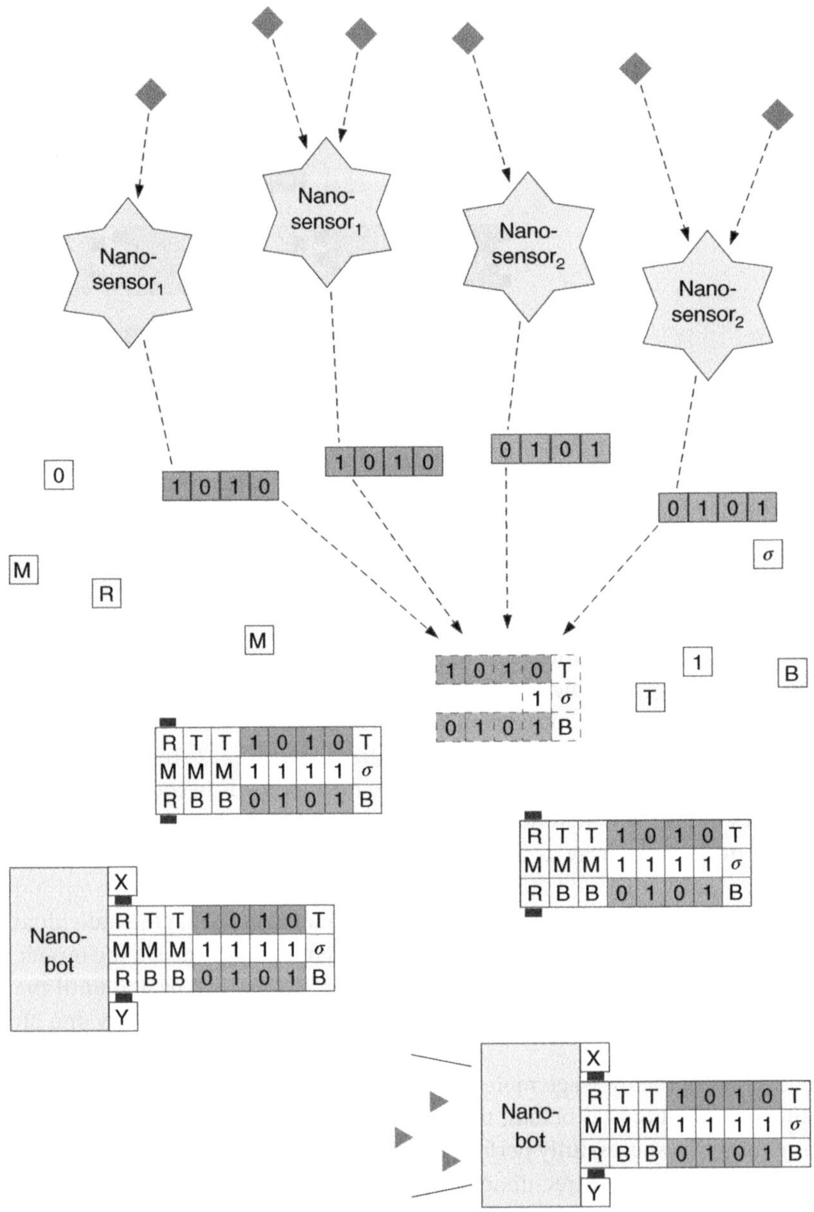

Figure 12.16 Model of the problems ADD as a nanonetwork. Nanosensors$_1$ and nanosensors$_2$ release two different input binary numbers upon detecting an event. This allows the already present input tiles and seed assemblies to execute the computation of an ADD. Upon complete assembly, nanobots bind the message and react to it. *Source*: Lau (2020)/with permission of Elsevier.

Figure 12.17 summarizes the result of 50 simulation runs in a histogram. The number of binding reactions required appears to be log-normally distributed. A bias toward higher numbers is observed. No message molecule was completed in under 8500 reactions.

The simulation runs have shown that the presented architecture is error prone. The top and bottom rows are likely to assemble early. Bindings in the middle are thus very prone to growth

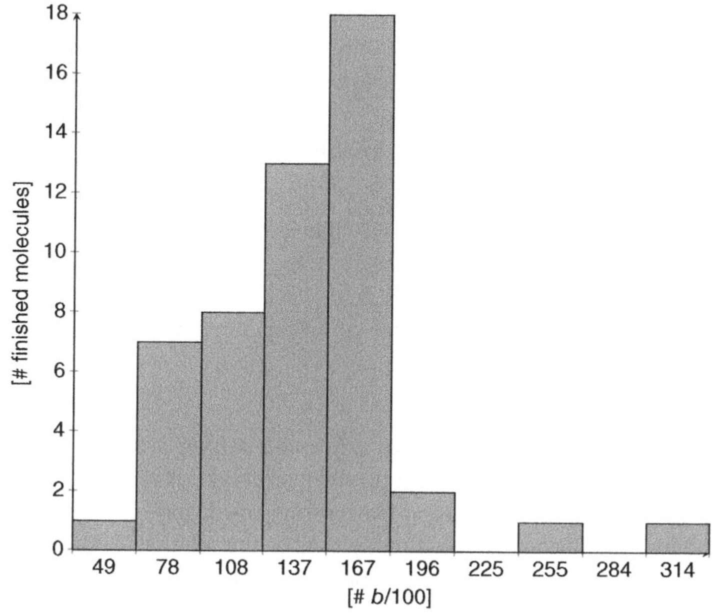

Figure 12.17 Result of 50 kTAM simulations of message molecule $\mathcal{M}_{Add}$. The histogram shows the number of complete molecules in relation to the required binding reactions *b*. *Source*: Lau (2020)/with permission of Elsevier.

errors. Only if the parameters G_{se} and G_{mc} are chosen very precisely, a correct composition can be guaranteed with a high degree of probability. This can be problematic in wet-lab experiments.

The assembly speed is limited due to the small interval of possible values for G_{se} and G_{mc} since tile detachments have to occur more frequently – if error reduction is a goal. After adjusting the parameters to a value that would correspond to $\tau = 3$ in the aTAM, there were seven permanent errors. However, these values only apply to exactly one simulated assembly at a time.

Overall, the number of glues used in different tile types must be minimized in order to prevent errors. This makes an error-free assembly process more likely. In addition, the assemblies should only provide a growth front in as few directions as possible at the same time. In the best case, each step is dependent on the previous one. The larger the target structure, the more likely the undesired interactions of intermediates become.

12.4.4 Solving Arbitrary Boolean Formulas

Now that we understand the basics of DNA-based nanonetworks and how they can perform computations, it is a good time to analyze their computational capabilities. In fact, tile-based computations are Turing-complete and are thus able to perform any kind of computation that a normal computer can. However, this Turing-completeness comes at a price of possibly thousands of unique tile types.

As a result, such tile-based Turing machines are usually not feasible at the nanoscale – at least in the near future. Due to physical limitations with DNA, the number of unique glues in a system is likely limited to around 10 000 (Patitz, 2014). Furthermore, many mathematical proofs claim that certain solutions exist, but rarely demonstrate such a system. For real-world applications, a *proof by demonstration* is much more desirable.

While it might be unclear how to create a small, fully functional Turing machine from DNA, it is indeed possible to create less powerful computational systems (Lau et al., 2020). One example would be circuits or the equally powerful Boolean formulas. They possess practically identical computational capabilities compared to a Turing machine. The only difference is that they cannot implement programs that run forever. Yet, for many practical purposes, programs that never terminate are not required.

Solving almost arbitrary decision problems is rather simple by using tiles. The following steps are necessary:

1. converting a computational problem into a Boolean formula Φ (if possible),
2. creating a truth table,
3. determining the *disjunctive normal form* (DNF) of a formula Φ,
4. minimization of the formula Φ using the Quine–McCluskey or an equally powerful method, and
5. designing a message molecule for each clause.

The first step is only possible for problems that terminate. While the *halting problem* states that it is not clear for all problems, there is a subset of many problems for which it is certain if and under which conditions they terminate. Furthermore, it is necessary to convert all problems into decision problems as only the results of those can be directly used at the nanoscale. Luckily, it is possible to convert functional problems and optimization problems into decision problems relatively easily.

The next step is the creation of a truth table for the given formula Φ. It is important to note that truth tables are of an exponential size, and it might thus be difficult to create DNA-based nanonetworks for more complex problems.

Next, the formula Φ has to be converted into DNF. The DNA just takes all entries of a truth table that evaluates to "1" and creates a *clause* for them. A clause combines all inputs with their respective truth values of that line of the truth table with the logical AND. All resulting clauses are then combined using the logical OR as it is sufficient when one of the clauses evaluates to "1" for the entire formula to be true. It is important to mention that there are more efficient procedures to arrive at the DNF of a formula, but the worst case always remains exponential in size. Optionally, but highly desirably, the DNA formula Φ′ can be further minimized using Karnaugh–Veitch maps or equivalent procedures (Karnaugh, 1953).

Finally, a message molecule is designed for every remaining clause. The process is illustrated in Figure 12.18. There are unique framework tiles σ_i, T_i, M_i, and B_i for every clause that serves as

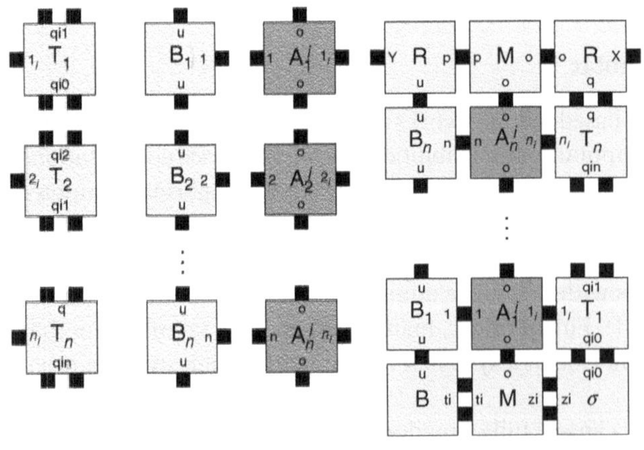

Figure 12.18 General procedure for creating tilesets from formulas in the DNF using the example of the truth Table 5.3. *Source*: Lau (2020)/with permission of Elsevier.

possible binding locations for inputs A_n^i. They further limit the growth of the message molecule to the length of the clauses.

The assembly process of these molecules functions analogous to the AND molecule and is illustrated in Figure 12.19. The entire molecule may only fully assemble including the receptor tiles if all tiles A_n^i are present. These have to be released by nanosensors under predefined conditions.

Once a single molecule is finished, the decision problem is regarded as true. One peculiarity of such systems is that it is unclear if a problem will *never* be true or if it is not true "yet." Logically speaking, a problem is either always false or false until it is true.

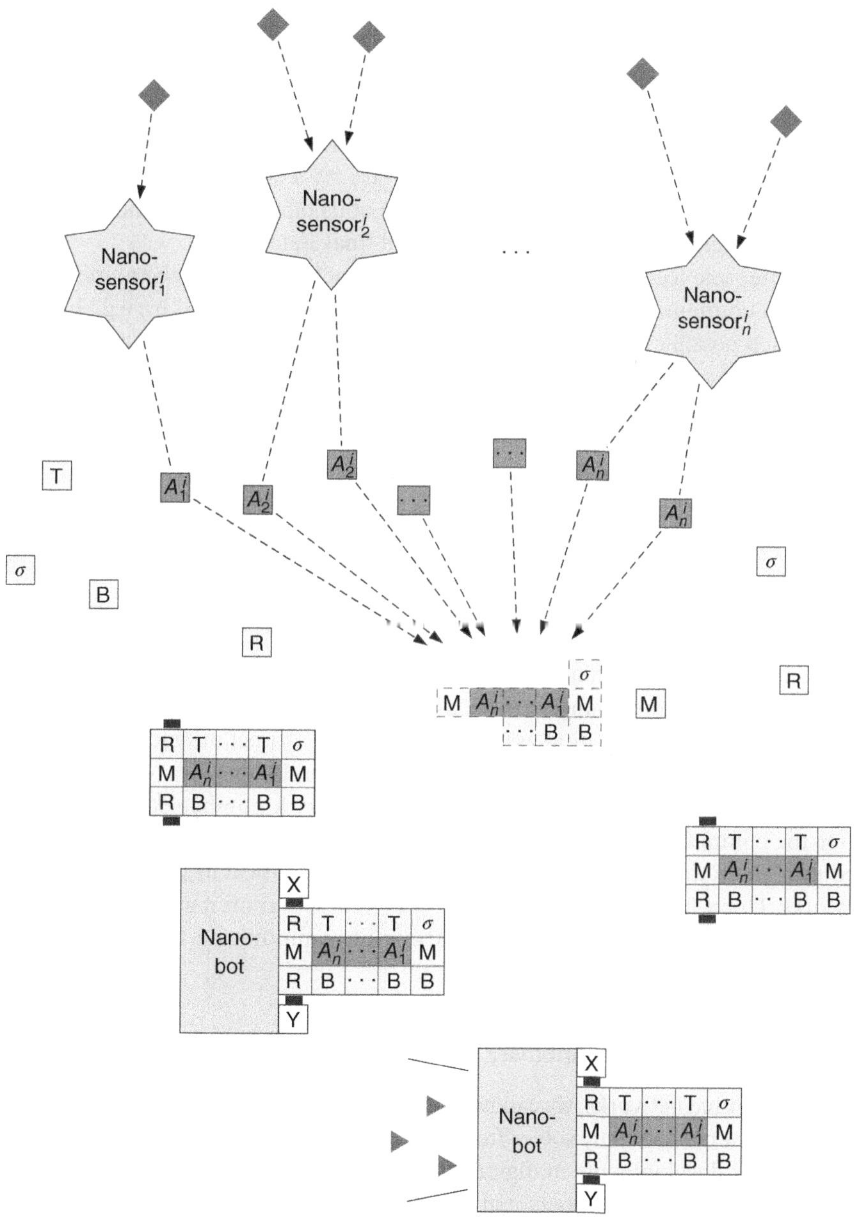

Figure 12.19 DNA-based nanonetwork for Boolean formulas. *Source*: Lau (2020)/with permission of Elsevier.

Finally, unlike regular logical systems, the principle of the excluded middle does not apply to tile-based self-assembly systems of that kind. Variables A_n^i that are true and false could theoretically coexist at the same time. While this is certainly unusual, it does not mean that such systems are not useful.

12.4.5 Solving Arbitrary Computations – Turing Networks

As a last somewhat generic contribution, we want to analyze how *specific* Turing machines may be emulated/implemented using tile-based self-assembly systems. While *universal* Turning machines may require thousands or hundreds of thousands of different tile types, the same is not true for Turing machines that solve specific problems. Here, the construction of such a system may require only a feasible number of different tile types while still, at least theoretically, capable of solving any computational task that is "small enough."

The construction principle itself is rather simple. The interested reader is encouraged to look back at Chapter 5 to refresh the knowledge of Turing machines. As a small reminder, a Turing machine is a tuple $M = (Q, \Sigma, \Gamma, \delta, q_0, q_\square)$ consisting of the state of its tapes/memory, the positions of the read/write heads, its program, and predefined initial and final states.

From theoretical computer science, we know that any Turing machine may also be represented as a *configuration graph*. A configuration graph consists of nodes that are labeled by, e.g., binary numbers that represent each possible state of a computation. The edges represent state transitions or computational steps that are valid given the program of such a machine. From that perspective, a computation of a Turning machine would be a *path* from the initial state to a final state.

This path is essentially a "state history" of the machine, and it is relatively simple to reconstruct the same history using tiles. The approach can be seen in Figure 12.20. At the top, we can see the state history of a specific Turing machine starting at the initial state q_0. From there, the state of the tape is changed bit by bit at each step in time until a final state $q_\square$ is reached.

At the bottom, we can see the same computation represented by rows of tiles with the same content as the Turning machine encoded in the markers and glues. There are specific tile types and glues that represent, for example, the read/write head position, which allows for the assembly of the next row. The content of the tape is then copied/propagated upward, and after a finite time, a final state is reached where no more changes to the assembly are possible. Just as before, it is possible to attach a receptor/ligand to the final assembly, allowing for direct interaction with the message molecule which encodes a successfully computed decision problem.

The entire nanonetwork can be seen in Figure 12.21. In this case, a number of nanosensors release the inputs for the computation, while all other tiles are always present in the communication channel/medium. The remaining architecture is more or less identical to all the previously presented nanonetworks. Upon full assembly, the result of the computation may be detected by nanobots that react by, e.g., releasing medication or starting to communicate the result to the macroscale by passing messages to a gateway.

12.4.6 Personalized Health Parameter Anomaly Detection

Based on the previously presented networks, we can now start to have a look at a more sophisticated DNA-based nanonetwork that conceptually solves frequent medical problems.

Extrapolating from current developments in digitalization and bioengineering, it is possible to enable diagnostic methods that work on a personal level and are therefore more inclusive and

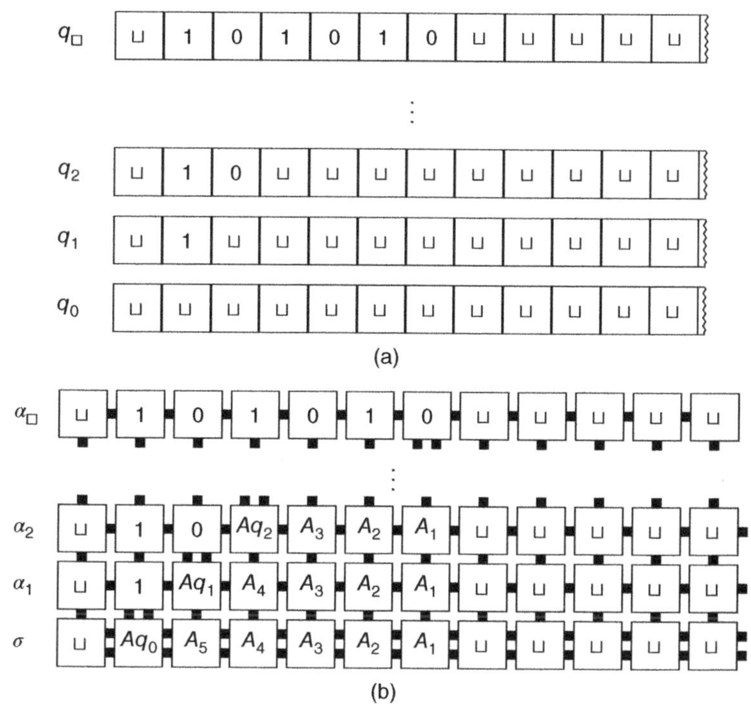

Figure 12.20 (a) Example tape content of a Turing machine in states $q_0, \ldots, q_\square$. (b) Example assembly that represents the configuration history of the Turing machine from (a). *Source*: Lau (2021b)/with permission of Elsevier.

precise. *Precision medicine* represents a paradigm shift in medicine, from generalized approaches to the treatment of a disease to strategies for the prevention, diagnosis, and therapy of diseases based on a person's unique characteristics (Duffy, 2015). The vision of medical nanonetworks perfectly fits the requirement, and they might be an important part of precision medicine (Mousa et al., 2020). The following DNA-based nanonetwork illustrates that nanodevices can help with detecting health parameter anomalies from a *personal norm* instead of a global average, taking precision medicine to the most personalized level possible.

In this paragraph, we demonstrate a DNA-based nanonetwork that measures a number of health parameters of a patient and stores them as tiles in the communication channel. Based on the measured data, the network can enter the second mode of operation, where it detects anomalies concerning the previously determined values. It can either immediately react to the measurements by releasing a previously stored medical payload or by communicating the results to an external device or person. We thus solve several problems at once and demonstrate that DNA-based nanonetworks are more powerful and versatile than classical medical tests in almost every way. They can compute, have memory, and can act completely autonomously if necessary.

Based on these capabilities, it is possible to design medication that only has an effect if the recipient is really sick. Another option is a class of medication that only applies if other health parameters are within a tolerable range or medication that only works if no harmful prior conditions apply. Yet, another use case is the design of a diabetes medication that is only triggered once blood sugar reaches critical levels.

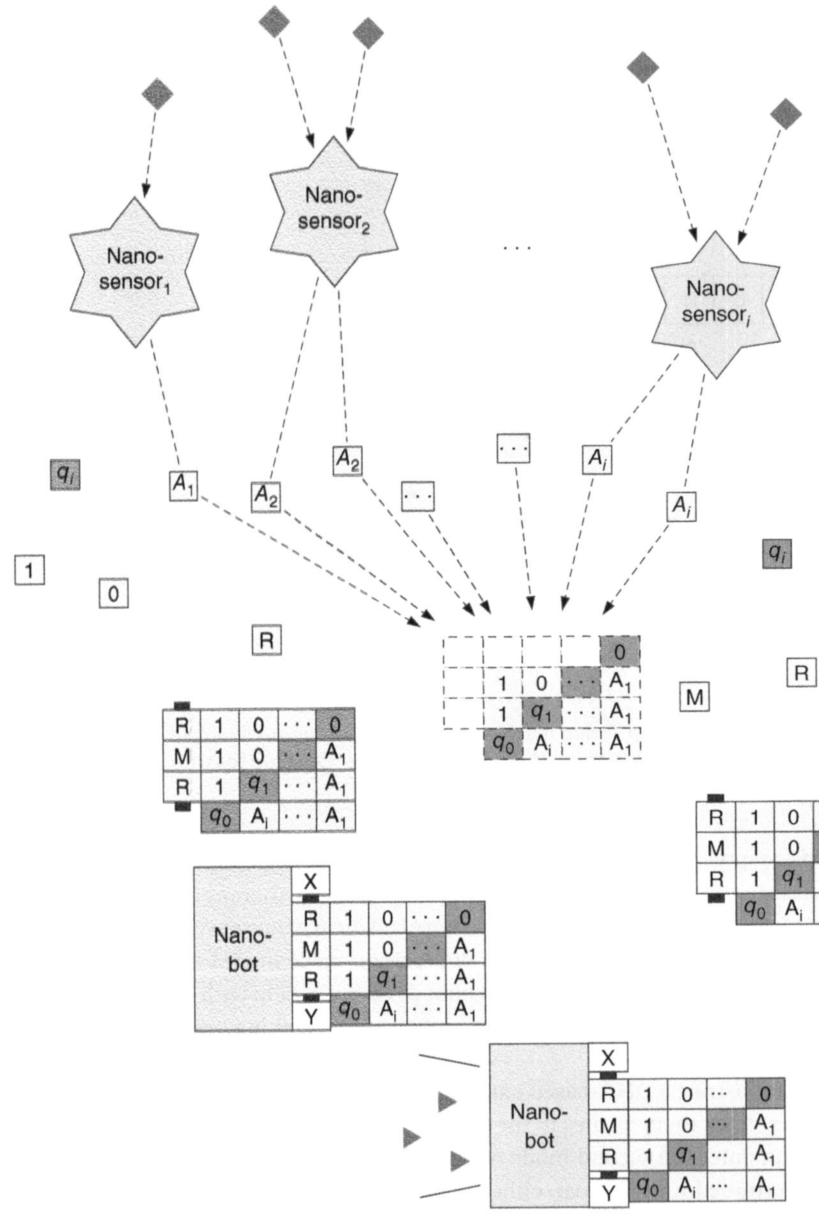

Figure 12.21 General reference architecture for nanonetworks based on Turing machines. It consists of $o \geq 0$ markers (orange rhomboids), $p \gg 1$ nanosensors, $q \geq 0$ message molecules, and $r \gg 1$ nanobots. Tiles are released upon detecting a previously defined event to initiate or continue the assembly process of the message molecules. The read/write-head position is indicated by the tiles q_0 to 0. The ligand "R" can only form once the top row fully assembled. Each row of tiles depends on the sufficient assembly of previous rows. *Source*: Lau et al. (2021b)/with permission of Elsevier.

12.4.7 Tile-based Anomaly Detection

While humans overall are similar, the normal ranges of various health parameters differ from person to person. For example, the blood pressure of humans who lift heavy weights is naturally elevated. However, this is usually not critical and just a bodily adaptation to weight training. For other people, the same value could be dangerously high, and blood pressure is just one of many examples. Generally, many health parameters follow a Gaussian distribution.

In classical medicine, medication is usually determined by comparison to a deviation from the population average, and the treatment follows a one-drug-fits-all model. While this can be a useful tool for the population average, it fails to account for individual variation. Especially people who deviate from the average cannot be reliably diagnosed. This basically affects all women as hormonal cycles could influence various bodily parameters. As a result, women are often excluded from clinical trials as it is more costly to account for the additional variables. Medication for women is usually obtained solely by examining men and extrapolated to women (Chilet-Rosell, 2014).

To solve this problem of generalization, we conceptualize a nanonetwork that first measures normal values for an individual and later detects anomalies in comparison. Thus, we present a functional nanonetwork design to solve this general problem while also allowing for immediate treatments. We conceptualize this network as a powerful medical test outside of human bodies but envision it to work inside living beings as well.

12.4.7.1 Phase 1

We divide the task of the nanonetwork into two main phases: first, measuring, and later monitoring and treatment. Figure 12.22 divides the first of the two phases into five subparts. The entire process starts once all tiles and the seed tile σ are present in the medium. Once started, nanosensor^1–nanosensor^n detect different health parameters and release specific tiles upon detection. This continues for a time t where message molecules of arbitrary length can form in part 3. Tiles R, B, and M are introduced and stop the assembly process by adding a receptor to the message molecule that prevents further assembly. The thus assembled message represents the concentration k of a measured parameter in unary coding.

The fully assembled messages can then bind to fitting boxes of maximum depth k like a key. Each of the opened boxes releases new seed tiles $k_>$ for phase 2. The boxes also release tiles that bind to all seed tiles $k - i_>$ and thus ensure the presence of only a single seed tile that represents a normal health parameter. Thus, the measured concentration is converted from unary coding to a single tile. It is also possible to release signal repeater tiles from the same boxes to amplify a weak signal. After this step, the network automatically enters phase 2.

12.4.7.2 Phase 2

Phase 2 is depicted in Figure 12.23. The initially released seed tiles $k_>$ begin forming message frameworks of length k plus a predefined threshold j. For the same time interval t, the messages are now allowed to fill the assembled frame before the medical test expires. If the messages fully assemble in the given time, this signifies the reaching of a critical, predefined threshold. The finalized messages can then bind to a different type of nanobot that either immediately reacts by releasing medication where it is needed or by communicating the measurements to the outside.

The functionality of the concentration counter can be seen in Figure 12.24 (a). A frame of height $k + j$ grows once seed tile $k_>$ starts the process. The tiles A_i have to fill the frame completely for the left part of the receptor R to form.

In the visionary scenario inside of the human body, the second phase has to be repeated continuously. It is worth mentioning that only the exceeding of a threshold may be detected. In DNA-based

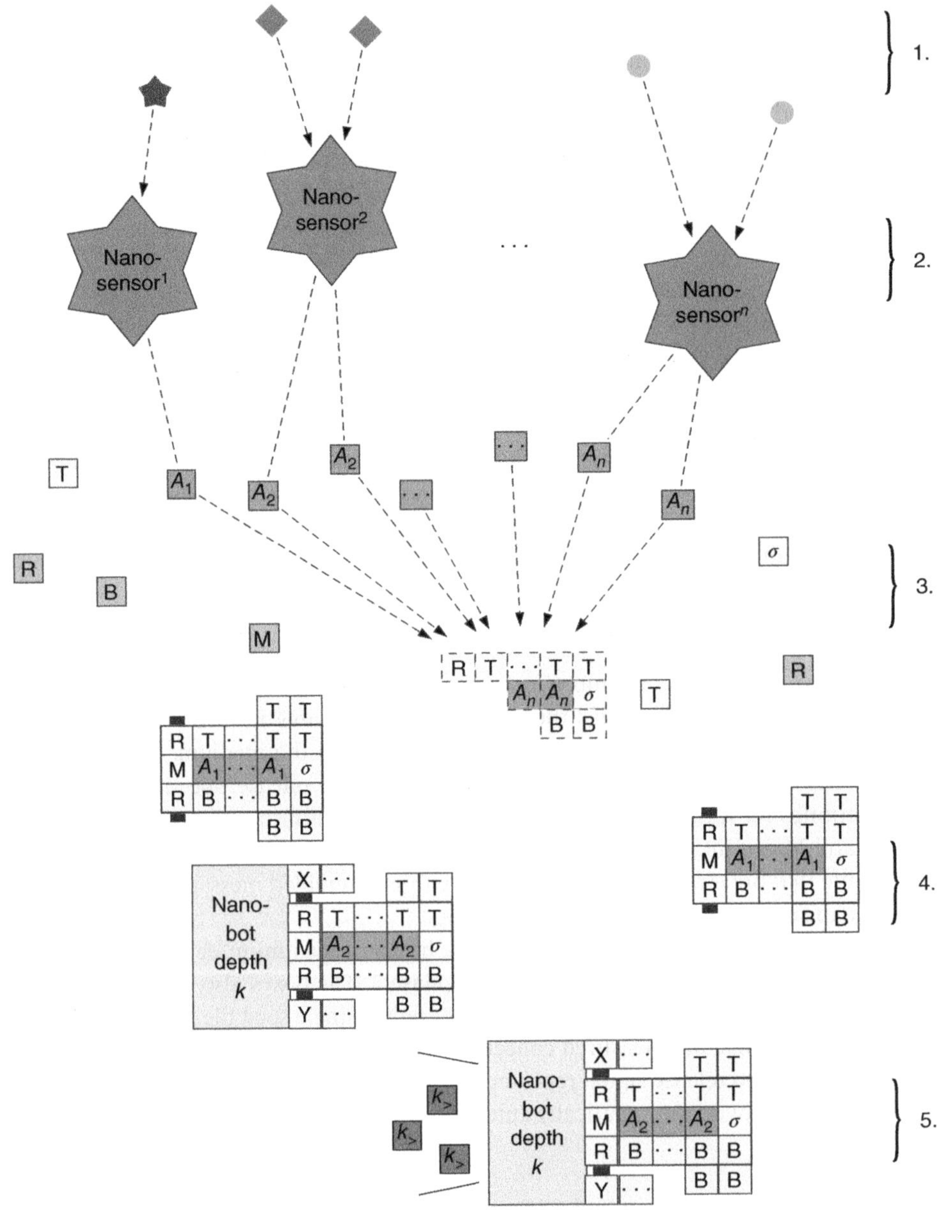

Figure 12.22 In phase 1, the purpose of the nanonetwork is to determine the normal range of body parameters 1−n. Nanosensor$_1^i$ counts the frequency of health parameters and releases tiles. In the medium, messages of arbitrary length form that are only stopped once a receptor tile R has been added externally. Now, the length of the message represents a unary number that stands for the normal range of parameter i over time t. The messages can now fully assemble and only bind to DNA-boxes of depth k or lower. Upon binding, the content of the boxes is released and now represents the threshold information as a single tile. *Source*: Lau et al. (2021b)/with permission of Elsevier.

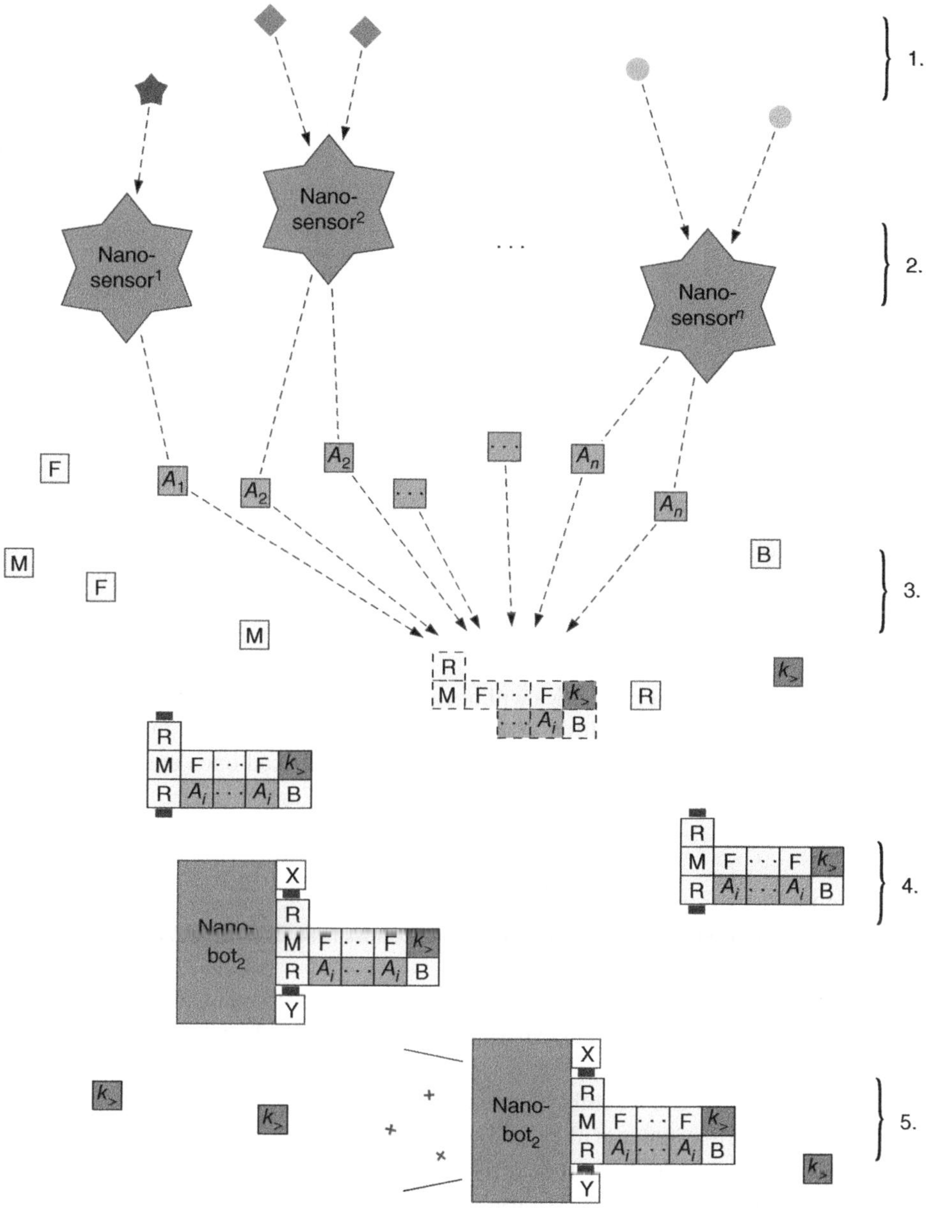

Figure 12.23 In phase 2, the purpose of the nanonetwork is to measure deviations from the previously determined normal values. The message molecules in the middle can only fully assemble if a threshold has been violated by a predefined value. Upon such an event, an appropriate nanobot is opened and the measurement communicated to an external party. Alternatively, the nanobots directly fix the anomaly. *Source*: Lau et al. (2021b)/with permission of Elsevier.

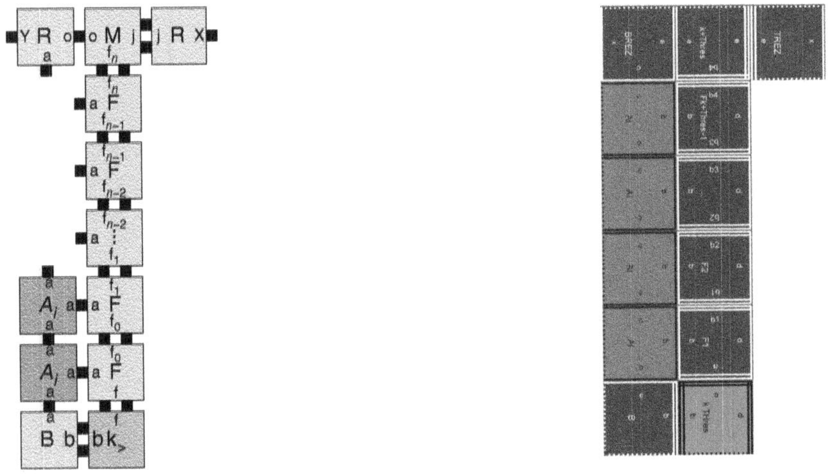

(a) (b)

Figure 12.24 (a) Assembly of a message molecule for the counting problem with a ligand. The frame structure demands up to k fitting tiles to bind before the ligand can be completed. (b) The result of the simulation of the same message molecule. (a) Conceptualization. (b) Simulation result. *Source*: Lau et al. (2021b)/with permission of Elsevier.

nanonetworks, we cannot tell apart if molecules will never finish or are *not yet* finished. Thus, we need a different approach to detect if too little of a health parameter is detected. One possibility is the engineering of a molecule with a finished receptor that gets blocked once enough tiles are present.

12.4.7.3 Evaluation

We use the both the ISU TAS from Patitz Patitz (2009) and the NetTAS software to evaluate the proposed assemblies. We created the molecules shown in Section 12.4.7 in the kTAM and 2HAM of the ISU TAS and the NetTAS.

We use the kTAM model to simulate the assembly processes under somewhat realistic conditions. The kTAM sensibly predicts wet-lab experiments, as shown in the work of Rothemund et al. (2004). DNA-based nanonetworks require precise environmental conditions, like temperature and monomer concentration, for the assembly process. However, it is possible to use DNA-based nanonetworks in controlled environments like a petri dish to analyze, for example, tissue samples.

Afterward, we analyze the tilesets with the 2HAM module. It tests all possible interactions between tiles and intermediate assemblies. We use the 2HAM to confirm the binding between nanobot receptors and message molecules and to investigate possible error sources of intermediate assemblies.

12.4.7.4 Realistic Simulation in the kTAM

We use the kTAM parameters $G_{se} = 10.4$ and $G_{mc} = 17.0$ to match the experiments from the work of Rothemund et al. (2004). G_{se} is the costs to break a binding, and G_{mc} stands for the monomer concentration. $\frac{G_{mc}}{G_{se}}$ roughly models the temperature parameter τ from the aTAM. $G_{mc} = 17.0$ model a tile concentration of 0.8 µM. This is a comparable concentration to the sodium in human blood (Lau et al., 2019). $G_{se} = 10.4$ further corresponds to a temperature of 32.7 °C. Lower values stand for higher temperatures and make binding reactions less likely. Each tile type occurs infinitely often

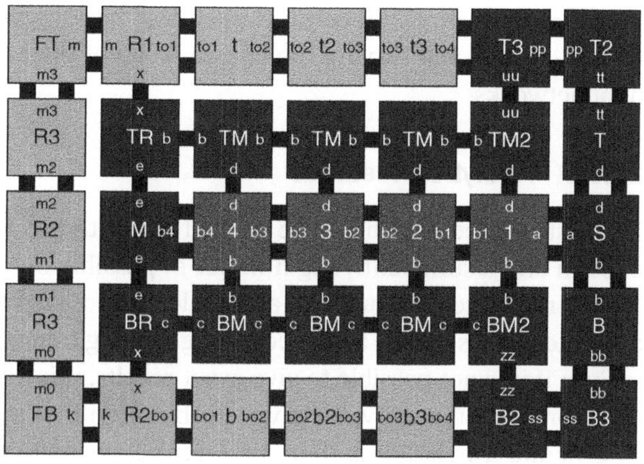

Figure 12.25 Approximation for the message molecules that count tiles in unary coding. Once finished, it is correctly bound to a fitting box, modeled as a receptor, of appropriate depth k. Any shorter message molecule cannot reach the receptor inside the tube to trigger the release of threshold tiles. *Source*: Lau et al. (2021b)/with permission of Elsevier.

in the medium with the same probability and distribution. These are reasonable assumptions for in vitro conditions but do not describe environments in the human body well. (Lau et al., 2019) offered some insights into already conducted wet-lab experiments.

The syntax of the simulator should be understood as follows: The text in the middle is the marker and serves the purpose of distinguishing tiles. The dotted line at the side means the np glue. A glue of strength one is represented with one line with a label next to it, and a glue of strength two is represented with two lines with a label.

Figure 12.24 (b) shows a fully assembled message molecule that counts the concentration of a health parameter in unary coding plus a threshold l. We chose a value of $k = 3$ and an additional threshold of $l = 2$. The dotted-line tile represents an arbitrary value k. The assembly finishes without growths or facet errors (Patitz, 2014).

12.4.7.5 Analysis in the 2HAM

In this section, the interaction of intermediate assemblies and receptors/message molecules is examined in the 2HAM. The simulations in the 2HAM show that the assembly process reaches a fixpoint upon which no additional intermediate assemblies can form, as displayed in Figure 12.25. Furthermore, the proposed message molecule from phase 1 binds to a nanobot without problems. It is not possible for the same molecule to bind to a deeper nanobot as the thick end of the message molecule blocks it from reaching the glues inside the receptor.

The proposed nanonetwork can measure a concentration value and persistently store the measured value as a tile. The network learns a "normal range" for a number of health parameters and can directly react upon exceeding a predefined range in phase 2. This approach is superior to regular medical tests as both memory and computational power can be utilized. The test is of a very general nature and can be adapted to various parameters or disease markers. This type of network thus theoretically allows us to solve yet unsolved medical problems that require state, which is difficult to implement in a feasible manner using tiles.

12.5 Verification Methods for Nanonetworks

As most research on all kinds of nanonetworks still happens on the level of ideas, concepts, and visions, it is necessary to evaluate the performance and correctness of novel ideas apart from wet-lab experiments. While actual wet-lab tests would obviously yield the most useful results, without functioning nanodevices, they are simply not possible. Yet, there are other tools available that are capable of verifying certain aspects of nanonetworks without those being available. In this section, we list and explain the current state of the art in evaluation and verification methods for nanonetworks.

12.5.1 Analytical Methods

The first method that should be applied before trying anything else is analytical. If a model is simple enough and a sufficiently big number of possible outcomes can be explored computationally, it is simply not necessary and not advisable to go a step further and simulate. In this case, the most accurate results can be achieved by, e.g., giving exact formulas.

Even if it is not possible to give exact results, it is often possible to derive upper or lower boundaries for the complexity of a given problem. In that case, analytical methods can be used as a tool to generate a hypothesis that can later be verified or falsified using simulation or even wet-lab experiments.

In the case of many electromagnetic or acoustic nanonetwork algorithms, analytical methods are often not applicable. They are *complex* systems where the many possible interactions between network participants create a state space that is infeasible to fully explore in most cases. Thus, it is better to simulate such networks in most cases.

For tile-based self-assembly, it is also not possible to derive exact results in most cases. While such systems might intuitively occur like simple stochastical processes, their inherent non-determinism makes them difficult if not impossible to predict. Yet, in some cases, it is indeed possible to gain some analytical insights due to the simple and linear design of the tiles involved.

Let us analyze this using the simple case of the previously introduced 4-bit AND message molecule. The expected value of the binding reactions required for a message molecule can be estimated analytically by determining the probability of correct binding of all tiles at suitable locations on the growth front for each assembly α_i of a given assembly sequence S. The summation gives the probability that a binding reaction is correct.

Now, for a desired probability Δ, it can be computed after how many binding reactions a correct binding in the expected value has taken place. Forming the sum over the expected values of all assemblies a_i results in the expected number of required binding reactions for a message molecule with a given assembly sequence S. The expected value can be estimated under the condition that for all assemblies a_i from S, it is true that $|a_i| \leq |a_{i+1}|$.

It is possible to give an upper bound for the expected value of the required binding reactions. For this, we assume that only one tile can correctly bind at a given location w of the growth front at any given time. The probability of this is $1/(|T| \cdot |w|)$. If we now solve the inequality $1 - ((|T| \cdot |w| - 1)/(|T| \cdot |w|))^p > \Delta$ for p, we get the required number of binding reactions for an assembly a_i from S. With 15 different tile types, $\Delta = 0.95$, and a growth front of size 4, this would correspond to 179 expected binding reactions. If you multiply p by $|S|$, you get the worst-case expected value of the required binding reactions of a message molecule for a given assembly sequence length $|S|$.

Since a potentially infinite number of assembly sequences exist for each message molecule and the actual assembly sequence is unknown due to the non-determinism of self-assembly systems, simulation is preferable to analytical consideration.

12.5.2 Complexity Analysis

Apart from classical approaches, the complexity theory methods introduced in Chapter 5 offer a variety of tools to better understand complex systems that exceed mere complicatedness.

Complexity theory aims at finding out certain properties about algorithms or models as accurately as possible. Even for systems where the general complexity is unclear, it is often possible to show that certain algorithms are only as difficult as another algorithm using *reductions*. The same methods are also applicable at the nanoscale for all types of networks and should be performed as one of the initial steps when coming up with a new model.

For electromagnetic or acoustic nanonetworks, there are well-established tools for analyzing the communication difficulty. In essence, there are two general ways to estimate that complexity:

1. counting messages, and
2. counting bits.

The easier but also more commonly used network complexity measure is counting the number of messages that are necessary to solve a given problem or implement an application. For this, we usually use the Landau-notation introduced in Chapter 5. We essentially analyze the network architecture and identify the aspect that grows the worst in comparison to the number of network participants. Note that the Landau notation usually uses the general "input" as a point of reference – the number of nanonodes is a *parameter* of the input that we know to be more useful. The interested reader is encouraged to have a look at (Downey and Fellows, 2013) for more information on the topic of *parameterized complexity*. Oftentimes, a problem might seem very difficult on the surface, but specifics of the input might make it a feasible problem nonetheless. Especially at the nanoscale this might be true as many of the available resources are limited.

Based on that, we simply count the messages that have to be sent over a network in the worst possible scenario or sometimes in an average scenario. Oftentimes, it would be necessary that all devices in a network forward a message, resulting in $O(n)$ messages, which would likely be infeasible at the nanoscale – even if the messages are small. At other times, all nanodevices have to exchange messages with all other nanodevices, resulting in $O(n^2)$ required messages. While this might be feasible for some macroscale networks, it is very likely not an option to use such algorithms at the nanoscale. The last class we discuss here is $O(2^n)$. This class describes the behavior of a network when broadcasting and even macroscale networks cannot handle this, let alone nanonetworks.

While counting messages already answers the most important questions of possible infeasibility, sometimes it is necessary to go a step further. In this case, we try to find out the minimum number of bits that have to be exchanged in order to solve a problem. Especially in resource-constrained nanonetworks, this might be necessary in some cases.

This approach is based on *information theory* and addresses the amount of information that may be contained in a bit string of a certain length. When using binary numbers to encode information, a bit string of length n may only assume a certain number of different states. In this case, starting from 0, it is possible to count all the way up to 2^n. If we want to gain insight into other network participants, we can exchange messages with them. For each bit we transmit, we can rule out *half of all possibilities*. As an example, a four-bit binary number has 16 options. As soon as we find out that the first bit is, e.g., a "1," there are only eight remaining options for the remaining bits. As a result, a message of length n contains $\log(n)$ information (Shannon, 1948).

Based on this simple principle, it is possible to derive upper and lower bounds for the number of bits that are necessary to solve a problem. Even using the best compression algorithm in the world or a flawless encoding scheme, it is not possible to use fewer bits than that.

For tile-based self-assembly systems, such tools are not as well known as most researchers use them for computational purposes only. In the following paragraphs, we introduce the *tile complexity* to estimate the difficulty of solving a given problem using message molecules. Instead of bits, we use tile types as basic units that contain information. In the following, we compare simple message molecules introduced here or in Chapter 5 with circuit classes. Circuits are interesting because they do not possess memory and are equivalent to Boolean formulas in their computational capabilities.

The complexity of various operations differs from computational model to computational model. The complexity-theoretical considerations from Chapter 5 only refer to circuits – other parameters are of interest for self-assembly systems. Table 12.8 shows a comparison of circuit implementations and message molecules. Often only upper bounds are known for the problems shown. These relate to the best-known implementation. A lower bound can be specified for the logical OR in self-assembly systems since only one tile is required for this.

In Chapter 5 and earlier in this chapter, the operations AND, ADD, and THRES were examined in more detail and modeled as nanonetworks. These problems correspond to circuit classes AC^0 (AND and ADD) and NC^1 (THRES), so they are rather simple operations. In self-assembly systems, the same operations require a tileset of linear size for AND, a tileset of constant size for ADD, and a tileset of greater than exponential size for THRES. This shows that simple circuit implementation does not necessarily mean that these problems can also be easily solved by self-assembly systems.

Most of the presented message molecules fall into the NC^1 class but sometimes have a very different tile complexity. In general, the tile complexity does not seem to be an appropriate measure of complexity for message molecules but it is rather used as a standard measure of the complexity

Table 12.8 Complexity of problems related to the input size n that are of interest for nanonetworks compared to circuits.

Problem	Space class	Assembly size	Tile complexity
ADD	AC^0	$O(n)$	$O(1)$
GEQ, LEQ, EQ	NC^1	$O(n)$	$O(n)$
SUB	AC^0	$O(n)$	$O(1)$
MULT	NC^1	$O(n^2)$	$O(1)$
DIV	NC^1	$O(n^2)$	$O(1)$
INC	AC^0	$O(n)$	$O(\log(n))$
AND	AC^0	$O(n)$	$O(n)$
OR	AC^0	$O(1)$	$O(1)$
XOR	AC^0	$O(n)$	$O(1)$
ODD, EVEN	AC^0	$O(1)$	$O(1)$
MIN, MAX	NC^1	—	—
THRES	NC^1	On	$O\binom{n}{k}$
IT-ADD	NC^1	$O(n^2)$	$O(1)$
MOD	NC^1	$O(n^2)$	$O(1)$
AVG	NC^1	—	—
MEDIAN	L	—	—
CUBE	—	$O(n^3)$	$O(\log n)$

Upper limits are given in each case. The tile complexity does not take the input into account. The size of an assembly corresponds to the number of tiles used (Lau, 2020).
Source: Lau et al. (2021b)/with permission of Elsevier.

of self-assembly computations. A large part of the required tiles in And and Thres goes back to the required frame structure, which provides a binding option for the input of the computation – the computation itself is still simple.

A sensible, more advanced approach is to determine the number of algorithmically relevant tile types or different glues. This is already the case with the Add problem since only the tileset of constant size that is involved in the assembly of the output is counted. Framework tile types and input are the same. In the case of the And problem, the complexity of the circuit class is equalized since only a constant tile complexity remains after deducting input and framework tile types. The problem Thres is also reduced with regard to its tile complexity if the procedure is identical. Only a linear number of algorithmically relevant tile types remains.

12.5.3 Simulation

While we already had several sections on specific simulation software, we never analyzed simulation as a method in general. Whenever analytical or theoretical methods reach their limits, either due to the size of the input, complexity, or missing options to simplify a problem, it is often beneficial to simulate as a next step. Simulation allows researchers to gain insight into systems that are otherwise hard to come by or even infeasible.

Overall, a simulation attempts to recreate all the relevant aspects of a physical system in a computer model. As reality is usually too complex to represent in computers without additional steps, simulation programmers often abstract and modularize. First, they usually focus on just a subset of relevant components to make the task computationally more feasible.

Next, researchers are usually interested in finding a solution to a much bigger problem but have to “cut corners” to make a simulation even possible. In reality, there are usually millions if not billions of components that interact with each other at any given time. If more than the interaction of a handful of parameters is desirable, the entire task would be computationally intractable. As a rule of thumb, the jump from two to three observed intertwined properties usually means that a problem is at least in the complexity class NP and thus very difficult to solve. In such cases, it is often necessary to focus on the different parameters individually and reassemble the partial results into a solution for the bigger problem. While this is not always possible, it is the best approach we have in many cases.

Next, any simulation has to make sure that outliers and other anomalies are taken care of. Natural systems typically behave in a somewhat predictable manner and results often follow a Gaussian distribution. The vast majority of systems thus behave in a similar way. However, there are outliers that can corrupt the interpretation of any simulation. It is usually necessary to repeat the same simulation many times to get an idea of how the system behaves on average. This is possible due to the *law of large numbers*.

The law of large numbers is a fundamental concept in probability theory and statistics. It states that as the number of trials or observations increases, the observed or empirical probability of an event will converge to its true or theoretical probability.

It is easy to understand this law using a series of coin flips. The probability of getting both heads or tails is 0.5. If we flip the coin 10 times, we may not observe an equal number of heads and tails. However, if we continue flipping the coin more and more times, the observed proportions of heads and tails will tend to get closer to the theoretical probabilities of 0.5 each. If we flip the coin 100 times, we might observe 51 heads and 49 tails. If we flip it 1000 times, we might see 502 heads and 498 tails. For very large numbers, the observed proportion of heads will converge more closely to 0.5.

We do the same thing with simulations where we try to identify the true average behavior of a system. However, as most systems we simulate are *complex*, it would be foolish to expect simple behaviors. Yet, a majority of simulation runs may still follow a somewhat stable pattern that might be useful for further research.

In the nanonetworking community, we need a number of different simulation tools to provide an answer to all the problems that arise in the community. Among them are simulators for networking questions, physical particle systems, self-assembly, protein folding, and many more. We have already talked about specific instances of simulation tools in before chapters and will not repeat ourselves here.

12.5.4 Organs-on-Chips

The logical next step in evaluating the performance of certain concepts and ideas is wet-lab experiments. As real-world systems are complex, some important factors have likely been overlooked in simulations and other previous analyses. Thus, it is necessary to test any new ideas using empirical experiments under conditions that are as close as possible to the intended application.

As most nanonetworks have medical use cases, wet-lab experiments are not always possible for ethical, political, or legal reasons. Any new medication must be tested thoroughly before the application on real humans or even animals is allowed. As a result, some researchers came up with organs-on-chips that we have briefly introduced in Chapter 7 (Huh et al., 2011).

Organs-on-chips try to replicate the conditions inside human bodies as realistically as possible. These chips are typically made from transparent materials, such as polymers, and contain hollow channels lined with living cells that mimic the specific functions of an organ.

The development of organs-on-chips is driven by the limitations of traditional cell cultures and animal models in accurately representing human physiology and predicting the response of human organs to drugs, diseases, and other interventions. Organs-on-chips offer a more physiologically relevant and personalized approach to studying human biology and advancing drug discovery and development.

Organs-on-chips utilize microfluidic systems, which involve precise control and manipulation of small fluid volumes within channels. These channels are designed to replicate blood vessels, airways, or other physiological structures found in organs. By flowing nutrient-rich media and other substances through the channels, researchers can recreate the dynamic microenvironment of the organ being modeled.

Cells derived from human organs are cultured on the chips to recreate the specific tissue types and functions. For example, liver-on-chip models may contain liver cells cultured in a way that mimics the liver's structure and function. The cells used are typically derived from human sources, such as induced pluripotent stem cells or primary cell cultures.

In addition to fluid flow and cell cultures, organs-on-chips can simulate the mechanical forces that organs experience. These forces include stretching, contraction, or fluid flow. By incorporating these mechanical forces, researchers can study the effects of physical stress on cells and tissues, providing insights into organ function and response.

Some advanced organs-on-chips technology allows for the integration of multiple organ models on a single platform. This enables the study of organ interactions, drug metabolism, and toxicity across different organs simultaneously. These interconnected systems provide a more comprehensive representation of human physiology and can help understand complex biological processes.

Organs-on-chips often include sensors and imaging techniques to monitor various parameters. These measurements provide real-time data on cell viability, metabolic activity, and gene expression. Monitoring these parameters allows researchers to assess the health and response of the cells and tissues, providing valuable insights into the functioning of organs and their responses to drugs or diseases.

The benefits of organs-on-chips are significant. They have the potential to reduce the reliance on animal testing as they provide a more human-relevant platform for studying diseases and evaluating drug efficacy and safety. Organs-on-chips may also enable personalized medicine approaches, allowing researchers to develop targeted therapies based on individual patient characteristics, once the technology reaches maturity.

12.5.5 Wet-lab

The last and most important step is the realistic test of a proposed application. For most nanonetwork concepts, such tests are currently not feasible as nanodevices simply do not exist. However, many of the components nanonetworks or nanodevices consist of may be tested individually to gain some insight. For example, molecular motors, energy sources, sensors, and actuators as well as point-to-point communication and many other components can be and in some cases already have been analyzed in wet-lab experiments.

Concerning holistic nanonetworks, only bacteria-based nanonetworks and DNA-based nanonetworks can be implemented given current technological capabilities. Acoustic, molecular, and electromagnetic nanonetworks lack the devices necessary and are mainly theoretical research subjects. That said, the communication aspects themselves may be simulated using larger machinery outside of the nanoscale.

As bacteria-based nanonetworks are relatively straightforward in their execution, we focus on DNA-based nanonetworks as a case study. (Rothemund et al., 2004) presented the results of a wet-lab experiment with DNA tiles. In the said experiment, an attempt was made to create a Sierpinski triangle from tiles. The Sierpinski triangle is the result of computing an XOR in tile-based self-assembly systems, and an example can be seen in Figure 12.26.

A tile concentration of 0.2 μM was used for each tile. The value corresponds to a quarter of the simulated concentration, which roughly corresponds to the salt concentration in human blood (0.14 μM). The results indicate that structures from 100 tiles can grow in 60–96 min with low error rates.

Such experiments offer great insights into the behaviors of DNA-based self-assembly, even if they simulate just one aspect of a DNA-based nanonetwork. Furthermore, from that and other experiments with tiles, we can extrapolate and estimate how other, similar systems might behave. As some

Figure 12.26 Image of a Sierpinski triangle that is the emergent result of computing an XOR in tile-based self-assembly systems. *Source*: Adapted from Rothemund et al. (2004).

message molecules from DNA-based nanonetworks only consist of 15–24 tiles, it can be expected that the error rate and the required assembly time are lower than in the given wet-lab experiment.

Yet, a comparison between simulation and wet-lab experiments is only of limited significance. With a tile concentration of 0.8 μM per tile, a message molecule AND was completed in 3000 binding reactions. In the wet-lab experiment, the desired structures were ready after one hour at a concentration of 0.2 μM.

Since the message molecules – in contrast to the 100 tiles of the wet-lab experiment – consist of only 15–24 tiles, it can be expected that they would be finalized more quickly. When extrapolating, we get about 10–16 minutes to complete the assemblies. In general, larger assemblies grow faster than small ones in the laboratory because the growth front contains more sites for binding. However, it must be considered that the growth front of the wet-lab experiment was very large, but the number of different tile types was small at 15–35.

In simulations, this initial situation leads to a small number of binding reactions being required. Such an effect is also to be expected in reality. However, it can be assumed that the total time for many tile types will hardly change if the absolute tile concentration is increased. This is a common practice when adding more tile types. Due to the increased total concentration, only more binding reactions will take place per unit of time.

The number of required nanorobots and nanosensors can be limited. In order to ensure that a message molecule can be created for each seed tile, the same number of input tiles and other inputs must be able to be distributed. If the tiles are placed in a DNA box with dimensions $42 \cdot 36 \cdot 36\,nm^3$ (Andersen et al., 2009), then 472 tiles of size $14.4 \cdot 4 \cdot 2\,nm^3$ Rothemund et al. (2004) can be calculated. If the presented squares of edge length 50 are folded into a box, 8 680 555 tiles could be accommodated in it.

Assuming equal distribution, 1.275×10^{10} DNA boxes would suffice to hold the required tiles with a media volume of 50 μl (Rothemund et al., 2004). If, on the other hand, a box with an edge length of 50 is used, it would be 694008 boxes. How the tiles are distributed on the nanosensors can be decided at will.

In the work of (Rothemund et al., 2004), the number of all required components was estimated upward, which is why this procedure is suitable for wet-lab experiments with message molecules.

Fully assembled message molecules behave like regular molecules used for communication. They are also subject to Brownian motion. For a detailed analysis of the behavior of molecules used for communication at the nanoscale, refer to the work of (Felicetti et al., 2014).

In summary, it is expected that message molecules and receptors can assemble correctly and interact meaningfully in a controlled environment such as a Petri dish.

12.6 Summary

In summary, there are several different ideas and concepts that may serve as a foundation for future directions the research field of nanonetworks might take. Some of them are strictly nature inspired, some take elements from natural systems into consideration, and others are completely man-made. Yet, all of the prevalent ideas and concepts are based on some foundation of already existing research.

No matter what type of nanonetwork we analyze in detail, all of them share certain constraints and environmental novelties/difficulties. Among them are dynamic and mobile network typologies, decentralization, self-organization, resource constraints, severely limited energy supplies, and unique applications. The majority of nanonetwork applications are currently envisioned to take place inside the human body to serve medical purposes like health monitoring or the direct treatment of diseases. Nanonetworks themselves are usually conceptually integrated into IoNTs to be part of a larger ecosystem.

Nature-inspired systems like bacteria-based nanonetworks may be fast to implement when it comes to first working prototypes. While they create good headlines and publicity for the research area, it is unclear what the limitations of such systems really are. It might be possible to modify the behavior of natural cells and bacteria to some extent to perform, e.g., computations or networking tasks, but once a certain threshold is exceeded, these organisms tend to die. This puts clear but yet unknown limits on the capabilities of such systems. Furthermore, it is likely still necessary to introduce some artificial components like gateways that aggregate measured data into such systems to create an interface to the macroscale.

Next to mainly natural systems, we have analyzed nanonetworks based on molecular communication. There are several different ways to implement those, ranging from, e.g., zinc ion concentrations to DNA-tiles as information carriers. All molecular communication networks have the following in common: They follow a partially nature-inspired communication approach based on the diffusion of small particles. Those work similarly to hormones and pheromones inside and outside the bodies of living creatures. There, it is mainly the type of particle that contains information and influences the organism in different ways.

In molecular communication networks based on simple particles, researchers often envision the concentration to encode bits. In DNA-based nanonetworks, however, there are different ways of encoding information that exceeds a mere concentration value. DNA molecules can also be understood as words from the alphabet $\{A, T, C, G\}$, which can contain much more information per particle but have the same communication latency. These networks are especially slow when it comes to transmitting bits, which might be a limiting factor. However, they could at least work given currently available technologies and have other advantages like biocompatibility.

The last group of nanonetworks we have discussed are mainly artificial, and this group includes both acoustic and electromagnetic nanonetworks. Many researchers (either directly or indirectly) hope that miniaturization continues as before and that near-future breakthroughs make nanodevices widely available. This would allow the networking community to utilize many of the already established tools without having to learn new paradigms.

Like all other nanonetworks, they consist of a very large number of nanodevices that communicate to the macroscale via a microscale gateway that aggregates and preprocesses data. One of the advantages is the speed of communication and the theoretically available bandwidth. Acoustic nanonetworks may communicate with the speed of sound in water of about 1500 m/s and electromagnetic nanonetworks transmit data at the speed of light in the terahertz band. Especially electromagnetic nanonetworks may have astounding bit rates but a very limited communication range due to molecular absorption.

While some of the presented networks *may* be tested under realistic conditions, few real experiments are actually conducted. Evaluation and verification usually happen on a theoretical level via analytical tools, complexity theory, or simulation. Only a handful of expensive wet-lab experiments are published every year, and they are usually far from complete nanonetwork implementations. For the time being, simulation is likely the best bet for reasons of feasibility and cost-effectiveness. The entire field is simply so young that it might take several years if not decades until industrial interest in the topic accelerates research progress and spawns real-world applications.

Yet, there are tremendous amounts of revolutionary and potentially paradigm-shifting ideas that might shape the future of how we think of communication and computation. The technological advancements of the near future will likely play out at the nanoscale where nanonetworks are a more than suitable paradigm to tackle the multitude of new challenges.

13

Ethical, Legal, and Social Issues

This chapter discusses various aspects that may influence the development and acceptance of nanonetworks that are not of a technical nature. No matter the technology, people tend to be cautious upon first introduction and require a number of positive experiences before they accept a new technology. This usually has to happen gradually as any big change all at once is usually experienced as a potential threat. Everything in our evolutionary history happened bit by bit and any deviation from that poses a felt risk.

This chapter discusses everything related to ethical, legal, and social issues that might arise in the future. We try to develop nanotechnologies that help people which requires frequent checking to see if the current developments and ideas are compatible with the current views in society.

We start this chapter with a short discourse on the evolutionary and psychological need for graduality. Next, we analyze potential environmental issues that nanonetworks might cause or face. We then discuss potential issues of biocompatibility and waste disposal of nanoscale materials. Pollution is a big issue, and it is necessary to avoid the accumulation of garbage that cannot even be seen by the naked eye.

Next, we analyze some issues related to politics and legal matters that have to be taken into account when designing potential new types of medication. There is an immense body of rules that vary from country to country that novel technologies and their development have to abide by.

Finally, we discuss potential dangers and catastrophic scenarios that science fiction literature proposed over the years. In many cases, these fears are not founded in real possibilities, but some of them might pose some valid concerns. In the end, we summarize the chapter and suggest possible future directions and investigate potential concerns as well as questions of acceptance.

13.1 The Process from Idea to Final Product

Just like natural processes, technology usually advanced gradually and improvements happen in small steps. The same is just as true for evolution as it is for the future development of nanonetworks – it is only that technology changes at a much faster rate. Furthermore, it is not only that such processes *naturally* develop slowly, most people also *expect* slow change. If researchers want their concepts to become reality, they have to slowly introduce them into the general awareness while simultaneously demonstrating their benefits. Negative publicity as with research into genetic modification should be avoided whenever possible.

We have already discussed various theoretical tools for evaluating nanonetwork concepts in Chapter 12. The next logical step is an empirical test where a previously generated hypothesis is tested under realistic conditions. As mentioned many times before, theoretical approaches have to

Nanonetworks: The Future of Communication and Computation, First Edition. Florian-Lennert A. Lau.

sacrifice accuracy and realism to enable feasible computer simulations. Thus, it is almost always necessary to test if the models reflect empirical observations somewhat accurately.

The first "real" step in a gradual process from a theoretical concept to the release of a product usually involves a number of Petri dish experiments where specific aspects of an often larger problem are analyzed in isolation. While Petri dishes are still very different from the target environment of, e.g. a human body, it is still possible to introduce a variety of new parameters that a simulation could not take into account. The new observations then either verify or falsify the hypothesis, and the models have to be adjusted until they somewhat accurately predict the behavior of the real world. Wet-lab experiments like that are usually expensive and not accessible for smaller universities or corporations, and it is nearly always more desirable to run a computer simulation.

Yet, after a number of iterations, the models are usually "good enough" to move on to the next phases of testing. The goal of this gradual process is to find a good balance between realism, costs, and sometimes politics/social influences. Many times, animal tests using, for example, lab rats are next in line. While this standard procedure usually works well for designing novel types of medication, rats and their environments are simply too different from human bodies. Unless researchers develop real nanoscale devices that could be tested inside of, e.g. a rat's circulatory system, it would be better to look for alternatives. While some concepts like those of bacteria-based nanonetworks exist and would likely function in the very near future, the new information gained from such experiments would likely be very limited and such an experiment would be "just" a proof of concept that might still generate some publicity.

One of the alternatives is tests inside fish tanks or more complex agricultural systems involving fish. Fish tanks provide a larger model of a mostly liquid environment that includes several biological components that might mimic a circulatory system closer. Furthermore, it is likely possible to use much bigger devices and still gain some insight into the multitude of possible influence factors we have previously ignored or generalized away in the models.

When all these steps have been somewhat successful and indicate a true benefit for, e.g. the field of medicine, the majority of research ambitions typically move away from the university context toward corporations that try to create marketable products. This phase presupposes that previous ambitions have led to the first prototypes of nanobots or nanonetworks that may now be refined until they meet medical regulations and political standards. Now, the typical process of testing and releasing new medicine starts. This process may include any combination of the previous steps, including testing on real people who typically suffer from a terminal illness. For those people, the potentially adverse byproduct of a poorly tested new technology is still better than the alternative.

Only when the new product successfully goes through all the previous steps will it reach maturity and be widely available on the market. Due to the inherent complexity and the potential political obstacles especially inside of the European Union, it might very well take decades until the presented process is finished. Yet, it might still be worth it if the new product provides benefits on the level of generality of a smartphone. After all, nanonetworks may not only perform a single function but offer advantages in a vast number of aspects of life.

13.2 Environment

As you can see, this rather long process involves a number of obstacles of physical, ethical, legal, and social nature. In the following paragraphs, we analyze various aspects of the target environment of nanodevices.

When discussing nanomachine deployment in humans, *environmental health and safety* (EHS) risks as well as ethical, legal and social issues should not be ignored. For assessing potential risks of new technologies, an evolutionary analysis can be beneficial (Dobzhansky, 2013). Everything a species was repeatedly exposed to during its evolutionary history is typically harmless or at least predictable (Anderson et al., 2016). For example, species adapt to exposure to certain substances and sometimes use them in their organisms as building blocks (food), while some of them even become essential (vitamins/essential fats). Particles that are less common or less common in high quantities often impose a significant risk to a person's health or may at least trigger unpredictable consequences.

In addition to that, there are several other environmental influences humans and other living creatures had to adapt to. As a rule of thumb, the more omnipresent some factor was, the bigger the possible impact. We now go through a list of evolutionary influence factors sorted according to significance. Please keep in mind that both the selection of factors as well as their order are subject to debate. However, it should be obvious that all of those factors can make or break the success of nanodevice or nanonetworks inside living organisms.

1. The first and likely most important factor is biochemical interactions. Understanding and leveraging biochemical interactions is crucial for medication development and nanonetwork design. The human body relies on intricate biochemical pathways for normal physiological processes. Medications or nanonetworks that can interact effectively with these pathways can affect specific molecular targets, modulate cellular processes, and achieve desired therapeutic effects. Designing drugs that mimic or inhibit specific biochemical reactions, such as enzyme inhibitors or receptor agonists/antagonists, can lead to more targeted and effective treatments. It is overall crucial to at least consider those biochemical factors that were stable for a long time in the evolutionary history of a species. That way it might be possible to anticipate potentially harmful or even lethal interactions between the organism and external substances.
2. The next important aspect is potential immune system responses. Considering the immune system response is vital when developing medications or nanonetworks. The immune system is responsible for recognizing and eliminating foreign substances or pathogens. Technically, this is a subclass of biochemical interactions that deserves its own section due to the complexity of the immune system. To avoid triggering adverse immune responses, medications or nanonetworks should be designed to minimize immune recognition. Strategies include developing surface coatings or modifications that prevent immune cell binding, designing stealth nanoparticles to evade immune surveillance, or incorporating immune modulatory agents to regulate immune responses. Ensuring compatibility with the immune system increases the chances of successful drug delivery and reduces the risk of immune-related side effects.
3. The next factor is (cellular) microenvironments. Different tissues and organs have unique cellular microenvironments, characterized by specific cell types, extracellular matrix composition, nutrient availability, and physical structures. Understanding these microenvironments is crucial for designing medications or nanonetworks that can effectively navigate and interact within them. For example, drug delivery systems can be engineered to respond to specific cellular cues, such as targeting receptors or enzymes expressed in target tissues, or utilizing nanoparticle size and surface properties that allow them to penetrate specific barriers or cell types. Adapting to cellular microenvironments enhances the targeted delivery of medications or nanonetworks and increases their therapeutic effectiveness.
4. The next factor is the environmental temperature. Maintaining optimal temperature conditions is important for medication development and nanorobot design. Extreme temperatures can

affect the stability, structure, and function of medications, nanonetworks, and especially DNA-based nanonetworks. They may cause denaturation of proteins, alter drug release kinetics, or compromise the integrity of networks. Therefore, ensuring that medications or nanonetworks can operate effectively within the physiological temperature range (around 36–37 °C) is crucial for their functionality and safety within the body.

5. On a similar scale compared to the temperature, the pH levels are also relevant. Different parts of the human body have varying pH levels, and medications or nanonetworks must be able to withstand and operate within these specific pH environments. For instance, the stomach has an acidic pH, while the blood has a slightly alkaline pH. Designing medications or nanonetworks that are pH-sensitive can enable controlled drug release or activation in targeted areas. pH-responsive drug delivery systems or nanonetworks can be designed to respond to specific pH conditions, such as using pH sensitive polymers that change their structure or solubility in response to pH variations, allowing for site-specific drug release.
6. Next come oxygen levels in and outside living creatures. Considering the oxygen levels within the body is important for medication development or nanorobot design. Different tissues and organs have varying oxygen concentrations. Medications or nanonetworks should be able to function effectively within these varying oxygen conditions to ensure their efficacy. Oxygen-sensitive drug delivery systems can be designed to release drugs in oxygen-deficient areas, such as solid tumors, where low oxygen levels are a characteristic feature. Additionally, oxygen sensors can be integrated into nanonetworks to detect and respond to local oxygen levels, guiding their behavior and drug release profiles.
7. Next come mechanical forces. Certain parts of the human body experience mechanical forces, such as compression, tension, or shear. This aspect requires much care as the human body is well adapted to various levels of physical activity that vary to extreme degrees in terms of the physical forces that act on the body. Medications or nanonetworks intended for use in mechanically stressed areas, such as joints or blood vessels, need to be able to withstand these forces without compromising their structural integrity or function. For example, drug-eluting stents used in cardiovascular interventions need to be flexible and capable of expansion without fracture. Nanonetworks designed for targeted drug delivery to joints should be able to withstand the mechanical stresses and movements associated with joint motion. Considering mechanical forces is crucial to ensure the durability and effectiveness of medications or nanonetworks in mechanically challenging environments.
8. Another important factor is metabolic processes. The human body undergoes complex metabolic processes, including the breakdown and transformation of drugs or foreign substances. Medications or nanonetworks should be designed to be metabolically stable and resist degradation to maintain their effectiveness and prolong their duration of action. The metabolism can influence drug efficacy, bioavailability, and toxicity. Understanding the metabolic pathways involved in the clearance of medications or nanonetworks and designing them with suitable chemical modifications or formulations can enhance their stability, bioavailability, and therapeutic potential.
9. Finally and possibly least importantly, we have electric and magnetic fields. The human body generates electric and magnetic fields naturally and is equally subjected to them. Medications or nanonetworks that interact with these fields can have specific applications. For example, in electrostimulation therapies, medications or nanonetworks may need to respond to electric fields to modulate nerve or muscle activity. In magnetically targeted drug delivery, medications or nanonetworks can be engineered with magnetic components to navigate and concentrate at specific locations under the influence of external magnetic fields. Designing medications or

nanonetworks to interface effectively with electric and magnetic fields enables targeted and controlled delivery, enhancing their therapeutic potential in these specific applications.

Next, we proceed along a hierarchy of generality and move toward more specific aspects. Now that we have a rough idea of what influence factors we have to expect, we can search for experimental values or insights from already existing areas of research. In this case, we can take especially experiential values from the medical field, as there is likely most data available. As living creatures are extraordinarily complex systems, we only demonstrate potential problems using a few key examples.

An event of special importance that shed light on the potentially very complex interactions inside the human body is the thalidomide tragedy. The thalidomide tragedy happened in the 1950s and 1960s when the development of new medication was less formalized. The medications called thalidomide or in Germany *Contergan* were initially prescribed against nausea in pregnant women but caused severe birth defects in the newborn children as a byproduct.

For the first time, it became clear that the human body is much more complex than initially expected and that unwanted side effects may happen more often than expected and must be avoided at all costs. It is not only about *what* substances are present but also about *when* they are administered. While this is already important for adults and especially for women with their hormonal complex menstrual cycle, it is even more important for fetuses and children. The simple presence of a drug at the wrong time can adversely affect the unborn life to sometimes extreme degrees. Fetuses need extremely precise environmental conditions for optimal growth.

The tragedy highlighted the critical importance of rigorous safety testing and evaluation during drug development. No matter what kind of medication is developed, unintended side effects happen all the time, sometimes years or decades later. For example, the famous Viagra pill was originally planned as heart medication and researchers only later repurposed the drug when they noticed said side effects.

However, the same is not only true for environments in the human body but also *outside* too. Whenever some substance acts inside of an environment that had no time to gradually adapt to it in an evolutionary process, unwanted side effects are to be expected. Normally, nature finds some kind of balance around major changes but that process takes time. In modern human history, there were and are several such events that took a long time for consequences to surface.

This includes fuels where lead has been added as a performance-enhancing agent to reduce engine knocking. However, it was later discovered that leaded gasoline emissions posed significant health and environmental hazards. Lead is a toxic metal, and prolonged exposure to lead can cause severe health issues, particularly affecting the nervous system. A similar crisis happened due to the CFC gas used in fridges and foam-blowing agents. The untested use led to the unintended interaction with the ozone layer which caused tremendous problems.

The most severe cases are likely the industrial use of radioactive materials where an entirely new type of health hazard has been discovered as well as the widespread use of plastics. In the first case, there was simply no evolutionary process that could have prepared life forms for the adverse effects of radiation and catastrophic consequences resulted whenever contaminated substances came into contact with regular environments or people. In the case of plastics, the effects started to surface much slower. As that class of material never existed in the evolutionary history of our planet, there were no organisms that could interact with the material. As a result, it was nearly indestructible and only breaks down into smaller and smaller pieces that harm especially sea creatures as garbage tends to accumulate in the oceans.

As you can see, the introduction of a completely novel material or drug can pose tremendous risks that have to be analyzed in depth before being made widely available. The catastrophes of the past have to be avoided at all costs. Unwanted side effects are the *normal case* and not a strange, rare, or unexplainable phenomenon. If something has an effect, it is almost guaranteed that there are also *side effects*.

13.2.1 Biocompatibility

Many of the previously mentioned points boil down to issues of *biocompatibility*. Biocompatibility describes the property of technology to minimize the harmful interaction with the biological target environment. Biocompatible technologies attempt to act in unison with organisms and minimize disruptiveness. Especially nanonetworks may reduce disruptiveness by precisely acting where intended without causing any unnecessary systemic effects.

Nanonetworks may tackle yet another problem that is currently unsolvable. Many substances have a healthy *range* in which they cause no harm to an organism and are oftentimes even beneficial. However, when that healthy range is not met or exceeded, harmful consequences would be the expected result. That healthy range varies from person to person, and nanonetworks might be able to individually determine that range and either inform the host or directly treat the problem.

The occurrence of some heavy metals in the human body is a suitable example. While zinc, copper, and iron are involved in several hundred processes in the body, too much of any of those typically leads to sometimes deadly diseases. The organs which are responsible for removing an excess of certain substances only adapted according to necessity. If there was never a need to get rid of copious amounts of iron from the blood, no evolutionary mechanism that deals with such problems can be expected to exist.

Under these circumstances, possible materials for in-body nanomachines are harshly restricted. Especially CMOS technologies might cause potential problems as they have been designed to function in different environments. As a result, they have to be modified to meet the new requirements.

One of the new and often overlooked problems would be *coating* we have mentioned in Chapter 3. Any object that is introduced into the human bloodstream automatically interacts with all other present substances, and in many cases, this leads to nanodevices being surrounded by proteins and fats. To still function properly, researchers have proposed to frequently stimulate, e.g. implants with ultrasound to "shake" the device sufficiently for coating proteins and fats to fall off.

While especially coating applies to *all* nanoobjects, DNA still appears to be a suitable building block as we have discussed before. It might even be possible to avoid parts of or the entire problem by "blocking" the open ends of DNA using weaker binding strands that may break off when a more suitable DNA sequence attempts a binding. Many of the previously mentioned issues likely do not occur at all. However, as DNA also encodes genetic information and potential to replicate, any DNA-based nanonetwork designer has to proceed with great care to avoid these issues.

13.3 Waste Disposal

All nanoscale devices or nanonetworks have an expected time to live where they may perform the desired function. Afterward, they may be considered as "garbage" and the remaining materials have to be dealt with. While there is no clear consensus on how such wastes may be removed from a living body, there are several ideas that we analyze in the following. That said, all of these are on the level of ideas and should be read with care. It might very well be that some or many of the presented methods do not work under realistic conditions.

1. The first and most obvious example is *avoidance* of the problem in the first place. Under normal conditions, the human body and that of other living creatures have a limited tolerance for foreign substances. If the concentration of the administered nanonetwork or swarm of nanodevices is sufficiently small, problems may be avoided in the first place. In many cases, it might even be a good idea to avoid the use of nanonetworks until there really is no other option to minimize potential risks. This procedure is akin to the limitations on X rays to avoid potential health hazards. In almost all cases, it is simply best to avoid the problem in the first place.
2. When it is not possible to avoid such problems altogether, the problem might be small enough for the human body to deal with on its own. The human body has various mechanisms including filter organs to eliminate waste, including the liver, kidneys, skin pores, spleen, and excretory system. If dysfunctional nanorobots are recognized as foreign entities, they may be broken down by enzymes or other biological processes and eliminated through urine, feces, sweat, or even the breath.

   However, it is important to consider the type of nanodevice and the materials they consist of. In the case of molecular nanodevices, there might be no problem as all the human body has the capability to deal with the resulting waste materials. The same might be true for DNA-based nanodevices. Filter organs like the spleen that naturally purify the blood may also be able to remove artificial DNA. As DNA is everywhere in all living creatures, it would be highly unlikely that the body does *not* possess such a mechanism.

   The most difficult case would be nanodevices on the basis of CMOS technologies or other heavy metals. While many of those elements naturally occur within the human body, the concentration is usually minuscule. If the natural thresholds are exceeded, some kind of *heavy metal poisoning* would likely be the result which might be life-threatening in some cases. While there are some methods to deal with them, like chelation therapy or phlebotomy, it can be very difficult to remove the, waste materials for good.
3. It would be much better if nanodevices were made of materials that do not pose such risks to create biodegradable nanodevices or nanonetworks. Designing nanorobots with biodegradable materials is a promising approach. These materials could be engineered to break down over time into harmless byproducts that can be metabolized or eliminated by the body. Researchers would need to select biocompatible materials and optimize their degradation rates to ensure safe and controlled breakdown. While there would still be waste materials, many of them can be harmless for the human body.
4. The next potential approach would be targeted drug therapies that counteract specific nanoscale wastes. Such medication might be able to neutralize dysfunctional nanorobots, accelerate the breakdown of such devices or assist the human immune system by increasing the detection probability. In the simplest case, such a drug must only remove the component that hides the nanodevices from the human immune system. In more complex scenarios, the drug of choice could attack the nanodevices directly while leaving the surrounding tissues unharmed. The resulting waste may then be removed by the usual bodily mechanisms that basically all of the presented methods have to use in some form or another.
5. In some cases, magnetic or electromagnetic removal might be possible. If the nanorobots possess magnetic properties or can be influenced by external electromagnetic fields, it may be possible to

guide them to specific areas for removal. Magnetic fields could be used to attract and manipulate the nanorobots, directing them toward regions where they can be extracted or where natural elimination processes are more effective. Some researchers have already demonstrated that it is possible to create nanodevices that may be moved via the application of an external magnetic field (Huang et al., 2016). The same mechanism may be used to gather nanoscale waste material in one area to remove it.

6. In cases where other methods are not viable, surgical intervention may be necessary. Minimally invasive techniques like endoscopy or laparoscopy could be employed to locate and extract dysfunctional nanorobots from specific areas within the body. These techniques utilize small incisions or natural body openings, minimizing tissue damage and promoting faster recovery. This approach is especially promising if the nanodevices have already been gathered in a small area. If this is not possible, it might be necessary to periodically and gradually remove parts of the blood until the concentration of waste materials reaches manageable levels. This idea is not new and has been used for hundreds of years if patients were suffering from too much iron in the blood (hemochromatosis/iron overload).
7. If simply removing parts of the blood is not feasible or not fast enough, it might be necessary to use other filter mechanisms or devices. It may be possible to develop specialized filtration devices that can selectively capture and remove dysfunctional nanorobots from the bloodstream or other bodily fluids. These devices could act as filters or traps to separate nanorobots based on their size, shape, or specific properties, preventing their circulation within the body. A similar approach would be dialysis, where the circulatory system is connected to an external filtration device to remove unwanted substances. This approach is often used in case of kidney failure, but it might be possible to adapt it for nanodevices too.
8. The last and possible most futuristic approach would be bioengineered micro-organisms. These micro-organisms could be designed to recognize and degrade or neutralize dysfunctional nanorobots, acting as natural eliminators or scavengers within the body. Similar approaches have already been tested with organisms that aid in degrading plastics or even nuclear waste, and it might be possible to design similar organisms to deal with nanobot waste materials (Rüthi et al., 2023). Unlike many of the previously presented approaches, this one might also be applicable outside of living creatures.

Please keep in mind that this is not an exhaustive list of possible approaches. Nonetheless, it should provide the interested reader with a good intuition on the possible ways to remove waste from the body or from the environment.

13.4 Politics and Legal Matters

Yet, other important aspects are political regulations and other legal matters that any new technological development has to abide by. Assuming that the proposed system may successfully complete the stages of simulation and in vitro testing, political, legal, and especially matters of social acceptance are still possible hindrances. International and national laws are still adapting to the upcoming technologies and even macroscale medical innovations are often hindered to a great degree by legal and political issues. For example, in Germany and the European Union, nanodevices and nanonetworks are most likely subject to the *German Medical Devices Act* (Johner et al., 2020) as many nanoscale applications are medical systems to provide measurements, diagnosis, and treatment. It is important to in mind that in other developed countries around the world, similar legislative issues and political barriers exist.

The *Medical Devices Act* implements the EU directives 93/42/EEC, 98/79/EEC, and 90/385/EEC. An ethics commission and a federal authority have to approve any new medical system before a clinical investigation is allowed. The system then receives a classification to reflect the level of risk according to Annex IX from Directive 93/42/EEC. As nanonetworks are potentially very invasive, may be long-lasting, and may bear not yet completely investigated risks of impairing human health (especially concerning long-term consequences), the highest risk classification Class III is to be expected.

The regulatory frameworks that restrict the development of new technologies play a crucial role in ensuring the safe development, production, and use of emerging technologies. However, when it comes to nanonetworks and nanodevices, there are unique challenges associated with adapting existing regulations to address their characteristics and potential risks.

Nanonetworks and nanodevices operate at the nanoscale, which presents novel characteristics and behaviors compared to macroscale materials and devices. At the nanoscale, materials may exhibit different chemical reactivity, increased surface area, and altered biological interactions. These unique properties can pose challenges for traditional regulatory frameworks that were primarily designed for larger-scale technologies. Adapting regulations to account for these specific characteristics becomes necessary to ensure adequate risk assessment and management.

Assessing the potential risks associated with nanonetworks and nanodevices requires a comprehensive understanding of their characteristics, behavior, and potential exposure pathways. However, due to the relatively recent emergence of nanotechnology, there may be gaps in scientific knowledge and standardized testing methods. Robust research and data collection are essential for evaluating the safety and environmental impact of these technologies. Regulators may face challenges in conducting thorough risk assessments and developing appropriate risk management strategies as the information available may be limited or still under development.

Regulatory frameworks often include systems for classifying and labeling products based on their potential risks. However, determining the appropriate classification and labeling requirements for nanonetworks and nanodevices can be challenging. The unique characteristics of nanomaterials and the lack of standardized definitions for nanotechnology-related terms make it difficult to establish clear criteria (Büther et al., 2017) (even though this book attempts to bridge that gap, among other things). This can result in inconsistencies in how nanonetworks and nanodevices are regulated and labeled across different jurisdictions, leading to confusion for both consumers and manufacturers.

Moreover, achieving international harmonization of regulations for nanonetworks and nanodevices is a significant challenge. Nanotechnology is a global field, and different countries or regions may have varying priorities, resource constraints, and legal frameworks. Lack of harmonization can result in barriers to trade, increased compliance costs, and difficulties for multinational companies operating in multiple jurisdictions. Bridging these gaps and establishing common regulatory approaches require collaboration among regulatory agencies, scientific experts, industry stakeholders, and policymakers.

Addressing these challenges requires continuous efforts. There must be initiatives focused on enhancing regulatory frameworks specifically tailored to nanotechnology, such as developing nanomaterial-specific regulations, guidelines for risk assessment, and standardized testing methods. Additionally, dialogue and cooperation between countries and international organizations are vital for promoting harmonization, sharing best practices, and aligning regulatory approaches for nanonetworks and nanodevices.

By addressing these challenges and adapting regulatory frameworks to account for the unique characteristics and potential risks of nanonetworks and nanodevices, it should be possible to foster the safe and responsible development and use of nanotechnology while ensuring public confidence in these emerging technologies.

13.5 Acceptance

With all of this in mind, it is important to also discuss the topic of public acceptance of new technologies. While it might be possible to create nanonetworks in the future, this does not automatically mean that they will be *popular*, even with all their potential advantages in mind. Humans are traditionally "conservative" and prefer stability over improvements unless the benefits are big enough or the process of introducing new technologies is slow and gradual. This attitude is rational from an evolutionary perspective as the vast majority of changes in behavior do not provide a lasting benefit for a species as a whole and the same trend can be observed when designing new products or technologies.

The public reception of potentially invasive medical nanotechnology must be considered and evaluated first. If it is rejected by society at large, it is less attractive to companies and might never be implemented at all – no matter how potentially beneficial it is. New technology has the potential to heavily influence everyday life after integration as most of us have experienced with the rise of smartphones. Through a change in society, new demands on technology arise, and the smartphone revolution produced a wide demand for high bandwidth internet access.

Nanonetworks may change the frequency of how often we have medical examinations as well as the job landscape of physicians overall. Yet, the societal consequences of continuous monitoring of body functions are ultimately unknown and subject to future research. Personal freedom may increase if medical treatment can be performed more independently from location and physician. Hunt and Mehta Hunt and Mehta (2013) investigated how nanotechnology is introduced, how it is perceived, and which actors are involved-including governments and how related sciences influence the framing.

The entire process could further be compared to the rise of genetic engineering in, e.g. Germany (Thiel, 2014), which is one of the "newer" studies on the topic. Genetic engineering has great potential to tackle many of the existing problems humans face – the potential might be even bigger than that of nanonetworks. Yet, the public view on the topic is overwhelmingly negative in Germany and other parts of the world. However, people are rarely able to provide reasonable or logical arguments to support their point of view.

While it is not entirely clear why the public perception of even *green* genetic engineering is bad, it is clear that such negative publicity could be a major obstacle for nanonetworks too. Identifying and discussing possible fears, risks, and problems with the general public during the entire process from concept to final product in advance would be immensely beneficial. Nanonetworks should be a tool that aids in increasing the standard of living and not yet another perceived danger.

Apart from that, there are only a handful of studies on the topic of acceptance. However, there are some case studies or risk analyses, like (Vierboom et al., 2008), that attempt to identify how people feel about (medical) nanodevices and for what reasons they would test out the technology. This study was carried out as early as 2008, but the results are still relevant today and the public perception of nanotechnology has not changed much.

To the best of my knowledge, this study is one of the biggest and most thorough surveys available. Overall, 1000 people participated. Of those, 52% were female and roughly 1/3 of all participants came from one of the three 15-year-long intervals starting at age 15 till 60. More importantly, the education level reflected the German average and no kind of diploma was over-represented.

Roughly 1/3 of all participants had never heard about the topic at all and had no prior knowledge. 150 people stated that they had heard something about the topic but could not remember where. Of those who did know a thing or two, most thought of nanotechnology in terms of miniaturization, paints, medicine, or interestingly the "lotus effect."

Figure 13.1 shows the results of a questionnaire with roughly 1000 participants on their approval of different nanotechnologies. The results are sorted by the overall approval from most at the left to least at the right. Most often, people agree that it would be a good idea to use nanotechnologies to improve the properties of materials, packaging, or textiles. However, the closer the questions came to applying nanotechnology to their own food or body, the more cautious people became. The only exception would be the improvement of teeth health which people generally approve of.

The study further tried to find out in whom people had the most trust when it came to advice on the topic. The results are displayed in Figure 13.2.

The results are sorted from "most trustworthy" to the left to "least trustworthy" to the right. The majority of people were willing to put a degree of trust in consumer rights organizations. That said, only 54% of people stated that they had full confidence in their judgment on nanotechnologies.

Next on the list are scientists working in the field. 47% of all participants stated that they had absolute faith in them, and 45% stated that they had at least some faith in their judgement. Only 8%

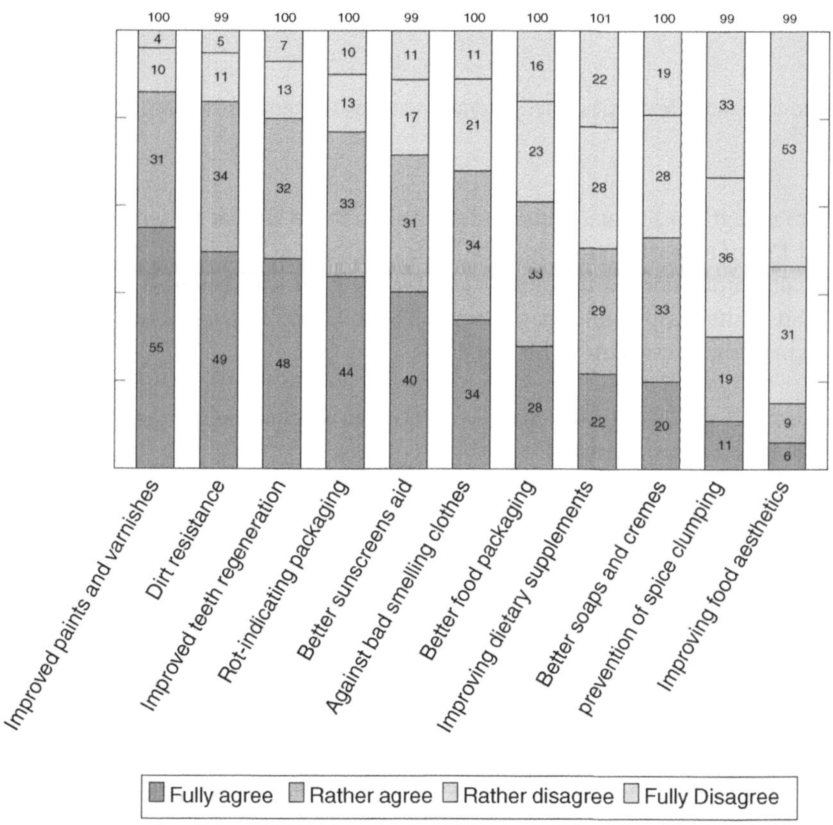

Figure 13.1 Results of a questionnaire on why people would consider the use of nanonetworks. *Source:* Vierboom et al. (2008)/BFR.

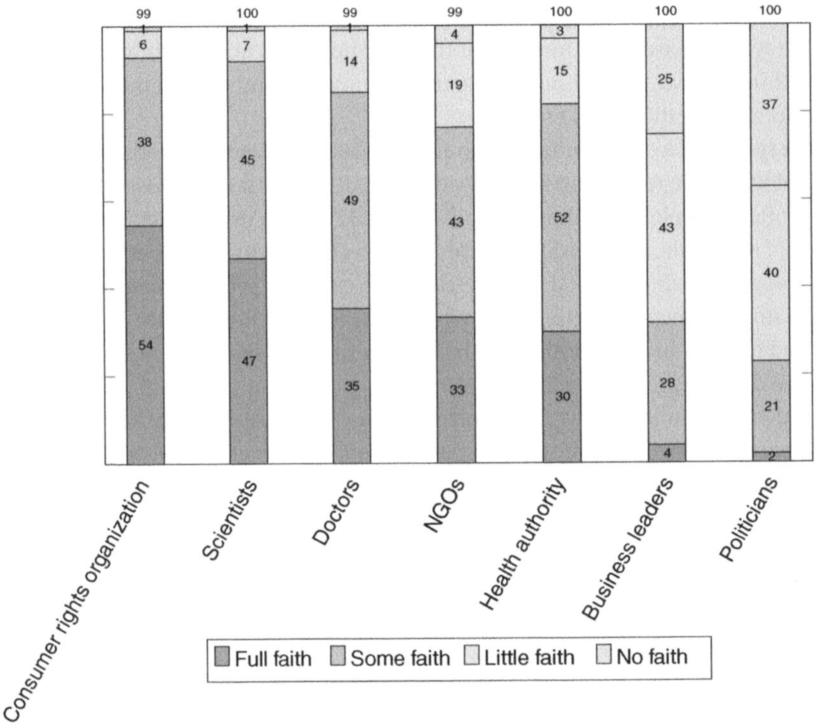

Figure 13.2 Readiness to try nanotechnologies when recommended by different institutions. *Source*: Vierboom et al. (2008)/BFR.

of all people had little to no faith in science. Interestingly, people had a similar amount of faith in doctors/physicians, but 15% of all participants doubted their judgment.

While NGOs and health protection authorities still had some approval, business leaders and government officials had limited approval at best. People had the least amount of confidence in politicians' abilities to correctly assess nanotechnologies.

The study further tried to figure out how people would react to possible future developments in nanotechnologies. Figure 13.3 shows the participant's reactions to various statements. The results are sorted from most approval to least approval.

Overall, people in Germany and likely the West as a whole are cautious of possible risks but are also interested in the potential of nanotechnology. Additionally, most people were confident that they could at least understand nanotechnologies sufficiently to make a judgment call. That said, most people were not afraid and not totally against novel developments either.

While the immediate focus of the study was not to assess the general opinion on medical nanotechnologies, the study nonetheless offers some insights. 41% of all participants saw the most potential of nanotechnologies in the medical field. The second item on the list, better environmental, protections, had only 25% approval. Only 4% of all participants saw some benefit in improving food. All in all, most people saw great potential for nanotechnologies in the medical field, even if their own bodies are affected. The same was definitely not true for any other area. It is to be expected that people would be even more willing to try out novel therapy routes if they are affected by a disease for which no conventional treatment exists. Especially in cancer therapy, this is already the case and that observation is very likely also valid when it comes to medical nanotechnologies.

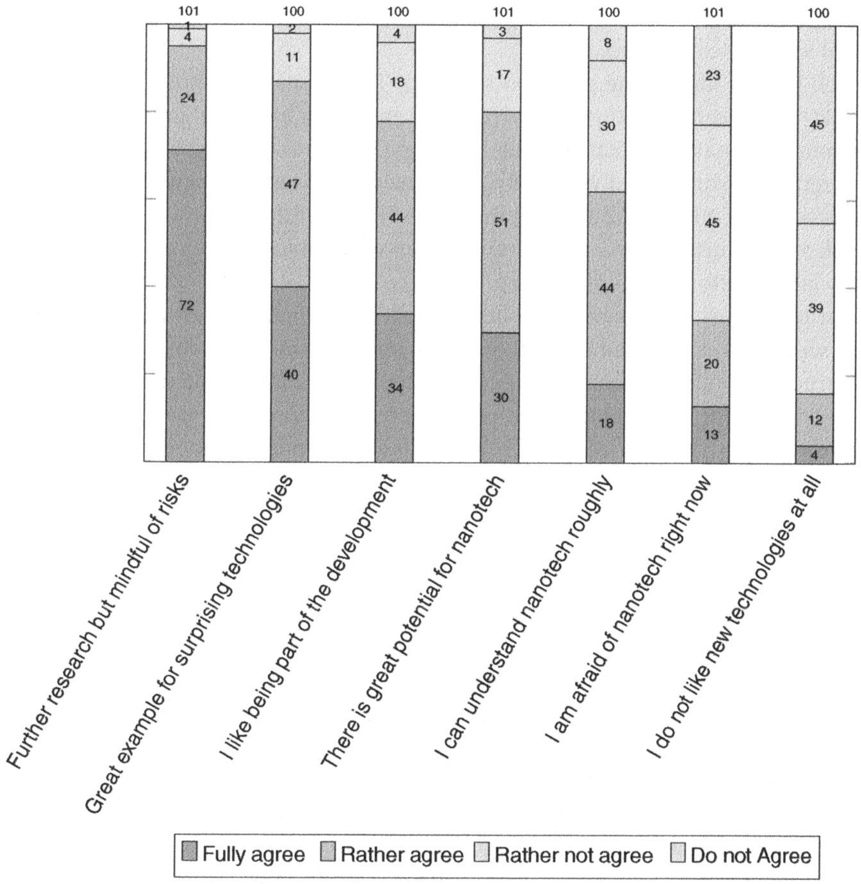

Figure 13.3 Readiness to support nanotechnologies in the future in different domains. The questions have been roughly translated/shortened. *Source*: Vierboom et al. (2008)/BFR.

13.6 Dangers and Fears

Finally, we want to discuss potential dangers, conspiracy theories, and even dystopic scenarios that many people came up with over the years. We start with a number of rather unrealistic scenarios, conspiracy theories, or even dystopian science fiction outlooks. As we have discussed many potential problems in detail before, we only briefly (re-)introduce them here.

- One of the likely most "popular" fears is that of an artificial intelligence takeover that sometimes includes nanobots. The idea has already been introduced in Chapter 2 and dates back to at least the 1960s. This scenario involves a superintelligent AI surpassing human intelligence and taking control, either by manipulating humans or by directly seizing power. The AI may perceive humans as a threat or decide to optimize the world in ways that are detrimental to humanity's well-being. This scenario might also include the use of nanodevices to influence humans even from the inside but is rather unrealistic as of right now. While there is a lot of talk about the possibility of an AI gaining consciousness as of 2023, it is rather unlikely to happen. In most cases, AIs are surprisingly simple extrapolative lookup tables that are not even Turing-complete – the impressive results of modern ANN stem from immense models and vast amounts of data and not really from any real learning capabilities.

- Another rather well-known idea is the gray goo scenario. This scenario, popularized by Eric Drexler in his book "Engines of Creation," involves self-replicating nanobots breaking free of control and consuming all matter on Earth, reducing everything into a gray goo of nanomaterial. The nanobots replicate exponentially, depleting resources and eventually causing ecological collapse. While this scenario might sound threatening, there is currently no known way to influence matter in such a way. Furthermore, the materials from which nanodevices are likely made are not that readily available to allow for such a spread.
- Yet, another scenario involves the use of nanobots in modern warfare. In a dystopian future, nanobots could be weaponized and used in warfare. These nanobots might be designed to target and destroy infrastructure, biological systems, or even specific individuals, causing widespread destruction and loss of life. The development of such weapons could lead to an escalation of conflicts and devastating consequences. While it is likely possible to weaponize nanodevices and the military has displayed an active interest in such matters, there are simpler ways to cause harm – at least in the near future. The restrictions current nanoscale systems are facing are usually just too impractical to offer any real benefit. To cause harm to others, there are better tools and the same is true for espionage.
- The last somewhat unrealistic fear stems from the possibility of nanobots to influence humans from inside, for example, via microchips in blood. Some conspiracy theorists envision nanobots infiltrating the human brain where they could potentially be used to control and manipulate individuals' thoughts, actions, and emotions. In this scenario, a centralized authority or malicious entity could gain control over people's minds, essentially creating a society of mind-controlled individuals stripped of free will. While the idea certainly invokes interest, the array of possible challenges to get simple microchips to work inside of the human bloodstream is simply too much. It might be possible to inject microchips into the bloodstream of people, but they would be rendered useless in a short time due to coating molecules. Furthermore, the array of different challenges we have discussed also applies here.

While many fears are likely founded on a lack of knowledge of what is and what is not possible, there are a few real dangers that could hardly be avoided and have to be researched in depth.

- A very realistic fear would be a potential loss of privacy. This is especially true as digital technologies are already used to learn as much as possible about users and potential customers. As of right now, improved targeted advertisement is luckily the main incentive to gather data about people. However, once the data is there, it could also be abused by authorities or criminal institutions. A recent and frightening example would be the Nazi Germany taking over other countries where they gained access to data of people living there including their ethnicity. Nanodevices as taught in this book could take the amount of gathered data another step further as they are often envisioned as a constant health monitoring tool. Nanobots, if misused, could be designed to infiltrate individuals' bodies without their consent, allowing for constant monitoring and surveillance of not only health data. This scenario involves a complete erosion of personal privacy as every action or even bodily function could be monitored by a centralized authority or malicious entity.
- Very related to that scenario is a potential "medical tyranny." In a dystopian future, nanobots intended for medical purposes could be exploited by a totalitarian regime. These nanobots might be used to control individuals' health, selectively cure or harm them, or serve as a means of population control. In the simplest and likely most immediately threatening case, the data nanobots have collected could be used to provide more data than people are comfortable with to health

insurance. If any problematic behavior is detected, insurance might get more expensive or is not provided at all. This scenario raises concerns about the abuse of power and the potential for medical oppression.

- Another medicine-related scenario concerns itself with potential new diseases and poisonings that might be caused by the nanobots themselves. Despite their potential for medical applications, nanobots could pose risks if they are not properly designed or controlled. Malfunctioning or rogue nanobots could inadvertently cause harm to human cells or tissues, leading to the development of diseases or severe health complications. In addition to that, there is currently no good solution for the waste material problem available and it is likely that some adverse side effects on health can be expected. As a result of the size, most nanoscale applications carry the risk of becoming a nanotoxin as a byproduct. Due to their small size, they could easily enter the blood through the lungs and may cause unintended problems there. Most nanoscale applications are designed for exactly one target environment, and the effects on any other environment are likely unknown and untested, but adverse interactions are to be expected. Whenever something is introduced to the body of a living creature that was not part of our evolutionary history, complications are to be expected.
- The last realistic problem would be further increased economic disparities. In that scenario, nanobots could exacerbate socioeconomic inequalities. Only the wealthy or privileged might have access to advanced nanobot technologies, leading to a stark division between those who can afford enhancements or medical treatments and those who cannot. This scenario could deepen existing societal disparities and create a divided world.

  While the development of such new technologies usually reaches all parts of society at some point, there is usually a phase where new products are extremely expensive. This is especially true for capitalist systems that *do* create immense benefits for everyone involved, but they usually come at the price of very unequal wealth distribution. It is important to keep in mind that humans are feeling-based creatures who cannot properly assess objective wealth and tend to be dissatisfied when people in their direct surroundings are richer than them. Sadly, capitalism has currently no answer to that problem of "emotional bias." As a result, this problem is nearly inevitable.

While this list is likely incomplete, it should give the interested reader a good intuition of the fears and dangers that advanced nanotechnologies might bring. All those fears have to be taken into account when trying to increase the rate of acceptance among the general public. In the end, novel types of medication should benefit all people and should not be seen as yet another tool of the "elites."

13.7 Summary

As we have seen, ethical, legal, and especially social matters should not be ignored in the process of developing technologies that have such a disruptive potential. It is crucial to involve people as much as possible in the process to gradually introduce new applications. As soon as something is introduced too fast without fully knowing the consequences, there is typically a backlash, as could be seen all over the world with Sars-Cov-2 vaccines. Products must thus be well tested especially concerning long-term consequences.

This is not only true for consequences concerning human health but also for the environment as a whole. New technologies and materials that have not emerged through an evolutionary process that allowed all other involved creatures to adapt to it carry a great risk. Even if the intended applications work as intended, there must still be ways to deal with the inevitable waste products of nanotechnologies. They tend to be very difficult to deal with as nanoscale materials are almost impossible to contain.

There are also political and legal problems that nanonetworks might face as most rules, laws, and guidelines have been developed for different types of medicine. As a result, it could be challenging to get a permit in regions like the EU. Their guidelines might be well suited to ensure the safety of their citizens for classical medications but are not suitable for nanotechnologies.

Finally, researchers and health officials alike must ensure that potential dangers and fears of the general population are respected and that everyone has a chance to educate themselves on the topic. People usually strongly dislike being forced to act in a certain way and rebel against that just for the sake of it. As a result, it is usually better to *offer* educational material but not force or overly advertise it. As soon as people suspect a loss of control or manipulation, an adverse emotional reaction is to be expected, even if the application in question would be beneficial. "Forcing" people for their own good is usually a dead end.

14

Conclusion

This chapter summarized the most important points and results of the entire book. We analyze the visions and open problems given all that we have learned in the previous chapters and attempt the predict future developments in the field. While accurate and most of all timely predictions are always subject to errors, sometimes it is indeed possible to predict certain trends, if not the specific temporal development.

In the end, we list the key messages that the interested reader should take away from this book. A new scientist should be properly prepared to face the challenges and misunderstandings that would otherwise lie ahead.

14.1 Summary

In this book, we have discussed vast parts of the area of nanoscale computation and communication as well as many required technologies. We first introduced the potential new reader to the area of nanonetworking by giving a short introduction to the topic, including some etymology, science fiction, and some novel example applications given the unique challenges at the nanoscale.

After that, we discussed the history of nanotechnologies in general, with a special focus on very early natural philosophy and manufacturing accuracy. We started in ancient India where a number of still prevalent ideas emerged and moved through ancient Greece toward the modern era. We then analyzed the historical advancements in tool development over the millennia and observed a seemingly exponential trend in miniaturization. As the human eye has a limited resolution, it was not possible to create devices that had details of less than a tenth of a millimeter in size.

However, the invention of ever more precise microscopes removed that limitation and allowed for further improvements that exceeded the limitations of prior advanced cultures of the past. We also discussed several microscope technologies that contributed to the development of modern technologies and are crucial for manipulating matter at very small scales. At the end of the chapter, we listed several examples of modern nanotechnologies to equip the interested reader with an intuition on what is and is not possible right now.

After that, we have further motivated the ongoing research into nanonetworks and nanoscale computation and communication applications in general. We started by illustrating how nanomaterials have already pervaded the modern industry. It is important that researchers are aware of the potential in various areas to attract funding and further advance the area as a whole. For that, we discussed potential applications in the military, geology, farming, and especially medicine. In medicine, there is great potential for improvement and there are many unsolved problems

Nanonetworks: The Future of Communication and Computation, First Edition. Florian-Lennert A. Lau.

where nanonetworks might offer a groundbreaking advantage and potential solutions that are unthinkable as of right now.

Next, we analyzed various different schemes of constructing nanoscale devices and nanonetworks specifically. While there are at least three overarching construction paradigms, not all of them are equally feasible to implement. For example, it is unclear if CMOS-based technologies may be scaled down further. Nature adaptation, however, is very likely to work but is subjected to different constraints and limitations.

Finally, we also discussed (programmable) self-assembly as a new paradigm for construction at the nanoscale. While we are currently not able to mass manipulate matter at a very small scale other than through photolithography, self-assembly does work and is already used as a construction mechanism via, e.g. special DNA molecules. In fact, DNA may be used for purposes of precise construction, computation, and even communication – all based on the same principles of self-assembly. We also presented the current state of the art in nanoscale construction. In short, while it is possible to create a variety of different nanostructures, the compatibility of various components is an unsolved problem. As we have mentioned numerous times, placing structures with a required precision down to the atom level is currently infeasible – at least on a meaningful scale.

We then discussed the topic of computation which enables most of the unique advantages compared to conventional medicine. With computation, it is suddenly possible to make conditional choices when monitoring or treating diseases. We have discussed several ways to encode state at the nanoscale and the corresponding models of computation. For each of those, we tried to estimate how feasible an implementation might currently be. In the end, self-assembly appears to be a suitable candidate to allow for nanoscale computation using already existing technologies. Finally, we have discussed possible ways of generating good or even optimal (collaborative) programs for nanonetworks. The task is surprisingly difficult and often computationally infeasible, yet "tricks" like lifting might render the problem feasible in at least *some* cases.

The last component of extraordinary importance for implementing nanonetworks is communication. Without the "trinity" of construction, computation, and communication, the foundation for nanonetworks would be missing.

Currently, there are at least three different ways of implementing communication in nanonetworks. The first and most familiar type is electromagnetic communication, which is basically a downscaled version of wireless communication as we know it. While EMC works well at the macroscale, the same is likely not true for the nanoscale where size limitations of antennas also limit the communication distance of nanodevices to around 2 mm.

An alternative to EMC might be acoustic communication that is not subject to the same type of molecular absorption. Acoustic communication uses mechanical waves as a medium for transmitting signals. While the message propagation happens at a much slower speed in, e.g. the bloodstream, it might still be a suitable solution. Yet, the devices are still often assumed to be downscaled versions of classical CMOS technologies, which might be unrealistic.

The last major type of communication is molecular communication. Molecular communication differs significantly from other communication media ad the information is now encoded using molecules instead of electromagnetic waves. While this approach severely limits the bandwidth and communication speed, it is extremely likely to work as the human body naturally used the approach for communication via hormones and pheromones. There are many possible ways to implement molecular communication, including DNA-based molecules that could also perform computations in the message in the channel, thereby bypassing potential space constraints in nanodevices.

In this context, we have also discussed potential means of achieving a multi-hop communication network, especially in the cases of EMC and acoustic communication. Molecular communication might be directly implemented as A-to-B communication without needing any hops in between. Finally, we have also discussed potential simulation software to validate or falsify hypotheses about the communication behavior in nanonetworks.

Another major influence factor on nanonetworks and especially their communication capability is their position in space and the various effects that influence it. Nanodevices are subject to several types of passive movement that affect any sufficiently small structure. The biggest influence would be the blood flow, which contributes the biggest change to physical positions.

Apart from that, nanodevices are also subject to diffusion. Diffusion/Brownian motion is usually influenced by the environmental temperature – the hotter it gets, the faster particles tend to move. In addition to that, the size of a structure also influences how affected they are by diffusion effects.

Apart from passive movement, there is also active movement, where a device manipulates its own position in space. Active movement is partially inspired by bacteria and their ability to move using flagella and motivated by the wish to use medication more locally in medicine. There are several ways to achieve active movement and a surprising number of researchers focus on nothing else. For example, it is possible to maneuver nanodevices using external magnetic fields, bubble propulsion, guide them along microtubules, or use other chemical properties to generate movement.

Once movement has taken place, another important step is to *localize* the nanodevices again, as they are usually envisioned to only become active when certain conditions are met. To localize nanodevices, there are several approaches like observing the concentration or age of information, trilateration, or proteome fingerprinting. All of them attempts to reduce the number of possible locations a nanodevice might be.

To process information based on their own location as well as manipulate it, nanodevices require both sensory and actuator components. Motors, like antennas, are special types of actuators/sensors or a combination of both.

In the most general sense, sensors are components that observe an environmental property and transform the observation into a format that machines or humans can use/understand. While there are countless sensor principles available, not all of them may be implemented using the restrictions at the nanoscale. For example, we have analyzed the most frequently used diagnostic methods to see if they could be used at the nanoscale. The result was that most procedures require large external devices to interpret or even read the results. However, some exceptions, e.g. based on antibodies could be used in nanonetworks.

Actuators fulfill the complementary function of sensors and manipulate an environmental property. In the simplest sense, actuators could be switches that start a reaction, dispensers that release medication, or antennas that manipulate electromagnetic fields. Without actuators, nanonetworks could not interact with their environment and would be limited to diagnostic purposes without any means to, e.g. treat diseases.

All actors and sensors, as well as many other components, require an energy supply to function. While nanodevices themselves are small, they still require a surprisingly large amount of energy to function properly. This is especially true for communication, and in WSN, it is estimated that about 95% or more of the energy is used to transmit information. In nanonetworks, this ratio is likely even worse and nanodevices may likely only transmit a single message every few hours. As a result, it is crucial to find especially network algorithms and routing protocols that minimize the required energy for communication.

If batteries are used, it might be the case that only a handful of messages could ever be sent per nanodevice before the energy runs out. In the case of ultra-capacitors, the amount of energy stored is even worse, albeit the recharging process is much faster.

An alternative to single-use batteries would be rechargeable energy storage that can be refilled using some kind of energy harvesting mechanism. For example, it might be possible to harvest sufficient amounts of energy from mechanical vibrations using the piezoelectric effect. Other publications propose the use of energy fuel cells to harvest energy from, e.g. sugar in the human body like some medical implants already do.

Especially medical applications require an additional timing component to attach a timestamp to the observed data. Medical data used for diagnosis is often only relevant for a time, and it matters a lot *when* an event occurred. It is often necessary to compare one measured parameter in a time-synchronized manner with others to be certain of a diagnosis. Without the capability to use absolute timestamps, this is very difficult.

Similarly, the process of synchronizing the internal clocks of nanodevices is not an easy task given the energy and computational constraints. One especially simple procedure to still achieve somewhat accurate clocks is based on fireflies that use nothing but an internal counter and observation of the blinking of neighbors to synchronize themselves.

However, this and many other network algorithms and heuristics are based on the availability of either pseudorandom or true random numbers. For this, we might be able to use pseudorandom generators or simply use a physical effect or environmental noise as a source of randomness.

The last potentially structurally relevant component we have discussed is security in nanonetworks. Security is of special importance as many of the proposed areas of application stem from the area of medicine, where great harm could be caused by adversaries that gain access to a system. While gaining physical or just physical access to a nanonetwork might be more difficult, the potential harm is greater and potentially deadly. One might still argue that it is possible to either gain a sample of blood or introduce additional nanostructures to a body via food, drinks, or even drugs if an attacker tries hard enough.

Implementing security varies according to the underlying communication paradigm used in the application. In the case of electromagnetic communication, it might be sufficient to use a lightweight version of already established algorithms and protocols. This is especially true for the gateway that likely aggregates in-body data and sends it wirelessly to external devices. As it is much bigger than nanodevices, regular algorithms should be applicable and even necessary, as the gateway is likely the first point an attacker might consider.

Inside nanonetworks themselves, sufficient security is difficult to achieve as resource constraints limit, e.g. the complexity of algorithms and the key lengths used. As a result, it might be necessary to look for alternative ways to conceal information and limit access to authorized parties.

In the case of molecular and DNA-based communication, classical procedures are likely completely infeasible as the additional overhead from security would increase message lengths manyfold. As a result, steganography might be a suitable alternative. In this case, we introduce sufficient "junk information" into the system so that the attacker has to invest resources to figure out if there is any message at all or if it is even relevant.

With all the previous information in mind, we were able to derive a number of holistic reference architectures for different types of nanonetworks. In the beginning, we discussed the differences and potential similarities between already existing MANETS or VANETS and nanonetworks. Some of the established algorithms might be usable at the nanoscale too, as demonstrated in BANs or swarm-based networks that are sometimes able to fully self-organize.

Next, we analyzed different types of nanonetworks, starting with acoustic, electromagnetic, molecular, and then DNA-based communication as a basis for networking. For each of them, we listed and specified as many relevant parameters as possible as well as likely/potential values for them. Among them are the number of participants, the communication range, the bandwidth, the construction material, and many more. Especially acoustic and electromagnetic nanonetworks are similar in nature and vary mainly in the communication technology used, as well as the parameters that are influenced by that.

The same is true for the reference architecture. Such nanonetworks are usually envisioned as in-body applications where a large number of devices perform predefined tasks inside of the body and communicate the results to a nearby gateway. This more powerful device aggregates and preprocesses the data before transmitting it from the microscale to the remaining parts of the IoNT outside.

Similarly, bacteria-based, molecular, and DNA-based nanonetworks are similar in character. In this case, direct physical contact between the molecules or bacteria is necessary for transmitting information. However, these networks also potentially introduce biological or hybrid components.

While they are likely more realistically implementable, they have a much lower data date and speed of communication. It might even be the case that only a handful of bits could be transmitted per minute, which would force network designers to reiterate many of the prevalent methods.

In terms of data aggregation and preprocessing, it is also necessary to have a more powerful gateway device that connects the nanoscale and the macroscale. In this case, the device must be able to gather or count molecular data without receptors being permanently blocked.

Finally, we have discussed several DNA-based nanonetwork architectures as this type of network is likely the least well known but potentially implementable using nothing but already available technologies. It is possible to solve many existing problems using DNA-based nanonetworks, including a distributed consensus, threshold breaches, and basic arithmetics, but also more complex procedures that can solve any mathematical function. In the end, we even presented a concept for a nanonetwork that tracks the normal ranges for individual health parameters and detects anomalies based on that. While such scenarios are complex, it is still good to know that they can at least work in theory.

In the end, we also discussed a number of techniques that can be used to verify the generated ideas and concepts in the research community. The first step is usually analytical methods, followed by a complexity analysis to put a new algorithm or application into the context of already existing ones. Next comes simulation as a tool for both the verification of hypotheses and to generate new ones that could be tested under more realistic circumstances like organ-on-chips or in wet-lab experiments.

The last content chapter of this book discussed the potential ethical, legal, social, and many other influences. We first reiterated the process from idea to final product to highlight the potential problems one might encounter in designing new systems.

Afterward, we analyzed many potential environmental interactions that might occur. This includes both the target function of a nanonetwork and side effects that are to be expected. This is especially true for problems of waste management and biocompatibility.

In the last part of the chapter, we tried to answer political questions and legal matters as well as the question of public acceptance. Just like genetically modified products, nanotechnologies tend to have a bad reputation and it is important to make the design process of such technologies as transparent as possible. Finally, we discussed potential dangers created by nanotechnologies.

14.2 The Future and Visions of Nanonetworks

One of the most difficult tasks is an attempt to predict future developments in an area of science. While it is *always* impossible to predict the future with any degree of certainty, the task of assigning a time horizon to potential developments is even harder. As a result, we will not even attempt this and stick to predictions on a very general level where it is hard to be wrong. We extrapolate past trends and only sort them into very broad categories of near-, middle-, and long-term visions that might become reality. While there is likely still a significant error, there is at least some chance to get things right.

14.2.1 Near Future

In recent years, significant advancements have been made in the field of nanonetworks and nanomedicine, paving the way for exciting near-future developments. We now explore six key areas that are poised to revolutionize targeted drug delivery and enhance healthcare through the integration of nanotechnology.

One of the most promising applications of nanonetworks in medicine is targeted drug delivery. Researchers have been exploring the use of, e.g. nanomotors, to transport drugs to specific locations within the body that are hard to reach. For example, poorly perfused areas of the body like cartilage, sinews, or bones could benefit tremendously from a higher concentration of medication. These nanomotors can be designed to navigate through the bloodstream and reach target sites, enabling precise delivery of therapeutic agents while minimizing side effects.

Furthermore, bacteria have been harnessed for their potential in nanonetworks, offering a unique approach to information processing and communication at the microscopic level. By engineering bacteria to communicate with each other and transmit signals, researchers aim to create bacterial nanonetworks that could be employed for various applications in medicine, such as monitoring and controlling microbial communities within the body to maintain health or combat infections. While these networks likely fulfill no immediate function, a proof of concept could be possible within the next 10–25 years.

Advancements in DNA nanotechnology have opened up possibilities for creating nanonetworks, e.g. within a Petri dish. DNA-based nanodevices can be programmed to interact and carry out specific tasks, forming intricate networks capable of information processing. These DNA nanonetworks hold immense potential for studying cellular processes, simulating biological systems, and investigating complex molecular interactions in a controlled environment. As all the necessary materials already exist and can be synthesized, it is highly likely that this type of nanonetwork will become a reality within 10 years with even more potential for future refinements and improvement.

Accurate localization and tracking of nanodevices within the body are crucial for effective nanomedicine applications. Researchers are exploring various localization techniques, such as using magnetic fields or ultrasound, to precisely monitor and guide nanomotors to their intended targets. Localization advancements will enhance the efficiency and reliability of targeted drug delivery and other nanomedicine interventions. Within the next 10–25 years, it is likely possible to precisely locate, e.g. the origin or metastases accurately.

As nanonetworks and nanomedicine continue to evolve, researchers have achieved isolated proofs of concept for individual nanonetwork components. These include nanosensors capable of detecting specific biomarkers, nanocarriers designed for drug encapsulation and release, and communication protocols for information exchange between nanomachines. These foundational advancements pave the way for integrating these components into fully functional nanonetworks that can operate synergistically to address complex medical challenges. While it is likely not possible to precisely combine all necessary nanonetwork components into holistic nanonetworks, the individual components will likely work within a generation.

Microbots, or tiny robots at the micrometer scale, hold great potential in nanomedicine due to their ability to perform medical functions within the body. Researchers are exploring and conceptualizing various microbot designs that combine mobility, sensing capabilities, and specific functionalities. These microbots can act as targeted drug delivery vehicles, perform minimally invasive surgeries, or assist in precise tissue engineering. While microbots are still roughly 1000 times bigger than nanodevices, it is still likely that the first functional microbots will be used within the next decade.

The near-future developments in nanonetworks and nanomedicine alone hold tremendous potential for revolutionizing targeted drug delivery and healthcare as a whole. From the use of simple nanomotors and bacterial nanonetworks to DNA-based networks and advanced localization techniques, these innovations are laying the foundation for more effective and personalized medical interventions. With isolated proofs of concept for various components and the emergence of the first useful microbots, the future of nanomedicine appears bright, offering exciting possibilities for improved patient outcomes and enhanced well-being.

14.2.2 Middle Future

Within two generations, even more advancements are to be expected and nanonetworks are poised to undergo significant advancements that could reshape healthcare and technology.

In the coming decades, we anticipate the emergence of the first real nanonetworks based on molecular communication. While their capabilities and specifications likely do not match those of established macroscopic networks, it is still possible to implement a variety of medical applications like that. By harnessing molecular messages, such as chemical gradients or signaling molecules, nanodevices will be able to exchange information and cooperate within complex networks. These nanonetworks will open up new avenues for collaborative drug delivery, precise sensing, and distributed computing at the nanoscale.

As nanotechnology progresses, its integration with agriculture holds immense potential for optimizing crop cultivation and food production. Nanonetworks could be deployed in farming applications to monitor soil conditions, detect plant diseases at an early stage, and precisely deliver nutrients or pesticides to specific plants. By leveraging nanosensors, nanomotors, and data-driven algorithms, nanonetwork farming applications could revolutionize the efficiency, sustainability, and yield of agricultural practices. It is highly likely that any nanonetwork technology is first tested in fish tanks or other farming systems before they are tested on actual humans.

While much of the focus in nanonetworks and nanomedicine has been on leveraging biological components as they are likely easier to use and implement, the mid-term future is likely to witness significant advancements in non-biological computation at the nanoscale. Non-biological computing elements, such as nanoelectronics, photonics, and quantum computing, might play an increasingly prominent role in nanonetworks. These non-biological components will enable faster, more powerful computations, as well as information processing. While those computational models will likely not meet the size demands of the nanoscale, the expected developments are nonetheless an important step toward further miniaturization.

In the next 25–50 years, we may further expect the first experimental tests of electromagnetic or acoustic nanonetwork prototypes using microbots. These networks involve miniature robots that are capable of communicating wirelessly using electromagnetic or acoustic waves. While these networks are not yet nanosized, these tests might still pave the way for the development of advanced swarm robotics and collaborative nanomedicine applications.

As we look ahead to the next 50 years, the potential advancements in nanonetworks and nanomedicine are awe-inspiring. From the emergence of the first real nanonetworks based on molecular communication to the application of nanotechnology in farming, the emergence of small-scale non-biological computational models, and the initial tests of EMC and acoustic nanonetworks with microbots, these developments will shape the future of healthcare and technology. It is through continuous innovation, interdisciplinary collaboration, and a long-term vision that we will unlock the transformative power of nanonetworks and harness their potential to improve human health and well-being.

14.2.3 Distant Future

Finally, we take a look at the distant future that the vast majority of us will not witness. In the next 100 years or more, nanonetworks and nanomedicine are expected to undergo groundbreaking developments that could reshape not only healthcare but also various other domains.

As nanotechnology continues to advance, we anticipate that the utilization of CNTs allow for the first nanonetworks as many researchers envision them right now. I personally think that much bigger challenges are involved that many scientists anticipate and therefore sort those advancements into the "distant future" category. CNTs possess exceptional electrical, mechanical, and thermal properties, making them ideal building blocks for nanodevices and nanonetworks, as soon as their precise placement in 3d is possible. By using electromagnetic communication paradigms, CNT-based nanonetworks could enable efficient, high-speed communication and computation at the nanoscale, paving the way for sophisticated applications in nanomedicine and beyond.

Not all of the developments will appear desirable to the majority of people. The long-term future of nanonetworks is likely to witness significant advancements in military applications. Nanotechnology could play a crucial role in enhancing soldier performance, protection, and medical care on the battlefield. Nanosensors embedded in military gear could provide real-time monitoring of vital signs, detect biochemical threats, and enable rapid response and treatment. Furthermore, nanodevices may be employed for advanced surveillance, autonomous drones, and intelligent warfare systems, ushering in a new era of military capabilities. Sadly, humans are especially creative when it comes to investigating the potential of new technologies for creating weapons.

Over the next century, breakthroughs in nanotechnology and nanomedicine are expected to enable the precise combination of all the previously presented nanoscale components into in-body nanonetworks. This integration will overcome challenges such as protein or fat coatings and other adverse interactions that currently hinder the functionality of nanodevices in the human body. Through advances in surface modification, bioengineering, and immune system interaction management, in-body nanonetworks will be able to navigate physiological barriers, deliver therapeutics with precision, and perform targeted interventions at the cellular level.

In the long-term future, the concept of a "medical singularity" could become a reality, where nanonetworks serve as universal tools against diseases. By optimizing nanotechnology, including targeted drug delivery, precise sensing, and molecular-level interventions, nanonetworks may be capable of addressing a near complete range of diseases and conditions. Advanced diagnostics, personalized treatments, and regenerative medicine facilitated by nanonetworks could revolutionize healthcare, leading to a future where previously incurable diseases become curable or at least manageable.

As nanonetworks evolve, they have the potential to become universal tools across various domains beyond medicine. The ability to communicate, compute, and interact at the nanoscale could revolutionize industries such as environmental monitoring, energy production and storage, information technology, and more. Nanonetworks may enable truly smart cities with interconnected infrastructure, highly efficient renewable energy systems, and advanced, potentially self-repairing nanoelectronics.

In summary, the long-term future of nanonetworks holds tremendous promise for transformative advancements. From true nanodevices based on CNTs and military applications to breakthroughs enabling in-body nanonetworks and the "medical singularity," nanotechnology will redefine healthcare, military capabilities, and various other domains. The potential universal applications of nanonetworks across industries will shape a future that is interconnected, technologically advanced, and marked by remarkable scientific achievements. I personally expect a shift that is akin to the rise of the personal computer and the internet if humanity can make it that far without major catastrophes or setbacks.

14.3 Key Message

In summary, nanonetworks hold immense potential to revolutionize life as we know it to achieve levels of convenience and carefreeness never before imaginable. To achieve this vision, holistic approaches and interdisciplinary collaboration are necessary as nanotechnologies affect all areas of science and daily life. At the nanoscale, even the simplest of tasks can pose unique challenges and it will require a huge effort to overcome those problems. Especially the compatibility of individual proposed solutions and the potentially adverse interactions with the environment have no satisfying solutions as of right now. While nature-inspired solutions are likely more compatible, they have a number of other potential problems too.

However, some words of warning are still necessary. While nanonetworks indeed carry this potential, objective improvements are usually not what the average person wants. As early as 2600 years ago, the Buddha already observed that humans are feeling-based creatures that are almost impossible to satisfy. Many of the problems nanonetworks address are in reality not rooted in a world that is against us or lackluster in various ways, but instead, the problem is to be found in our own attitude of craving. No matter how much wealth and convenience we accumulate, we can be absolutely certain that it is never enough and that people will look for various scapegoats if something is not going their way. After a while, the tremendous benefits are the "new normal" and many people will only see problems, no longer having perspective on the good things – sadly, this behavior is already very common nowadays, where people take wealth for granted and start undermining its foundation. No matter how good something initially feels or how much relief it brings, humans are not capable of experiencing the same amount of pleasure from it twice. We simply "adapt" to it and need more the next time – this flaw is sadly built into human nature and was very useful in our evolutionary past but lost its usefulness with the emergence of civilization. Yet, we still have to consider this at all times when developing novel technologies. While this might paint a rather grim picture, the *objective* improvements nanonetworks will bring are still worth it, even though it might not "feel like it" anymore once people get used to them.

Bibliography

351 Pleasant Street and Suite B. #442 Northampton MA. Federal Spending: Where Does the Money Go, 2022. URL https://www.nationalpriorities.org/budget-basics/federal-budget-101/spending/.

Sergi Abadal, Eduard Alarcón, Albert Cabellos-Aparicio, Max C. Lemme, and Mario Nemirovsky. Graphene-enabled wireless communication for massive multicore architectures. *IEEE Communications Magazine*, 51(11):137–143, 2013.

Sergi Abadal, Tie-Jun Cui, Tony Low, and Julius Georgiou. Programmable metamaterials for software-defined electromagnetic control: Circuits, systems, and architectures. *IEEE Journal on Emerging and Selected Topics in Circuits and Systems*, 10(1):6–19, 2020.

Acharya Buddharakkhita Gotama The Buddha. *Dhammapada Verse 204*. Buddhist Publication Society, 1987a. URL https://www.buddhanet.net/pdf_file/scrndhamma.pdf.

Leonard Max Adleman. Molecular computation of solutions to combinatorial problems. *Nature*, 369:40, 1994.

Ali Aghebati-Maleki, Sanam Dolati, Majid Ahmadi, Amir Baghbanzhadeh, Milad Asadi, Ali Fotouhi, Mehdi Yousefi, and Leili Aghebati-Maleki. Nanoparticles and cancer therapy: Perspectives for application of nanoparticles in the treatment of cancers. *Journal of Cellular Physiology*, 235(3):1962–1972, 2020. doi: https://doi.org/10.1002/jcp.29126. URL https://onlinelibrary.wiley.com/doi/abs/10.1002/jcp.29126.

Babak Ahmadi, Kristian Kersting, Martin Mladenov, and Sriraam Natarajan. Exploiting symmetries for scaling loopy belief propagation and relational training. *Machine Learning*, 92(1):91–132, 2013.

Aiwok. English: Venus of Galgenberg made of green serpentine 30,000 years ago, April 2010. URL https://commons.wikimedia.org/wiki/File:Venus_vom_Galgenberg.JPG.

I. F. Akyildiz and J. M. Jornet. The internet of nano-things. *IEEE Wireless Communications*, 17(6):58–63, December 2010. ISSN 1536-1284.

I.F. Akyildiz, W. Su, Y. Sankarasubramaniam, and E. Cayirci. Wireless sensor networks: A survey. *Computer Networks*, 38(4):393–422, 2002. ISSN 1389-1286. doi: 10.1016/S1389-1286(01)00302-4. URL https://www.sciencedirect.com/science/article/pii/S1389128601003024.

Ian F. Akyildiz, Won-Yeol Lee, Mehmet C. Vuran, and Shantidev Mohanty. Next generation/dynamic spectrum access/cognitive radio wireless networks: A survey. *Computer Networks*, 50(13):2127–2159, 2006.

Ian Fuat Akyildiz, Fernando Brunetti, and Cristina Blázquez. Nanonetworks: A new communication paradigm. *Computer Networks*, 52(12):2260–2279, August 2008. ISSN 1389-1286. doi: 10.1016/j.comnet.2008.04.001.

Nanonetworks: The Future of Communication and Computation, First Edition. Florian-Lennert A. Lau.

Ian Fuat Akyildiz, Josep Miquel Jornet, and Massimiliano Pierobon. Nanonetworks: A new frontier in communications. *Communications of the ACM*, 54(11): 84–89, November 2011. ISSN 0001-0782. doi: 10.1145/2018396.2018417.

Abdullah Alabdulatif, Navod Neranjan Thilakarathne, Zaharaddeen Karami Lawal, Khairul Eahsun Fahim, and Rufai Yusuf Zakari. Internet of Nano-Things (IONT): A comprehensive review from architecture to security and privacy challenges. *Sensors*, 23(5):2807, 2023.

Faisal Aldaye and Hanadi Sleiman. Modular access to structurally switchable 3D discrete DNA assemblies. *Journal of the American Chemical Society*, 129(44):13376–13377, 2007.

Catherine Alix-Panabieres. Perspective: The future of liquid biopsy. *Nature*, 579(7800):S9, 2020.

Eric Allender, David A. Mix Barrington, Tanmoy Chakraborty, Samir Datta, and Sambuddha Roy. Planar and grid graph reachability problems. *Theory of Computing Systems*, 45(4):675–723, 2009.

Apostolos Almpanis, Christophe Corre, and Adam Noel. Agent based modeling of the rhizobiome with molecular communication and game theory. In *Proceedings of the 6th Annual ACM International Conference on Nanoscale Computing and Communication, NANOCOM 2019*, Dublin, Ireland, September 25–27, 2019, pages 20:1–20:7, 2019. doi: 10.1145/3345312.3345476.

Carme Alvarez and Raymond Greenlaw. A compendium of problems complete for symmetric logarithmic space. *Computational Complexity*, 9(2):123–145, 2000. ISSN 1420-8954.

Paolo Amato, Massimo Masserini, Giancarlo Mauri, and Gianfranco Cerofolini. Early-stage diagnosis of endogenous diseases by swarms of nanobots: An applicative scenario. In M. Dorigo, et al., *Swarm Intelligence. ANTS 2010*. Lecture Notes in Computer Science, vol. 6234, pages 408–415. Springer-Verlag, Berlin, Heidelberg, 2010. ISBN 978-3-642-15461-4.

Islamshah Amlani, Alexei Orlov, Gregory Snider, Craig Lent, and Gary Bernstein. Realization of a functional cell for quantum-dot cellular automaton. *Science*, 277:928–930, August 1997.

Ebbe Andersen, Mingdong Dong, Morten M. Nielsen, Kasper Jahn, Ramesh Subramani, Wael Mamdouh, Monika M. Golas, Bjoern Sander, Holger Stark, Cristiano L. P. Oliveira, et al. Self-assembly of a nanoscale DNA box with a controllable lid. *Nature*, 459(7243):73–76, 2009. ISSN 0028-0836. doi: 10.1038/nature07971.

Christopher Anderson, Elizabeth J. Clarke, Adam P. Arkin, and Christopher A. Voigt. Environmentally controlled invasion of cancer cells by engineered bacteria. *Journal of Molecular Biology*, 355(4):619–627, 2006.

D. Sydor Anderson, M. J. Sydor, P. Fletcher, and A. Holian. Nanotechnology: The risks and benefits for medical diagnosis and treatment*Journal of Nanomedicine & Nanotechnology*, 7(4):e143, 2016.

David Angeli. A tutorial on chemical reaction network dynamics. *European Journal of Control*, 15:398–406, May 2009. doi: 10.3166/ejc.15.398-406.

Francis Anthony. *Panacea aurea: sive tractatus duo de ipsius auro potabili*. Froben, 1618.

Udi Apsel and Ronen I. Brafman. Extended lifted inference with joint formulas. In *UAI-11 Proceedings of the 27th Conference on Uncertainty in Artificial Intelligence*, pages 74–83. AUAI Press, 2011.

Thierry Arrabal, Dominique Dhoutaut, and Eugen Dedu. Efficient multi-hop broadcasting in dense nanonetworks. In *2018 IEEE 17th International Symposium on Network Computing and Applications (NCA)*, pages 1–9. IEEE, 2018a.

Thierry Arrabal, Dominique Dhoutaut, and Eugen Dedu. Efficient density estimation algorithm for ultra dense wireless networks. In *2018 27th International Conference on Computer Communication and Networks (ICCCN)*, pages 1–9. IEEE, 2018b.

Thierry Arrabal, Florian Büther, Dominique Dhoutaut, and Eugen Dedu. Congestion control by deviation routing in electromagnetic nanonetworks. In *6th ACM International Conference on*

Nanoscale Computing and Communication 2019 (ACM NanoCom'19), Dublin, Ireland, September 2019. ISBN 978-1-4503-6897-1/19/09. doi: 10.1145/3345312.3345465.

Ilja C. W. Arts and Peter C. H. Hollman. Polyphenols and disease risk in epidemiologic studies. *The American Journal of Clinical Nutrition*, 81(1):317S–325S, 2005.

Carl A. Ascoli. Could mutations of SARS-CoV-2 suppress diagnostic detection? *Nature Biotechnology*, 39(3):274–275, 2021.

Baris Atakan, Ozgur Akan, and Sasitharan Balasubramaniam. Body area nanonetworks with molecular communications in nanomedicine. *IEEE Communications Magazine*, 50(1):28–34, January 2012. ISSN 0163-6804. doi: 10.1109/MCOM.2012.6122529.

Monique Augustine Vigu Axelos and Marcel Van de Voorde. *Nanotechnology in Agriculture and Food Science*. Applications of Nanotechnology. Wiley, 2017. ISBN 9783527697731. URL https://books.google.de/books?id=RtZRDgAAQBAJ.

Sasitharan Balasubramaniam and Pietro Lio'. Multi-hop conjugation based bacteria nanonetworks. *IEEE Transactions on Nanobioscience*, 12(1):47–59, 2013. ISSN 15361241. doi: 10.1109/TNB.2013.2239657.

David R. Barbero and Samuel D. Stranks. Functional single-walled carbon nanotubes and nanoengineered networks for organic-and perovskite-solar-cell applications. *Advanced Materials*, 28(44):9668–9685, 2016.

Matteo Bassetti, Tobias Welte, and Richard G. Wunderink. Treatment of gram-negative pneumonia in the critical care setting: Is the beta-lactam antibiotic backbone broken beyond repair? *Critical Care*, 20(1):1–9, 2015.

Suddhasatwa Basu. *Fuel Cell Science and Technology*. Springer, 2007.

Ray Baughman, Anvar A. Zakhidov, and Walt A. De Heer. Carbon nanotubes – the route toward applications. *Science*, 297(5582):787–792, 2002.

Florent Becker, Éric Rémila, and Nicolas Schabanel. Time optimal self-assembly for 2D and 3D shapes: The case of squares and cubes. In Ashish Goel, Friedrich C. Simmel, and Petr Sosík, editors, DNA Computing: 14th International Meeting on DNA Computing, DNA 14, Prague, Czech Republic, June 2–9, 2008. Revised Selected Papers, pages 144–155. Springer-Verlag, Berlin, Heidelberg. ISBN 978-3-642-03076-5. doi: 10.1007/978-3-642-03076-5_12.

Richard Bellman. A Markovian decision process. *Journal of Mathematics and Mechanics*, 6(5):679–684, 1957.

Steven Bellovin. Frank miller: Inventor of the one-time pad. *Cryptologia*, 35(3):203–222, 2011.

Philipp Bende, Florian Lau, and Stefan Fischer. Error-Resistant scaling of Three-Dimensional nanoscale shapes on the basis of DNA-Tiles. In *6th ACM International Conference on Nanoscale Computing and Communication 2019 (ACM NanoCom'19)*, Dublin, Ireland, September 2019.

Yaakov Benenson, Binyamin Gil, Uri Ben-Dor, Rivka Adar, and Ehud Shapiro. An autonomous molecular computer for logical control of gene expression. *Nature*, 429(6990):423–429, May 2004. ISSN 0028-0836.

Dadi Bi, Apostolos Almpanis, Adam Noel, Yansha Deng, and Robert Schober. A survey of molecular communication in cell biology: Establishing a new hierarchy for interdisciplinary applications. *IEEE Communications Surveys & Tutorials*, 23(3):1494–1545, 2021.

A. Bicen and Ian Fuat Akyildiz. End-to-end propagation noise and memory analysis for molecular communication over microfluidic channels. *IEEE Transactions on Communications*, 62(7):2432–2443, July 2014. doi: 10.1109/TCOMM.2014.2323293.

Gerd Binnig, H. Fuchs, Ch. Gerber, H. Rohrer, E. Stoll, and E. Tosatti. Energy-dependent state-density corrugation of a graphite surface as seen by scanning tunneling microscopy. *Europhysics Letters*, 1(1):31, 1986.

Jeffrey J. Birac, William B. Sherman, Jens Kopatsch, Pamela E. Constantinou, and Nadrian C. Seeman. Architecture with GIDEON, a program for design in structural DNA nanotechnology. *Journal of Molecular Graphics and Modelling*, 25(4):470–480, 2006.

Avrim L. Blum and Ronald L. Rivest. Training a 3-node neural network is NP-complete. *Neural Networks*, 5(1):117–127, 1992.

Nicolas Boillot, Dominique Dhoutaut, and Julien Bourgeois. Going for large scale with nano-wireless simulations. In *Proceedings of the 2nd Annual International Conference on Nanoscale Computing and Communication*, pages 1–2, 2015.

Tobias Braun. Circuit minimization in quantum-dot cellular automata. Master thesis, Universität zu Lübeck, 2017.

Tanya Braun and Ralf Möller. Parameterised queries and lifted query answering. In *IJCAI-18 Proceedings of the 27th International Joint Conference on Artificial Intelligence*, pages 4980–4986. IJCAI Organization, 2018.

Tanya Braun, Stefan Fischer, Florian Lau, and Ralf Möller. Lifting DecPOMDPs for nanoscale systems – a work in progress. In *10th International Workshop on Statistical Relational AI at the 1st International Joint Conference on Learning and Reasoning*, 2021. URL https://arxiv.org/abs/2110.09152.

Robert Brijder. Computing with chemical reaction networks: a tutorial. *Natural Computing*, 18(1):119–137, March 2019. ISSN 1572-9796. doi: 10.1007/s11047-018-9723-9.

Hans D. Bruhn, Ralf Junker, Heiner Schäfer, and Stefan Schreiber. *LaborMedizin: Indikationen, Methodik und Laborwerte Pathophysiologie und Klinik*. Schattauer Verlag, 2011.

Yuriy Brun. Arithmetic computation in the tile assembly model: Addition and multiplication. *Theoretical Computer Science*, 378(1):17–31, 2007.

Stephen Bush, Janet Paluh, Guiseppe Piro, Vijay Rao, Venkatesha Prasad, and Andrew Eckford. Defining communication at the bottom. *IEEE Transactions on Molecular, Biological and Multi-Scale Communications*, 1(1):90–96, March 2015.

Kaspar David Buss. Circuit minimization in quantum-dot cellular automata with genetic algorithms. Master thesis, Universität zu Lübeck, 2017.

Florian Büther, Florian Lau, Marc Stelzner, and Sebastian Ebers. A formal definition for nanorobots and nanonetworks. In *The 17th International Conference on Next Generation Wired/Wireless Advanced Networks and Systems + The 10th conference on Internet of Things and Smart Spaces*, St. Petersburg, Russia, September 2017. Springer. doi: 10.1007/978-3-319-67380-6.20.

Florian Büther, Immo Traupe, and Sebastian Ebers. Hop count routing: A routing algorithm for resource constrained, Identity-Free medical nanonetworks. In *5th ACM International Conference on Nanoscale Computing and Communication 2018 (ACM NanoCom'18)*, Reykjavik, Iceland, September 2018. ISBN 978-1-4503-5711-1/18/09. doi: 10.1145/3233188.3233193.

Angela Sara Cacciapuoti, Jessica Illiano, Michele Viscardi, and Marcello Caleffi. Quantum internet: The dawn of the quantum paths. In *Proceedings of the 9th ACM International Conference on Nanoscale Computing and Communication*, pages 1–2, 2022.

Cadnano, 2017. URL http://cadnano.org/welcome.

Wenshan Cai, Uday K. Chettiar, Alexander V. Kildishev, and Vladimir M. Shalaev. Optical cloaking with metamaterisal, nature photonics. *Nature Photonics*, 1:224–227, 2007.

Called Biface of San Isidro, 2009. URL https://commons.wikimedia.org/wiki/File:Bifaz_de_San_Isidro_(M.A.N._1942-101-4-4723)_01.jpg.

Sebastian Canovas-Carrasco, Antonio-Javier Garcia-Sanchez, and Joan Garcia-Haro. A nanoscale communication network scheme and energy model for a human hand scenario. *Nano Communication Networks*, 15:17–27, 2018. ISSN 1878-7789. doi: 10.1016/j.nancom.2018.01.005. URL http://www.sciencedirect.com/science/article/pii/S1878778917300868.

K. Mani Chandy and Leslie Lamport. Distributed snapshots: Determining global states of distributed systems. *ACM Transactions on Computer Systems (TOCS)*, 3(1):63–75, 1985.

Ho-Lin Chen and Ashish Goel. Error free self-assembly using error prone tiles. In *Proceedings of the 10th International Meeting on DNA Based Computers*, pages 274–283, 2004.

Ho-Lin Chen and Ashish Goel. Error free self-assembly using error prone tiles. In *Proceedings of the 10th International Conference on DNA Computing, DNA'04*, 2005.

Junghuei Chen and Nadrian Charles Seeman. Synthesis from DNA of a molecule with the connectivity of a cube. *Nature*, 350(6319):631–633, 1991.

Yu-Ao Chen, Qiang Zhang, Teng-Yun Chen, Wen-Qi Cai, Sheng-Kai Liao, Jun Zhang, Kai Chen, Juan Yin, Ji-Gang Ren, Zhu Chen, et al. An integrated space-to-ground quantum communication network over 4,600 kilometres. *Nature*, 589(7841):214–219, 2021.

Qi Cheng, Gagan Aggarwal, Michael H. Goldwasser, Ming-Yang Kao, Robert Schweller, and Pablo Moisset de Espanés. Complexities for generalized models of self-assembly. *SIAM Journal on Computing*, 34:1493–1515, 2005.

Elisa Chilet-Rosell. Gender bias in clinical research, pharmaceutical marketing, and the prescription of drugs. *Global Health Action*, 7(1):25484, 2014.

Andrew Chiu, George Davida, and Bruce Litow. Division in logspace-uniform NC1. *RAIRO – Theoretical Informatics and Applications*, 35(3):259–275, 2001.

Arkadiusz Chworos, Isil Severcan, Alexey Koyfman, Patrick Weinkam, Emin Oroudjev, Helen Hansma, and Luc Jaeger. Building programmable jigsaw puzzles with RNA. *Science*, 306(5704):2068–2072, 2004.

Arthur Charles Clarke. *The Next Tenants*. Tales from the White Hart. 1957. URL http://www.isfdb.org/cgi-bin/title.cgi?56730.

Peter Clote and Evangelos Kranakis. *Boolean Functions and Computation Models*. Texts in Theoretical Computer Science. An EATCS Series. Springer-Verlag, Berlin, Heidelberg, 1 edition, 2002.

Luis Cobo and Ian F. Akyildiz. Bacteria-based communication in nanonetworks. *Nano Communication Networks*, 1(4):244–256, 2010.

COMSOL Multiphysics. Introduction to COMSOL Multiphysics®. Burlington, MA, 9(2018):32, 1998 [accessed February].

Stephen Cook and Pierre McKenzie. Problems complete for deterministic logarithmic space. *Journal of Algorithms*, 8(3):385–394, 1987. ISSN 0196-6774. doi: 10.1016/0196-6774(87)90018-6. URL http://www.sciencedirect.com/science/article/pii/0196677487900186.

Serge Cosnier, Alan Le Goff, and Michael Holzinger. Towards glucose biofuel cells implanted in human body for powering artificial organs. *Electrochemistry Communications*, 38:19–23, 2014.

Flaviu Cristian. Probabilistic clock synchronization. *Distributed Computing*, 3: 146–158, 1989.

G. T. Csanady. Turbulent diffusion: Elementary statistical theory and atmospheric applications. In *Turbulent Diffusion in the Environment*. Geophysics and Astrophysics Monographs, vol. 3, pages 46–81. Springer, Dordrecht, 1973.

Charles Darwin. *On the Origin of Species*. John Murray, 1859.

Didier Descouens. English: Chopper – Different views of the same specimen, November 2010. URL https://commons.wikimedia.org/wiki/File:Galet_MHNT_PRE.2009.0.200.1.jpg.

Dominique Dhoutaut, Thierry Arrabal, and Eugen Dedu. Bit simulator, an electromagnetic nanonetworks simulator. In *Proceedings of the 5th ACM International Conference on Nanoscale*

Computing and Communication, NANOCOM '18, pages 1–6. New York, NY, USA, 2018. Association for Computing Machinery. ISBN 9781450357111. doi: 10.1145/3233188.3233205.

Didia. *Deutsch: Der widderköpfige Schöpfergott Chnum formt das göttliche Kind Ihy (Horus/König) auf der Töpferscheibe und die Göttin Isis verleiht ihm Leben, indem sie ihm die Hierogylyphen für Leben (Anch) entgegenstreckt.* Aus dem Mammisi des Trajan in Dendera, 2014. URL https://commons.wikimedia.org/wiki/File:Chnum-ihy-isis.jpg.

Marc van Dijk and Alexandre M. J. J. Bonvin. 3D-dart: A DNA structure modelling server. *Nucleic Acids Research*, 37(Suppl 2):W235–W239, 2009.

DNA Lab: Xgrow tile assembly simulator, 2003. *URL* http://www.dna.caltech.edu/Xgrow/.

Theodosius Dobzhansky. Nothing in biology makes sense except in the light of evolution. *The American Biology Teacher*, 75(2):87–91, 2013.

Brian S. Donahue and R. F. Abercrombie. Free diffusion coefficient of ionic calcium in cytoplasm. *Cell Calcium*, 8(6):437–448, 1987.

Shawn M. Douglas, Hendrik Dietz, Tim Liedl, Björn Högberg, Franziska Graf, and William M. Shih. Self-assembly of DNA into nanoscale three-dimensional shapes. *Nature*, 459(7245):414–418, 2009a.

Shawn M. Douglas, Adam H. Marblestone, Surat Teerapittayanon, Alejandro Vazquez, George M. Church, and William M. Shih. Rapid prototyping of 3D DNA-origami shapes with Cadnano. *Nucleic Acids Research*, 37(15):5001–5006, 2009b.

Shawn Douglas, Ido Bachelet, and George Church. A logic-gated nanorobot for targeted transport of molecular payloads. *Science*, 335(6070):831–834, 2012.

Rod G. Downey and Michael R. Fellows. Fixed-parameter tractability and completeness I: Basic results. *SIAM Journal on Computing*, 24(4):873–921, 1995.

Rodney G. Downey and Michael R. Fellows. *Fundamentals of Parameterized Complexity*, volume 4. Springer, 2013.

Mildred S. Dresselhaus, Gene Dresselhaus, and Peter C. Eklund. *Science of Fullerenes and Carbon Nanotubes: Their Properties and Applications.* Elsevier, 1996.

Falko Dressler and Stefan Fischer. Connecting in-body nano communication with body area networks: Challenges and opportunities of the Internet of Nano Things. *Nano Communication Networks*, 6:29–38, 2015.

David J. Duffy. Problems, challenges and promises: perspectives on precision medicine. *Briefings in Bioinformatics*, 17(3):494–504, August 2015. ISSN 1467-5463. doi: 10.1093/bib/bbv060.

David J. Duffy. Problems, challenges and promises: perspectives on precision medicine. *Briefings in Bioinformatics*, 17(3):494–504, 2016.

Daniel Dykhuizen. Species numbers in bacteria. *Proceedings. California Academy of Sciences*, 56(6 Suppl 1):62–71, June 2005. ISSN 0068-547X. URL https://www.ncbi.nlm.nih.gov/pmc/articles/PMC3160642/.

Sebastian Ebers. Technologien für Fahrzeug-Ad-Hoc-Kommunikation. PhD thesis, Universität zu Lübeck, February 2016.

Andrew W. Eckford. Nanoscale communication with Brownian motion. In *2007 41st Annual Conference on Information Sciences and Systems*, pages 160–165. IEEE, 2007.

Albert Einstein. Über einen die erzeugung und verwandlung des lichtes betreffenden heuristischen gesichtspunkt, 1905.

Albert Einstein, Boris Podolsky, and Nathan Rosen. Can quantum-mechanical description of physical reality be considered complete? *Physical Review*, 47 (10):777, 1935.

Masayuki Endo, Kumi Hidaka, Takayuki Kato, Keiichi Namba, and Hiroshi Sugiyama. DNA prism structures constructed by folding of multiple rectangular arms. *Journal of the American Chemical Society*, 131(43):15570–15571, 2009.

Paula Enzian, Astrid Link, Christian Schell, Carina Malich, and Ramtin Rahmanzadeh. Light-induced permeabilization of liposomes. In Tayyaba Hasan, editor, *17th International Photodynamic Association World Congress*, volume 11070, pages 118–123. International Society for Optics and Photonics, SPIE, 2019. doi: 10.1117/12.2526071.

Paula Enzian, Christian Schell, Astrid Link, Carina Malich, Ralph Pries, Barbara Wollenberg, and Ramtin Rahmanzadeh. Optically controlled drug release from light-sensitive liposomes with the new photosensitizer 5,10-DiOH. *Molecular Pharmaceutics*, 17(8):2779–2788, 2020. ISSN 1543-8384. doi: 10.1021/acs.molpharmaceut.9b01173.

Alper Erturk and Daniel J. Inman. *Piezoelectric Energy Harvesting*. John Wiley & Sons, 2011.

Evologics. Acoustic Modems | EvoLogics, 2000. URL https://evologics.de/acoustic-modems.

Bengt Fagrell, Arnost Fronek, and Marcos Intaglietta. A microscope-television system for studying flow velocity in human skin capillaries. *American Journal of Physiology-Heart and Circulatory Physiology*, 233(2):H318–H321, 1977.

N. Farsad, H. B. Yilmaz, A. Eckford, C. B. Chae, and W. Guo. A comprehensive survey of recent advancements in molecular communication. *IEEE Communications Surveys & Tutorials*, 18(3):1887–1919, February 2016. ISSN 1553-877X.

Luca Felicetti, Mauro Femminella, and GianLuca Reali. A simulation tool for nanoscale biological networks. *Nano Communication Networks*, 3(1):2–18, 2012. ISSN 1878-7789. doi: 10.1016/j.nancom.2011.09.002. URL http://www.sciencedirect.com/science/article/pii/S1878778911000524.

Luca Felicetti, Mauro Femminella, Gianluca Reali, Paolo Gresele, and Marco Malvestiti. Simulating an in vitro experiment on nanoscale communications by using BiNS2. *Nano Communication Networks*, 4(4):172–180, 2013. ISSN 1878-7789. doi: 10.1016/j.nancom.2013.08.003. URL http://www.sciencedirect.com/science/article/pii/S1878778913000471.

Luca Felicetti, Mauro Femminella, Gianluca Reali, Paolo Gresele, Marco Malvestiti, and John Daigle. Modeling CD40-based molecular communications in blood vessels. *IEEE Transactions on NanoBioscience*, 13(3):230–243, September 2014. ISSN 1536-1241. doi: 10.1109/TNB.2014.2340134.

Adolf Fick. Ueber diffusion. *Annalen der Physik*, 170(1):59–86, 1855.

Alex S. Fraser. Simulation of genetic systems by automatic digital computers I. Introduction. *Australian Journal of Biological Sciences*, 13:484–491, January 1957.

Robert Freitas. Current status of nanomedicine and medical nanorobotics. *Journal of Computational and Theoretical Nanoscience*, 2(1):1–25, 2005. ISSN 1546-1955.

Robert Freitas Jr,. *Nanomedicine, Volume I: Basic Capabilities*. Landes Bioscience, Georgetown, Texas, April 1999.

Tsu Ju Fu and Nadrian Charles Seeman. DNA double-crossover molecules. *Biochemistry*, 32(13):3211–3220, 1993.

Kenichi Fujibayashi and Satoshi Murata. Thermodynamic simulations of DNA tile self-assembly. In *Proceedings of the 2nd International Conference on Nano-Networks*, Nano-Net '07, pages 20:1–20:5. ICST (Institute for Computer Sciences, Social-Informatics and Telecommunications Engineering), Brussels, Belgium, 2007. ISBN 978-963-9799-10-3. URL http://dl.acm.org/citation.cfm?id=1459290.1459317.

Anahi Gajardo, Andre Moreira, and Eric Goles. Complexity of Langton's ant. *Discrete Applied Mathematics*, 117(1–3):41–50, 2002.

Laura Galluccio, Tommaso Melodia, Sergio Palazzo, and Enrico Santagati. Medium access control and rate adaptation for ultrasonic intrabody sensor networks. *IEEE/ACM Transactions on Networking*, 23(4):1121–1134, August 2015. ISSN 1063-6692. doi: 10.1109/TNET.2014.2316675.

Cyril Gavoille, Christian Glacet, Nicolas Hanusse, and David Ilcinkas. On the communication complexity of distributed name-independent routing schemes. In *International Symposium on Distributed Computing*, pages 418–432. Springer, 2013.

Cody Geary, Paul Wilhelm Karl Rothemund, and Ebbe Andersen. A single-stranded architecture for cotranscriptional folding of RNA nanostructures. *Science*, 345(6198):799–804, 2014.

Marcel Gehrke, Tanya Braun, and Ralf Möller. Lifted temporal maximum expected utility. In *Proceedings of the 32nd Canadian Conference on Artificial Intelligence, Canadian AI 2019*. Springer, 2019a.

Marcel Gehrke, Tanya Braun, Ralf Möller, Alexander Waschkau, Christoph Strumann, and Jost Steinhäuser. Lifted maximum expected utility. In *Artificial Intelligence in Health*, pages 131–141. Springer, 2019b.

Regine Geyer. Bloodvoyagers: Conceptional design and first implementation of a body simulator for the realistic modelling of the work environment of medical nanobots. Master thesis, Universität zu Lübeck, 2017.

Regine Geyer, Marc Stelzner, Florian Büther, and Sebastian Ebers. Bloodvoyagers: Simulation of the work environment of medical nanobots. In *Proceedings of the 5th ACM International Conference on Nanoscale Computing and Communication*, NANOCOM '18, New York, NY, USA, 2018a. Association for Computing Machinery. ISBN 9781450357111. doi: 10.1145/3233188.3233196.

Regine Geyer, Marc Stelzner, Florian Büther, and Sebastian Ebers. BloodVoyagerS – Simulation of the work environment of medical nanobots. In *5th ACM International Conference on Nanoscale Computing and Communication 2018 (ACM NanoCom'18)*, Reykjavik, Iceland, September 2018b. URL https://dx.doi.org/10.1145/3233188.3233196.

Siavash Ghavami, Farshad Lahouti, and Ali Masoudi-Nejad. Modeling and analysis of abnormality detection in biomolecular nano-networks. *Nano Communication Networks*, 3(4):229–241, 2012.

John G. Gibbs and Y.-P. Zhao. Autonomously motile catalytic nanomotors by bubble propulsion. *Applied Physics Letters*, 94(16):163104, 2009.

Jorge Torres Gómez, Ketki Pitke, Lukas Stratmann, and Falko Dressler. Age of information in molecular communication channels. *Digital Signal Processing*, 124:103108, 2022.

Buddha Gotama. *The Dhammapada*. Nanoparticles, Approximately 500 BCE, 600-520 BC, n.d.

Thomas Graham. Xxvii. On the law of the diffusion of gases. *The London, Edinburgh, and Dublin Philosophical Magazine and Journal of Science*, 2(9):175–190, 1833.

M. Gregori and I. F. Akyildiz. A new nanonetwork architecture using flagellated bacteria and catalytic nanomotors. *IEEE Journal on Selected Areas in Communications*, 28(4):612–619, May 2010. ISSN 0733-8716. doi: 10.1109/JSAC.2010.100510.

Maria Gregori, Ignacio Llatser, Albert Cabellos-Aparicio, and Eduard Alarcón. Physical channel characterization for medium-range nanonetworks using flagellated bacteria. *Computer Networks*, 55(3):779–791, 2011.

Gregory Bowman. Folding@home, October 2000. URL https://foldingathome.org/gregory-bowman-phd/.

Lov K. Grover. A fast quantum mechanical algorithm for database search. In *Proceedings of the 28th Annual ACM Symposium on Theory of Computing*, pages 212–219, 1996.

Dalia Grybauskaite. World's Smallest Nativity Scene Can Only be Viewed with a Microscope, was Created with the Help of 3D Printers – TechEBlog, 2018. URL https://www.techeblog.com/smallest-nativity-scene-microscope-3d-printed/.

Zhen Gu, Alex Aimetti, Qun Wang, Tram Dang, Yunlong Zhang, Omid Veiseh, Hao Cheng, Robert Langer, and Daniel Anderson. Injectable nano-network for glucose-mediated insulin delivery. *ACS Nano*, 7(5):4194–4201, 2013.

Ertan Gul, Baris Atakan, and Ozgur B. Akan. NanoNS: A nanoscale network simulator framework for molecular communications. *Elsevier Nano Communication Networks*, 1(2):138–156, June 2010. doi: 10.1016/j.nancom.2010.08.003.

Aydin Guney, Baris Atakan, and Özgür Akan. Mobile ad hoc nanonetworks with collision-based molecular communication. *IEEE Transactions on Mobile Computing*, 11(3):353–366, 2012. doi: 10.1109/TMC.2011.53.

Riccardo Gusella and Stefano Zatti. The accuracy of the clock synchronization achieved by TEMPO in Berkeley UNIX 4.3 BSD. *IEEE Transactions on Software Engineering*, 15(7):847–853, 1989.

Mohamed Ahmed Hail. Named data networking for the Internet of Things. PhD thesis, University of Luebeck, April 2018.

P. Hamalainen, T. Alho, M. Hannikainen, and T. D. Hamalainen. Design and implementation of low-area and low-power AES encryption hardware core. In *9th EUROMICRO Conference on Digital System Design – Architectures, Methods and Tools (DSD 2006)*, pages 577–583, Dubrovnik, Croatia, August 2006. IEEE. doi: 10.1109/DSD.2006.40.

Dongran Han, Suchetan Pal, Jeanette Nangreave, Zhengtao Deng, Yan Liu, and Hao Yan. DNA origami with complex curvatures in three-dimensional space. *Science*, 332(6027):342–346, 2011.

Ismo Hänninen and Jarmo Takala. *Arithmetic Design on Quantum-Dot Cellular Automata Nanotechnology*. Springer-Verlag, Berlin, Heidelberg, 2008. ISBN 978-3-540-70550-5. doi: 10.1007/978-3-540-70550-5_6.

Yudong Hao, Martin Kristiansen, Ruojie Sha, Jens Birktoft, Carina Hernandez, Chengde Mao, and Nadrian Charles Seeman. A device that operates within a self-assembled 3D DNA crystal. *Nature Chemistry*, 9:824–827, 2017.

Nor Zaidi Haron and Said Hamdioui. Why is CMOS scaling coming to an end? In *2008 3rd International Design and Test Workshop*, pages 98–103, December 2008. doi: 10.1109/IDT.2008.4802475.

Noha Hassanin, Maha M. Elkishki, and Laila H. Fawzy. Trace elements and their relation to diabetes mellitus and obesity. *Journal of Recent Advances in Medicine*, 2:128–132, January 2021. doi: 10.21608/jram.2020.46094.1093. URL https://jram.journals.ekb.eg/article_123247.html.

Wendi Rabiner Heinzelman, Anantha Chandrakasan, and Hari Balakrishnan. Energy-efficient communication protocol for wireless microsensor networks. In *Proceedings of the 33rd Annual Hawaii International Conference on System Sciences*, page 10. IEEE, 2000.

Tad Hogg and Robert Freitas Jr,. Acoustic communication for medical nanorobots. *Elsevier Nano Communication Networks*, 3(2):83–102, June 2012. ISSN 1878-7789. doi: 10.1016/j.nancom.2012.02.002.

Robin Holliday. Molecular aspects of genetic exchange and gene conversion. *Genetics*, 78(1):273, 1974.

Steven Holtzen, Todd Millstein, and Guy Van den Broeck. Generating and sampling orbits for lifted probabilistic inference. In *UAI-19 Proceedings of the 35th Conference on Uncertainty in Artificial Intelligence*, pages 1–10. AUAI Press, 2019.

Matthias Homeister. *Quantum Computing Verstehen*. Springer, 2008.

John Joseph Hopfield. Kinetic proofreading: A new mechanism for reducing errors in biosynthetic processes requiring high specificity. *Proceedings of the National Academy of Sciences of the United States of America*, 71(10):4135–4139, October 1974. ISSN 0027-8424. URL http://www.ncbi.nlm.nih.gov/pmc/articles/PMC434344/. 4530290[pmid].

Farah Hoteit, Dominique Dhoutaut, Winston K. G. Seah, and Eugen Dedu. Dynamic ring-based forwarder selection to improve packet delivery in ultra-dense nanonetworks. In *Proceedings of the*

9th ACM International Conference on Nanoscale Computing and Communication, NANOCOM '22, New York, NY, USA, 2022. Association for Computing Machinery. ISBN 9781450398671. doi: 10.1145/3558583.3558849.

Hen-Wei Huang, Mahmut Selman Sakar, Andrew J. Petruska, Salvador Pane, and Bradley J. Nelson. Soft micromachines with programmable motility and morphology. *Nature Communications*, 7:12263, July 2016. URL http://dx.doi.org/10.1038/ncomms12263.

Dongeun Huh, Geraldine A. Hamilton, and Donald E. Ingber. From 3D cell culture to organs-on-chips. *Trends in Cell Biology*, 21(12):745–754, 2011.

Willem H. Hundsdorfer, Jan G. Verwer, and W. H. Hundsdorfer. *Numerical Solution of Time-Dependent Advection-Diffusion-Reaction Equations*, volume 33. Springer, 2003.

Geoffrey Hunt and Michael Mehta. *Nanotechnology: "Risk, Ethics and Law"*. Routledge, 2013.

Gavriel Iddan, Gavriel Meron, Arkady Glukhovsky, and Paul Swain. Wireless capsule endoscopy. *Nature*, 405(6785):417–417, 2000.

Sumio Iijima. Helical microtubules of graphitic carbon. *Nature*, 354(6348):56, 1991.

Neil Immerman and Susan Landau. The complexity of iterated multiplication. *Information and Computation*, 116(1):103–116, 1995. ISSN 0890-5401.

Ivan Ivanov, Tanja Vidaković-Koch, and Kai Sundmacher. Recent advances in enzymatic fuel cells: Experiments and modeling. *Energies*, 3(4):803–846, 2010.

Arthur M. Jaffe. The millennium grand challenge in mathematics. *Notices of the AMS*, 53(6):652–660, 2006.

Birgit Jenner, Johannes Köbler, Pierre McKenzie, and Jacobo Torán. Completeness results for graph isomorphism. *Journal of Computer and System Sciences*, 66(3):549–566, 2003. ISSN 0022-0000.

L. Jensen, J. Bangsbo, and Y. Hellsten. Effect of high intensity training on capillarization and presence of angiogenic factors in human skeletal muscle. *The Journal of Physiology*, 557(2):571–582, 2004.

Yubing Jian, Bhuvana Krishnaswamy, Caitlin M. Austin, Ozan Bicen, Jorge Perdomo, Sagar Patel, Ian Fuat Akyildiz, Craig Forest, and Raghupathy Sivakumar. nanoNS3: Simulating bacterial molecular communication based nanonetworks in Network Simulator 3. In *Proceedings of the 3rd ACM International Conference on Nanoscale Computing and Communication*, NANOCOM'16, pages 17:1–17:7. ACM, September 2016. ISBN 978-1-4503-4061-8. doi: 10.1145/2967446.2967464.

Laiming Jiang, Yang Yang, Ruimin Chen, Gengxi Lu, Runze Li, Di Li, Mark S. Humayun, K. Kirk Shung, Jianguo Zhu, Yong Chen, et al. Flexible piezoelectric ultrasonic energy harvester array for bio-implantable wireless generator. *Nano Energy*, 56:216–224, 2019.

Liuyi Jin, Lihua Zuo, Zhipei Yan, and Radu Stoleru. Nanocommunication-based impermeable region mapping for oil reservoir exploration. In *NANOCOM '19: Proceedings of the Sixth Annual ACM International Conference on Nanoscale Computing and Communication*, pages 1–7, September 2019.

Christian Johner, Matthias Hölzer-Klüpfel, and Sven Wittorf. *Basiswissen medizinische Software: Aus-und Weiterbildung zum certified professional for medical software*. dpunkt.verlag, 2020.

J. M. Jornet. A joint energy harvesting and consumption model for self-powered nano-devices in nanonetworks. In *2012 IEEE International Conference on Communications (ICC)*, pages 6151–6156, June 2012.

Josep Miquel Jornet and Ian Fuat Akyildiz. Graphene-based plasmonic nano-antenna for terahertz band communication in nanonetworks. *IEEE Journal on Selected Areas in Communications*, 31(12):685–694, December 2013. doi: 10.1109/JSAC.2013.SUP2.1213001.

Josep Miquel Jornet and Ian F Akyildiz. Femtosecond-long pulse-based modulation for terahertz band communication in nanonetworks. *IEEE Transactions on Communications*, 62(5):1742–1754, 2014.

Neville Kallenbach, Rong-Ine Ma, and Nadrian Charles Seeman. An immobile nucleic acid junction constructed from oligonucleotides. *Nature*, 305(5937):829, 1983. ISSN 0028-0836.

Peter Karlson and Adolf Butenandt. Pheromones (ectohormones) in insects. *Annual Review of Entomology*, 4(1):39–58, 1959.

Maurice Karnaugh. The map method for synthesis of combinational logic circuits. *Transactions of the American Institute of Electrical Engineers, Part I: Communication and Electronics*, 72(5):593–599, 1953.

Max Kaussow. Modeling and simulation of DNA-based nanonetworks. Master thesis, Universität zu Lübeck, 2022.

Hiroshi Kawano. Complete reconfiguration algorithm for sliding cube-shaped modular robots with only sliding motion primitive. In *2015 IEEE/RSJ International Conference on Intelligent Robots and Systems (IROS)*, pages 3276–3283, 2015.

Yonggang Ke, Shawn Douglas, Minghui Liu, Jaswinder Sharma, Anchi Cheng, Albert Leung, Yan Liu, William Shih, and Hao Yan. Multilayer DNA origami packed on a square lattice. *Journal of the American Chemical Society*, 131(43):15903–15908, 2009a.

Yonggang Ke, Jaswinder Sharma, Minghui Liu, Kasper Jahn, Yan Liu, and Hao Yan. Scaffolded DNA origami of a DNA tetrahedron molecular container. *Nano Letters*, 9(6):2445–2447, 2009b.

Anthony Kenny. *A New History of Western Philosophy*. OUP Oxford, 2012.

S. Kerzenmacher, J. Ducrée, R. Zengerle, and F. Von Stetten. Energy harvesting by implantable abiotically catalyzed glucose fuel cells. *Journal of Power Sources*, 182(1):1–17, 2008.

V. K. Khanna. *Nanosensors: Physical, Chemical, and Biological*. Series in Sensors. CRC Press, 2016. ISBN 9781439827130. URL https://books.google.de/books?id=8RjOBQAAQBAJ.

Kwanoh Kim, Xiaobin Xu, Jianhe Guo, and D. L. Fan. Ultrahigh-speed rotating nanoelectromechanical system devices assembled from nanoscale building blocks. *Nature Communications*, 5(1):3632, 2014.

Brian Kirby, Jason Campbell, Burak Aksak, Padmanabhan Pillai, James Hoburg, Todd C. Mowry, and Seth Copen Goldstein. Catoms: Moving robots without moving parts. In *Proceedings of the National Conference on Artificial Intelligence*, volume 20, page 1730. AAAI Press; MIT Press, Menlo Park, CA; Cambridge, MA; London; 1999, 2005.

Donald Ervin Knuth. Big omicron and big omega and big theta. *ACM SIGACT News*, 8(2):18–24, 1976.

Hideki Kobayashi, Mads Kaern, Michihiro Araki, Kristy Chung, Timothy S. Gardner, Charles R. Cantor, and James J. Collins. Programmable cells: interfacing natural and engineered gene networks. *Proceedings of the National Academy of Sciences of the United States of America*, 101(22):8414–8419, 2004.

David Kriesel. *Ein kleiner Überblick über Neuronale Netze*, 2007. URL erhältlich auf http://www.dkriesel.com.

Harry W. Kroto, A. W. Allaf, and S. P. Balm. C_{60}: Buckminsterfullerene. *Chemical Reviews*, 91(6):1213–1235, 1991.

Pawel Kulakowski, Kamil Solarczyk, and Krzysztof Wojcik. Routing in fret-based nanonetworks. *IEEE Communications Magazine*, 55(9):218–224, 2017.

Tanmay Kulkarni and Gymama Slaughter. Enzymatic glucose biofuel cell and its application. *Journal of Biochips & Tissue Chips*, 5:1000111, July 2015. doi: 10.4172/21530777.1000111.

Mehmet Şükrü Kuran, H. Birkan Yilmaz, Tuna Tugcu, and Ian Fuat Akyildiz. Interference effects on modulation techniques in diffusion based nanonetworks. *Nano Communication Networks*, 3(1):65–73, 2012. doi: 10.1016/j.nancom.2012.01.005.

Thomas LaBean, Erik Winfree, and John H. Reif. Experimental progress in computation by self-assembly of DNA tilings, 2000.

Leslie Lamport. Time, clocks, and the ordering of events in a distributed system. *Communications of the ACM*, 21(7):558–565, July 1978. ISSN 0001-0782.

Leslie Lamport. How to make a correct multiprocess program execute correctly on a multiprocessor. *IEEE Transactions on Computers*, 46(7):779–782, 1997.

Leslie Lamport. Time, clocks, and the ordering of events in a distributed system. In *Concurrency: the Works of Leslie Lamport*, pages 179–196, 2019.

Edmund Landau. *Handbuch der lehre von der verteilung der primzahlen*. volume 01. Leipzig B.G. Teubner, 1909.

James Lathrop, Jack Lutz, and Scott Summers. Strict self-assembly of discrete sierpinski triangles. In S. Barry Cooper, Benedikt Löwe, and Andrea Sorbi, editors, *Computation and Logic in the Real World: Third Conference on Computability in Europe*, CiE 2007, Siena, Italy, June 18–23, 2007. *Proceedings*, pages 455–464. Springer-Verlag, Berlin, Heidelberg. ISBN 978-3-540-73001-9. doi: 10.1007/978-3-540-73001-9_47.

Florian-Lennert Adrian Lau. DNA-basierte Nanonetzwerke. PhD Thesis, Universität zu Lübeck, 2020.

Florian Lau, Florian Büther, and Bennet Gerlach. Computational requirements for nano-machines: There is limited space at the bottom. In *4th ACM International Conference on Nanoscale Computing and Communication*, ACM NanoCom'17, pages 11:1–11:6, Washington DC, USA, September 2017. ACM. doi: 10.1145/3109453.3109458. in press.

Florian Lau, Florian Büther, Regine Geyer, and Stefan Fischer. Computation of decision problems within messages in DNA-tile-based molecular nanonetworks. *Nano Communication Networks*, 2019. ISSN 1878-7789. doi: 10.1016/j.nancom.2019.05.002. URL http://www.sciencedirect.com/science/article/pii/S1878778919300018.

Florian-Lennert Lau, Regine Wendt, and Stefan Fischer. Solving generic decision problems by in-message computation in DNA-based molecular nanonetworks. In *15th International Conference on Body Area Networks*, BodyNets '20, Cyberspace, 2020. ICST. Event-place: Cyberspace due to Corona.

Florian Lau, Regine Wendt, and Stefan Fischer. DNA-based molecular communication as a paradigm for multi-parameter detection of diseases. In *8th ACM International Conference on Nanoscale Computing and Communication 2021 (ACM NanoCom'21)*, Virtual Conference, September 2021a. ACM.

Florian-Lennert Adrian Lau, Regine Wendt, and Stefan Fischer. Efficient in-message computation of prevalent mathematical operations in DNA-based nanonetworks. *Nano Communication Networks*, 28:100348, June 2021b. ISSN 1878-7789. doi: 10.1016/j.nancom.2021.100348. URL http://www.sciencedirect.com/science/article/pii/S1878778921000090.

Florian-Lennert Lau, Tanya Braun, Ralf Möller, and Stefan Fischer. Using decPOMDPcoms to holistically model and program nanodevices and emergent nanonetworks. In *9th ACM International Conference on Nanoscale Computing and Communication 2022 (ACM NanoCom'22)*, Barcelona Catalunya Spain, October 2022a.

Florian-Lennert Lau, Bennet Gerlach, Regine Wendt, and Stefan Fischer. Towards personalized precision medicine using DNA-Based molecular communication networks. In *9th ACM International Conference on Nanoscale Computing and Communication 2022 (ACM NanoCom'22)*, Barcelona Catalunya Spain, October 2022b.

Bryden Le Bailly. Computing: Nothing more than DNA. *Nature Nanotechnology*, September 2016. ISSN 1748-3387. URL http://dx.doi.org/10.1038/nnano.2016.173. Research Highlights.

Ji Youn Lee, Soo-Yong Shin, Tai Hyun Park, and Byoung-Tak Zhang. Solving traveling salesman problems with DNA molecules encoding numerical values. *Biosystems*, 78(1):39–47, 2004. ISSN 0303-2647. doi: 10.1016/j.biosystems.2004.06.005. URL http://www.sciencedirect.com/science/article/pii/S0303264704001157.

Stanislaw Lem. *The Invincible*. Wydawnictwo MON, 1964.

Stanislaw Lem. *Pokój na Ziemi*. Harcourt Brace, 1984.

Filip Lemic, Sergi Abadal, Wouter Tavernier, Pieter Stroobant, Didier Colle, Eduard Alarcón, Johann Marquez-Barja, and Jeroen Famaey. Survey on terahertz nanocommunication and networking: A top-down perspective, 2019. URL https://arxiv.org/abs/1909.05703.

Filip Lemic, Sergi Abadal, Wouter Tavernier, Pieter Stroobant, Didier Colle, Eduard Alarcón, Johann Marquez-Barja, and Jeroen Famaey. Survey on terahertz nanocommunication and networking: A top-down perspective. *IEEE Journal on Selected Areas in Communications*, 39(6):1506–1543, 2021.

Eric Lerner. The essentials of Buddha Dhamma in meditative practice, 1995. URL https://www.accesstoinsight.org/lib/authors/khin/wheel231.html.

Zetang Li and Zhong Lin Wang. Air/liquid-pressure and heartbeat-driven flexible fiber nanogenerators as a micro/nano-power source or diagnostic sensor. *Advanced Materials*, 23(1):84–89, 2011.

Xiaojun Li, Xiaoping Yang, Jing Qi, and Nadrian Charles Seeman. Antiparallel DNA double crossover molecules as components for nanoconstruction. *Journal of the American Chemical Society*, 118(26):6131–6140, 1996.

Hong Li, Chuan Xu, Navin Srivastava, and Kaustav Banerjee. Carbon nanomaterials for next-generation interconnects and passives: Physics, status, and prospects. *IEEE Transactions on Electron Devices*, 56(9):1799–1821, 2009.

Jinxing Li, Berta Esteban-Fernández de Ávila, Wei Gao, Liangfang Zhang, and Joseph Wang. Micro/nanorobots for biomedicine: Delivery, surgery, sensing, and detoxification. *Science Robotics*, 2(4):eaam6431, 2017.

Suping Li, Qiao Jiang, Shaoli Liu, Yinlong Zhang, Yanhua Tian, Chen Song, Jing Wang, Yiguo Zou, Gregory J. Anderson, Jing-Yan Han, Yung Chang, Yan Liu, Chen Zhang, Liang Chen, Guangbiao Zhou, Guangjun Nie, Hao Yan, Baoquan Ding, and Yuliang Zhao. A DNA nanorobot functions as a cancer therapeutic in response to a molecular trigger in vivo. *Nature Biotechnology*, 36:258 EP –, February 2018. doi: 10.1038/nbt.4071.

Anthony Liekens and Chrisantha Fernando. Turing complete catalytic particle computers. In Fernando Almeida e Costa, Luis Mateus Rocha, Ernesto Costa, Inman Harvey, and Antonio Coutinho, editors, *Advances in Artificial Life*, pages 1202–1211, 2007. Springer-Verlag, Berlin, Heidelberg. ISBN 978-3-540-74913-4.

Chenxiang Lin, Yan Liu, Sherri Rinker, and Hao Yan. DNA tile based self-assembly: Building complex nanoarchitectures. *ChemPhysChem*, 7(8): 1641–1647, 2006.

Pik Kwan Lo, Pierre Karam, Faisal Aldaye, Christopher McLaughlin, Graham Hamblin, Gonzalo Cosa, and Hanadi Sleiman. Loading and selective release of cargo in DNA nanotubes with longitudinal variation. *Nature Chemistry*, 2(4):319–328, 2010.

Bruce E. Logan, Bert Hamelers, René Rozendal, Uwe Schröder, Jürg Keller, Stefano Freguia, Peter Aelterman, Willy Verstraete, and Korneel Rabaey. Microbial fuel cells: Methodology and technology. *Environmental Science & Technology*, 40(17):5181–5192, 2006.

Lucie A. Low, Christine Mummery, Brian R. Berridge, Christopher P. Austin, and Danilo A. Tagle. Organs-on-chips: Into the next decade. *Nature Reviews Drug Discovery*, 20(5):345–361, 2021.

Xiaojun Ma, Masoud Hashempour, Lei Wang, and Fabrizio Lombardi. Manufacturing yield of QCA circuits by synthesized DNA self-assembled templates. In *Proceedings of the 20th Symposium on Great Lakes Symposium on VLSI*, GLSVLSI '10, pages 275–280, New York, NY, USA, 2010. ACM. ISBN 978-1-4503-0012-4. doi: 10.1145/1785481.1785546.

Tasnuva Mahjabin, Alina Olteanu, Yang Xiao, Wenlin Han, Tieshan Li, and Wei Sun. A survey on DNA-based cryptography and steganography. *IEEE Access*, 11:16423–116451, 2023.

Heimo Mairbäurl. Anpassung des erythrozytären sauerstofftransports an belastung und höhe. *Jahrbuch 2014 Österreichischer Gesellschaft für Alpin- und Höhenmedizin*, pages 161–185, January 2014.

Angela Mammana, Gregory Carroll, Jetsuda Areephong, and Ben Feringa. A chiroptical photoswitchable DNA complex. *The Journal of Physical Chemistry B*, 115(40):11581–11587, 2011. doi: 10.1021/jp205893y. PMID: 21879715.

Radu Marculescu, Umit Y. Ogras, Li-Shiuan Peh, Natalie Enright Jerger, and Yatin Hoskote. Outstanding research problems in NoC design: System, microarchitecture, and circuit perspectives. *IEEE Transactions on Computer-Aided Design of Integrated Circuits and Systems*, 28(1):3–21, 2008.

Abraham Maslow. A theory of human motivation. *Psychological Review*, 50(4):370, 1943.

Maurice Walshe Gotama The Buddha. *The Long Discourses of the Buddha, Sutta 1*. Wisdom Publication, 1987b. URL https://suttacentral.net/dn2/en/.

Maurice Walshe Gotama The Buddha. *The Long Discourses of the Buddha, Sutta 2*. Wisdom Publication, 1987c. URL https://suttacentral.net/dn2/en/.

Edward McCluskey Jr,. Minimization of Boolean functions. *Bell system Technical Journal*, 35(6):1417–1444, 1956.

Alfred J. Menezes, Paul C. van Oorschot, and Scott A. Vanstone. *Handbook of Applied Cryptography*, page 810. CRC Press, Boca Raton, FL, 1996.

Deyev Sergei Mikhailovich and E. N. Lebedenko. Modern technologies for creating synthetic antibodies for clinical application. *Acta Naturae*, 1(1):32–50, 2009.

Brian Milch, Luke S. Zettelmoyer, Kristian Kersting, Michael Haimes, and Leslie Pack Kaelbling. Lifted probabilistic inference with counting formulas. In *AAAI-08 Proceedings of the 23rd AAAI Conference on Artificial Intelligence*, pages 1062–1068. AAAI Press, 2008.

Kao Ming-Yang and Vijay Ramachandran. DNA self-assembly for constructing 3D boxes. In Peter Eades and Tadao Takaoka, editors, *Algorithms and Computation: 12th International Symposium*, ISAAC 2001 Christchurch, New Zealand, December 19–21, 2001 Proceedings, pages 429–441. Springer-Verlag, Berlin, Heidelberg, 2001. ISBN 978-3-540-45678-0. URL https://doi.org/10.1007/3-540-45678-3_37.

Marc Z. Miskin, Alejandro J. Cortese, Kyle Dorsey, Edward P. Esposito, Michael F. Reynolds, Qingkun Liu, Michael Cao, David A. Muller, Paul L. McEuen, and Itai Cohen. Electronically integrated, mass-manufactured, microscopic robots. *Nature*, 584(7822):557–561, 2020.

Shahram Mohrehkesh, Michele Weigle, and Sajal Das. *Energy Harvesting in Nanonetworks*, pages 319–347. Springer International Publishing, 2017. ISBN 978-3-319-50688-3.

George Edward Moore. Cramming more components onto integrated circuits. *Electronics*, 38(8):114–117, 1965.

Michael Moore, Akihiro Enomoto, Tadashi Nakano, Tatsuya Suda, Atsushi Kayasuga, Hiroaki Kojima, Hitoshi Sakakibara, and Kazuhiro Oiwa. Simulation of a molecular motor based communication network. In *Proceedings of the 1st International Conference on Bio Inspired Models of Network, Information and Computing Systems*, BIONETICS '06, New York, NY, USA, 2006. ACM. ISBN 1-4244-0463-0. doi: 10.1145/1315843.1315867.

Michael John Moore, Tatsuya Suda, and Kazuhiro Oiwa. Molecular communication: Modeling noise effects on information rate. *IEEE Transactions on Nanobioscience*, 8(2):169–180, 2009.

Tayana Morkel, Jan H. P. Eloff, and Martin S. Olivier. An overview of image steganography. In *ISSA*, volume 1, pages 1–11, 2005.

Shaker A. Mousa, Raj Bawa, and Gerald F. Audette. *The Road from Nanomedicine to Precision Medicine*. CRC Press, 2020.

Wolf-Dieter Müller-Jahncke, Christoph Friedrich, and Ulrich Meyer. *Arzneimittelgeschichte*. Wiss. Verlagsgesellschaft, 2005.

Kary Mullis. The polymerase chain reaction (nobel lecture). *Angewandte Chemie International Edition in English*, 33(12):1209–1213, 1994.

Douglas B. Murphy. *Fundamentals of Light Microscopy and Electronic Imaging*. John Wiley & Sons, 2002.

Mohammad Aiman Mustaffa, Faiz Arith, Nur Syamimi Noorasid, Mohd Shahril Izuan Mohd Zin, Kok Swee Leong, Fara Ashikin Ali, Ahmad Nizamuddin Muhammad Mustafa, and Mohd Muzafar Ismail. Towards a highly efficient ZnO based nanogenerator. *Micromachines*, 13(12):2200, 2022.

Tadashi Nakano. Molecular communication: A 10 year retrospective. *IEEE Transactions on Molecular, Biological and Multi-Scale Communications*, 3(2):71–78, June 2017. doi: 10.1109/TMBMC.2017.2750148.

Aniruddh Nath and Pedro Domingos. A language for relational decision theory. In *Proceedings of the 6th International Workshop on Statistical Relational Learning*, 2009.

Katja Nau. Winzige riesen in unserem alltag, 2016. URL https://www.bmbf.de/publikationen/?P=1979.

Harold C. Neu. The crisis in antibiotic resistance. *Science*, 257(5073):1064–1073, 1992.

Mai Linh Edith Nguyen. Analyse und evaluation von simulatoren für die kommunikation in nanonetzwerken. Master thesis, Universität zu Lübeck, 2017.

Alec Nielsen, Bryan S. Der, Jonghyeon Shin, Prashant Vaidyanathan, Vanya Paralanov, Elizabeth A. Strychalski, David Ross, Douglas Densmore, and Christopher A. Voigt. Genetic circuit design automation. *Science*, 352(6281):aac7341, 2016.

Sami Nummelin, Juhana Kommeri, Mauri A Kostiainen, and Veikko Linko. Evolution of structural DNA nanotechnology. *Advanced Materials*, 30(24):1703721, 2018.

Dmytro Nykypanchuk, Mathew Maye, Daniel Van Der Lelie, and Oleg Gang. DNA-guided crystallization of colloidal nanoparticles. *Nature*, 451(7178):549–552, 2008.

Rachel K. O'Reilly, Andrew J. Turberfield, and Thomas R. Wilks. The evolution of DNA-templated synthesis as a tool for materials discovery. *Accounts of Chemical Research*, 50(10):2496–2509, 2017. doi: 10.1021/acs.accounts.7b00280. PMID: 28915003.

Frans A. Oliehoek and Christopher Amato. *A Concise Introduction to Decentralised POMDPs*. Springer, 2016.

Giacomo Oliveri, Douglas H. Werner, and Andrea Massa. Reconfigurable electromagnetics through metamaterials – a review. *Proceedings of the IEEE*, 103(7):1034–1056, 2015.

Norman Packard and Stephen Wolfram. Two-dimensional cellular automata. *Journal of Statistical Physics*, 38(5):901–946, March 1985. ISSN 1572-9613. doi: 10.1007/BF01010423.

Jing Pan, Tae-Gon Cha, Feiran Li, Haorong Chen, Nina Bragg, and Jong Choi. Visible/near-infrared subdiffraction imaging reveals the stochastic nature of DNA walkers. *Science Advances*, 3:e1601600, January 2017. doi: 10.1126/sciadv.1601600.

Sung Yong Park, Abigail Lytton-Jean, Byeongdu Lee, Steven Weigand, George Schatz, and Chad Mirkin. DNA-programmable nanoparticle crystallization. *Nature*, 451(7178):553–556, 2008.

Aaron N. Parks, Alanson P. Sample, Yi Zhao, and Joshua R. Smith. A wireless sensing platform utilizing ambient RF energy. In *2013 IEEE Topical Conference on Biomedical Wireless Technologies, Networks, and Sensing Systems*, pages 154–156. IEEE, 2013.

Lorenzo Pasotti, Susanna Zucca, and Paolo Magni. 23 – Modelling for synthetic biology. In Ewart Carson and Claudio Cobelli, editors, *Modelling Methodology for Physiology and Medicine*, pages 545–564. Elsevier, Oxford, 2nd edition, 2014. ISBN 978-0-12-411557-6. doi: 10.1016/B978-0-12-411557-6.00023-9. URL https://www.sciencedirect.com/science/article/pii/B9780124115576000239.

Jonathan Pate, Felix Zamora, Scott M. D. Watson, Nicholas G. Wright, Benjamin R. Horrocks, and Andrew Houlton. Solution-based DNA-templating of sub-10 nm conductive copper nanowires.

Journal of Materials Chemistry C, 2(43):9265–9273, October 2014. ISSN 2050-7534. doi: 10.1039/C4TC01632G. URL https://pubs.rsc.org/en/content/articlelanding/2014/tc/c4tc01632g.

Matthew Patitz. Simulation of self-assembly in the abstract tile assembly model with ISU TAS. *CoRR*, abs/1101.5151, 2009. URL http://arxiv.org/abs/1101.5151.

Matthew Patitz. An introduction to tile-based self-assembly and a survey of recent results. *Natural Computing*, 13(2):195–224, 2014. ISSN 1572-9796. doi: 10.1007/s11047-013-9379-4.

Matthew J. Patitz. Webtas, Online, Zugriff 30. Oktober 2022. URL http://self-assembly.net/software/WebTAS/WebTAS-latest/.

Karp Patrick. Efficient nanonetworks for frequent medical problems. Master thesis, Universität zu Lübeck, 2022.

Omkar Y. Pawar, Snehal L. Patil, Rahul S. Redekar, Sharad B. Patil, Sooman Lim, and Nilesh L. Tarwal. Strategic development of piezoelectric nanogenerator and biomedical applications. *Applied Sciences*, 13(5):2891, 2023.

Dieter Perl, Uwe Mueller, Udo Heinemann, and Franz Schmid. Two exposed amino acid residues confer thermostability on a cold shock protein. *Nature Structural & Molecular Biology*, 7(5):380, 2000.

William Wesley Peterson and Daniel T. Brown. Cyclic codes for error detection. *Proceedings of the IRE*, 49(1):228–235, 1961.

M. Pierobon and Ian Fuat Akyildiz. A physical end-to-end model for molecular communication in nanonetworks. *IEEE Journal on Selected Areas in Communications*, 28(4):602–611, May 2010. ISSN 0733-8716. doi: 10.1109/JSAC.2010.100509.

Giuseppe Piro, Luigi Alfredo Grieco, Gennaro Boggia, and Pietro Camarda. Nano-Sim: Simulating electromagnetic-based nanonetworks in the network simulator 3. In *Proceedings of the 6th International ICST Conference on Simulation Tools and Techniques*, SimuTools '13, pages 203–210, Cannes, France, March 2013. ICST (Institute for Computer Sciences, Social-Informatics and Telecommunications Engineering). ISBN 978-1-4503-2464-9. doi: 10.4108/icst.simutools.2013.251699. URL http://dl.acm.org/citation.cfm?id=2512734.2512762.

Max Planck. Zur theorie des gesetzes der energieverteilung im normalspektrum, pages 237–245. Berlin, 1900.

Plato. The republic, 375 BC, n.d. URL https://en.wikipedia.org/wiki/Republic_(Plato).

Werner Pluta. Microsoft stellt automatischen DNA-speicher vor, 2019. URL https://www.golem.de/news/speichertechnik-microsoft-stellt-automatischen-dna-speicher-vor-1903-140188.html.

David Poole. First-order probabilistic inference. In *IJCAI-03 Proceedings of the 18th International Joint Conference on Artificial Intelligence*, pages 985–991. IJCAI Organization, 2003.

Michael C. Potter. Electrical effects accompanying the decomposition of organic compounds. *Proceedings of the Royal Society of London. Series B, Containing Papers of a Biological Character*, 84(571):260–276, 1911.

PUB FIPS. 197. Advanced Encryption Standard (AES), National Institute of Standards and Technology, US Department of Commerce, November 2001, 2009. URL http://csrc.nist.gov/publications/fips/fips197/fips-197.pdf.

Willard Quine. The problem of simplifying truth functions. *The American Mathematical Monthly*, 59(8):521–531, 1952.

Marianne Reibold, P. Paufler, A. A. Levin, W. Kochmann, N. Pätzke, and D. C. Meyer. Carbon nanotubes in an ancient Damascus sabre. *Nature*, 444(7117):286, 2006.

Marcos I. Restrepo, Oriol Sibila, and Antonio Anzueto. Pneumonia in patients with chronic obstructive pulmonary disease. *Tuberculosis and Respiratory Diseases*, 81(3):187–197, 2018.

Michael F. Reynolds, Alejandro J. Cortese, Qingkun Liu, Zhangqi Zheng, Wei Wang, Samantha L. Norris, Sunwoo Lee, Marc Z. Miskin, Alyosha C. Molnar, Itai Cohen, et al. Microscopic robots with onboard digital control. *Science Robotics*, 7(70):eabq2296, 2022.

Vincent Rijmen and Joan Daemen. Advanced encryption standard. In *Proceedings of Federal Information Processing Standards Publications, National Institute of Standards and Technology*, pages 19–22, 2001.

John W. Romanishin, Kyle Gilpin, and Daniela Rus. M-blocks: Momentum-driven, magnetic modular robots. In *2013 IEEE/RSJ International Conference on Intelligent Robots and Systems*, pages 4288–4295. IEEE, 2013.

P. W. K. Rothemund. Design of DNA origami. In *ICCAD-2005. IEEE/ACM International Conference on Computer-Aided Design, 2005*, pages 471–478, 2005. doi: 10.1109/ICCAD.2005.1560114.

Paul Wilhelm Karl Rothemund. Folding DNA to create nanoscale shapes and patterns. *Nature*, 440(7082):297–302, March 2006. ISSN 0028-0836. doi: 10.1038/nature04586.

Paul Wilhelm Karl Rothemund and Erik Winfree. The program-size complexity of self-assembled squares (extended abstract). In *Proceedings of the Thirty-second Annual ACM Symposium on Theory of Computing*, STOC '00, pages 459–468. ACM, 2000. ISBN 1-58113-184-4. doi: 10.1145/335305.335358.

Paul Wilhelm Karl Rothemund, Nick Papadakis, and Erik Winfree. Algorithmic self-assembly of DNA Sierpinski triangles. *PLOS Biology*, 2(12), 12 2004. doi: 10.1371/journal.pbio.0020424.

Florian Rudnitzki, Susanne Feineis, Ramtin Rahmanzadeh, Elmar Endl, Johanna Lutz, Jürgen Groll, and Gereon Hüttmann. siRNA release from gold nanoparticles by nanosecond pulsed laser irradiation and analysis of the involved temperature increase. *Journal of Biophotonics*, 11(9):e201700329, 2018. doi: 10.1002/jbio.201700329.

Stuart Russell and Peter Norvig. *Artificial Intelligence: A Modern Approach*. Pearson, 2020.

Joel Rüthi, Mattia Cerri, Ivano Brunner, Beat Stierli, Michael Sander, and Beat Frey. Discovery of plastic-degrading microbial strains isolated from the alpine and arctic terrestrial plastisphere. *Frontiers in Microbiology*, 14: 1178474, 2023.

Walter John Savitch. Relationships between nondeterministic and deterministic tape complexities. *Journal of Computer and System Sciences*, 4(2):177–192, 1970. ISSN 0022-0000.

Kim Kristin Scharringhausen. Nanosensorik im menschlichen körper: Entwicklung eines systemkonzepts für nanogeräte im blutkreislauf. Master thesis, Universität zu Lübeck, 2018.

Nadrian Charles Seeman. Nucleic acid junctions and lattices. *Journal of Theoretical Biology*, 99(2):237–247, 1982. ISSN 0022-5193. doi: 10.1016/0022-5193(82)90002-9. URL http://www.sciencedirect.com/science/article/pii/0022519382900029.

Nadrian Charles Seeman and Hanadi Sleiman. DNA nanotechnology. *Nature Reviews Materials*, 3(1):17068, 2018.

Sven Seuken and Shlomo Zilberstein. Memory-bounded dynamic programming for DEC-POMDPs. In *IJCAI-07 Proceedings of the 20th International Joint Conference on Artificial Intelligence*, pages 2009–2015. IJCAI Organization, 2007.

Claude E. Shannon. A mathematical theory of communication. *Bell System Technical Journal*, 27(3), 1948. URL http://dblp.uni-trier.de/db/journals/bstj/bstj27.html.

Sunil Singhal, Shuming Nie, and May Wang. Nanotechnology applications in surgical oncology. *Annual Review of Medicine*, 61:359–373, 2010.

Maria Staiano, Angela Pennacchio, Antonio Varriale, Alessandro Capo, Adelia Majoli, Clotilde Capacchione, and Sabato D'Auria. Enzymes as sensors. In *Methods in Enzymology*, volume 589, pages 115–131. Elsevier, 2017.

Mark Staples, Karen Daniel, Michael Cima, and Robert Langer. Application of micro- and nano-electromechanical devices to drug delivery. *Pharmaceutical Research*, 23(5):847–863, 2006. ISSN 1573-904X.

Fabian Steinmetz and Bernd-Christian Renner. Taking LoRA for a dive: CSS for low-power acoustic underwater communication. In *2022 6th Underwater Communications and Networking Conference (UComms)*, pages 1–5. IEEE, 2022.

Marc Stelzner and Immo Traupe. FCNN: Location awareness based on a lightweight hop count routing body coordinates concept. In *6th ACM International Conference on Nanoscale Computing and Communication 2019 (ACM NanoCom'19)*, Dublin, Ireland, September 2019. ISBN 978-1-4503-6897-1. doi: 10.1145/3345312.3345493.

Marc Stelzner, Falko Dressler, and Stefan Fischer. Function centric networking: An approach for addressing in in-body nano networks. In *3rd ACM International Conference on Nanoscale Computing and Communication*, NANOCOM'16, New York, NY, USA, September 2016a. ACM. ISBN 978-1-4503-4061-8.

Marc Stelzner, Florian Lau, Katja Freundt, Florian Büther, Mai Linh Nguyen, Cordula Stamme, and Sebastian Ebers. Precise detection and treatment of human diseases based on nano networking. In *11th International Conference on Body Area Networks*, Turin, Italy, December 2016b. EAI.

Ke Sun, Teng-Sing Wei, Bok Yeop Ahn, Jung Yoon Seo, Shen J. Dillon, and Jennifer A. Lewis. 3D printing of interdigitated Li-Ion microbattery architectures. *Advanced Materials*, 25(33):4539–4543, 2013.

Szymon Szott, Katarzyna Kosek-Szott, Piotr Gawłowicz, Jorge Torres Gómez, Boris Bellalta, Anatolij Zubow, and Falko Dressler. Wi-Fi meets ML: A survey on improving IEEE 802.11 performance with machine learning. *IEEE Communications Surveys & Tutorials*, 24(3):1843–1893, July 2022. ISSN 1553-877X. doi: 10.1109/COMST.2022.3179242.

Abdelmajid Taibi, Antoine Durant, Valeria Loscri, Anna Vegni, and Luigi Spada. Controlling light by curvilinear metasurfaces. In *NANOCOM '19: Proceedings of the 6th Annual ACM International Conference on Nanoscale Computing and Communication*, pages 1–6, September 2019.

Tatsuo Takaishi, Akira Numata, Ryouji Nakano, and Katsuhiko Sakaguchi. Approach to high efficiency diesel and gas engines. *Mitsubishi Heavy Industries Review*, 45(1):21–24, 2008.

Marco S. L. Tang, Simon Chi-Chin Shiu, Maia Godonoga, Yee-Wai Cheung, Shaolin Liang, Roderick M. Dirkzwager, Andrew B. Kinghorn, Lewis A. Fraser, Jonathan G. Heddle, and Julian A. Tanner. An aptamer-enabled DNA nanobox for protein sensing. *Nanomedicine: Nanotechnology, Biology and Medicine*, 14(4):1161–1168, 2018. ISSN 1549-9634. doi: 10.1016/j.nano.2018.01.018. URL https://www.sciencedirect.com/science/article/pii/S1549963418300297.

Thanissaro Bhikkhu Gotama The Buddha. The connected discourses of the Buddha, Sutta 2.26, 1997. URL https://www.accesstoinsight.org/tipitaka/sn/sn02/sn02.026.than.html.

The Biggest Industries in the United States – worldatlas.com, 1994. URL https://www.worldatlas.com/articles/which-are-the-biggest-industries-in-the-united-states.html.

The DNA and Natural Algorithms Group. The Xgrow simulator, 2003. URL http://www.dna.caltech.edu/Xgrow/ [Accessed: 04-May-2018].

Bhikkhu Nanamoli Thera. *The Life of the Buddha*. The Buddhist Publication Society, Kandy, 1972. URL https://store.pariyatti.org/life-of-the-buddha.

Manuel Thiel. *Grüne Gentechnik in Deutschland: Einstellungen der Bevölkerung*, volume 4. ibidem-Verlag/ibidem Press, 2014.

Ageliki Tsioliaridou, Christos Liaskos, Sotiris Ioannidis, and Andreas Pitsillides. Corona: A coordinate and routing system for nanonetworks. In *Second Annual International Conference on Nanoscale*

Computing and Communication, NANOCOM' 15, pages 18:1–18:6, New York, NY, USA, 2015. ACM. ISBN 978-1-4503-3674-1.

A. Tsioliaridou, C. Liaskos, E. Dedu, and S. Ioannidis. Packet routing in 3D nanonetworks: A lightweight, linear-path scheme. *Nano Communication Networks*, 12:63–71, 2017. ISSN 1878-7789. doi: 10.1016/j.nancom.2017.01.001. URL http://www.sciencedirect.com/science/article/pii/S1878778916300898.

Alan Mathison Turing. On computable numbers, with an application to the Entscheidungsproblem. *Proceedings of the London Mathematical Society*, 2(1):230–265, 1937.

Mathias Uhlén, Linn Fagerberg, Björn M. Hallström, Cecilia Lindskog, Per Oksvold, Adil Mardinoglu, Åsa Sivertsson, Caroline Kampf, Evelina Sjöstedt, Anna Asplund, et al. Tissue-based map of the human proteome. *Science*, 347:1260419, January 2015. ISSN 1095-9203. doi: 10.1126/science.1260419.

Guy Van den Broeck, Nima Taghipour, Wannes Meert, Jesse Davis, and Luc De Raedt. Lifted probabilistic inference by first-order knowledge compilation. In *IJCAI-11 Proceedings of the 22nd International Joint Conference on Artificial Intelligence*, pages 2178–2185. IJCAI Organization, 2011.

Vamsi Vankamamidi, Marco Ottavi, and Fabrizi Lombardi. Tile-based design of a serial memory in QCA. In *15th ACM Great Lakes Symposium on VLSI*, GLSVLSI '05, pages 201–206, New York, NY, USA, 2005. ACM. ISBN 1-59593-057-4.

V. C. Venkatesh and Sudin Izman. *Precision Engineering*. McGraw-Hill Education, 2007.

Jacques Versieck. Trace elements in human body fluids and tissues. *Critical Reviews in Clinical Laboratory Sciences*, 22:97–184, 1985. ISSN 1040-8363. doi: 10.3109/10408368509165788. URL https://pubmed.ncbi.nlm.nih.gov/3891229/.

Carl Vierboom, Ingo Härlen, and Johannes Simons. Wahrnehmung der nanotechnologie in der bevölkerung. In *Repräsentativerhebung und morphologisch-psychologische Grundlagenstudie*. Bundesinstitut für Risikobewertung, Berlin, 2008.

Heribert Vollmer. *The NC Hierarchy*, pages 107–171. Springer-Verlag, Berlin, Heidelberg, 1999. ISBN 978-3-662-03927-4.

Sumeet Walia, Charan M. Shah, Philipp Gutruf, Hussein Nili, Dibakar Roy Chowdhury, Withawat Withayachumnankul, Madhu Bhaskaran, and Sharath Sriram. Flexible metasurfaces and metamaterials: A review of materials and fabrication processes at micro-and nano-scales. *Applied Physics Reviews*, 2(1):011303, 2015.

Frank Walsh and Sasitharan Balasubramaniam. Reliability and delay analysis of multihop virus-based nanonetworks. *IEEE Transactions on Nanotechnology*, 12(5):674–684, 2013. ISSN 1536125X. doi: 10.1109/TNANO.2013.2268389.

Walus Group. QCADesigner, 2009.

Hao Wang. *Dominoes and the Aea Case of the Decision Problem*, pages 218–245. Springer Netherlands, Dordrecht, 1990. ISBN 978-94-009-2356-0. doi: 10.1007/978-94-009-2356-0_11.

Xudong Wang. Piezoelectric nanogenerators – Harvesting ambient mechanical energy at the nanometer scale. *Nano Energy*, 1(1):13–24, 2012. ISSN 2211-2855.

Joseph Wang and Wei Gao. Nano/microscale motors: Biomedical opportunities and challenges. *ACS Nano*, 6(7):5745–5751, 2012. doi: 10.1021/nn3028997. PMID: 22770233.

Zhong Lin Wang and Jinhui Song. Piezoelectric nanogenerators based on zinc oxide nanowire arrays. *Science*, 312(5771):242–246, 2006.

Zhong Lin Wang and Wenzhuo Wu. Nanotechnology-enabled energy harvesting for self-powered micro-/nanosystems. *Angewandte Chemie International Edition*, 51(47):11700–11721, 2012.

Silja Wang, Gereon Huttmann, Zhenxi Zhang, Alfred Vogel, Reginald Birngruber, Shifalika Tangutoori, Tayyaba Hasan, and Ramtin Rahmanzadeh. Light-controlled delivery of monoclonal antibodies for targeted photoinactivation of Ki-67. *Molecular Pharmaceutics*, 2015. ISSN 1543-8384. doi: 10.1021/acs.molpharmaceut.5b00260.

Tiesheng Wang, Meisam Farajollahi, Yeon Sik Choi, I-Ting Lin, Jean E. Marshall, Noel M. Thompson, Sohini Kar-Narayan, John D. W. Madden, and Stoyan K. Smoukov. Electroactive polymers for sensing. *Interface focus*, 6 (4):20160026, 2016. https://creativecommons.org/licenses/by/4.0/.

Xiao Wang, Ruojie Sha, Martin Kristiansen, Carina Hernandez, Yudong Hao, Chengde Mao, James Canary, and Nadrian Charles Seeman. An organic semiconductor organized into 3D DNA arrays by "bottom-up" rational design. *Angewandte Chemie*, 129(23):6545–6548, 2017.

James Dewey Watson and Francis Harray Crick. A structure for deoxyribose nucleic acid. *Nature*, 171:737–738, April 1953. URL http://www.nature.com/nature/dna50/watsoncrick.pdf.

H. Wegele, L. Müller, and J. Buchner. Hsp70 and hsp90 – a relay team for protein folding. In *Reviews of Physiology, Biochemistry and Pharmacology*, volume 151, pages 1–44. Springer-Verlag, Berlin, Heidelberg, 2004.

Bryan Wei, Mingjie Dai, and Peng Yin. Complex shapes self-assembled from single-stranded DNA tiles. *Nature*, 485(7400):623–626, 2012.

Regine Wendt and Stefan Fischer. MEHLISSA – a medical holistic simulation architecture for nanonetworks in humans. In *7th ACM International Conference on Nanoscale Computing and Communication 2020 (ACM NanoCom'20)*, USA, September 2020.

Regine Wendt, Florian-Lennert Lau, Lena Unger, and Stefan Fischer. *Proteome Fingerprinting as a Localization Scheme for Nanobots*, New York, NY, USA, 2023, ACM. ISBN 9798400700347. doi: 10.1145/3576781.3608728.

Avi Wigderson. The complexity of graph connectivity. In *International Symposium on Mathematical Foundations of Computer Science*, pages 112–132. Springer, 1992.

Wikimedia Commons. File:Mao-DX-schematic-2.svg – Wikimedia Commons, the free media repository, 2019a. URL https://commons.wikimedia.org/w/index.php?title=File:Mao-DX-schematic-2.svgoldid=346087234 [Accessed: 17-May-2019].

Wikimedia Commons. File:Holliday junction coloured.png – Wikimedia Commons, the free media repository, 2019b. URL https://commons.wikimedia.org/w/index.php?title=File:Holliday_junction_coloured.pngoldid=346086825 [Accessed: 17-May-2019].

Wikipedia. Xenobot – Wikipedia, the Free Encyclopedia, 2023. URL http://en.wikipedia.org/w/index.php?title=Xenobotoldid=1154384490 [Accessed: 06-September-2023] This file is licensed under the Creative Commons Attribution 4.0 International license. You are free: to share – to copy, distribute and transmit the work to remix – to adapt the work Under the following conditions: attribution – You must give appropriate credit, provide a link to the license, and indicate if changes were made. You may do so in any reasonable manner, but not in any way that suggests the licensor endorses you or your use.

Franz Wilhelmstötter, 2021. Jenetics. URL http://jenetics.io/.

Christof Windeck. Amd Ryzen 3000: Am 7.7. kommt die 7-nm-CPU – heise online, 2019. URL https://www.heise.de/newsticker/meldung/AMD-Ryzen-3000-Am-7-7-kommt-die-7-nm-CPU-4312622.html.

Erik Winfree and Renat Bekbolatov. Proofreading tile sets: Error correction for algorithmic self-assembly. In J. Chen and J. Reif, editors, *DNA Computing. DNA 2003*. Lecture Notes in Computer Science, volume 2943, pages 126–144. Springer-Verlag, Berlin, Heidelberg, 2003.

Erik Winfree, Furong Liu, Lisa Wenzler, and Nadrian Charles Seeman. Design and self-assembly of two-dimensional DNA crystals. *Nature*, 394(6693): 539–544, 1998.

Stephen Wolfram. Computation theory of cellular automata. *Communications in Mathematical Physics*, 96(1):15–57, March 1984. ISSN 1432-0916. doi: 10.1007/BF01217347.

World Health Organization. Antimicrobial resistance global report on surveillance: 2014 summary. Technical report, World Health Organization, 2014.

Rico Wysocki. Efficient algorithms for creating cubes in DNA self-assembly systems. Bachelor thesis, Universität zu Lübeck, 2019.

Ming Xie. *Fundamentals of Robotics: Linking Perception to Action*. Series in Machine Perception and Artificial Intelligence. World Scientific Publishing, 2003. ISBN 9789812383358.

Xtile index, 2009. URL http://www.guptalab.org/xtile/.

Cuiping Yao, Florian Rudnitzki, Gereon Hüttmann, Zhenxi Zhang, and Ramtin Rahmanzadeh. Important factors for cell-membrane permeabilization by gold nanoparticles activated by nanosecond-laser irradiation. *International Journal of Nanomedicine*, 12:5659–5672, 2017. doi: 10.2147/IJN.S140620.

Jejoong Yoo, Chen-Yu Li, Scott Michael Slone, Christopher Maffeo, and Aleksei Aksimentiev. A practical guide to molecular dynamics simulations of DNA origami systems. In *DNA Nanotechnology*, pages 209–229. Springer, 2018.

Syuhei Yoshino, Takeo Miyake, Takeo Yamada, Kenji Hata, and Matsuhiko Nishizawa. Molecularly ordered bioelectrocatalytic composite inside a film of aligned carbon nanotubes. *Advanced Energy Materials*, 3(1):60–64, 2013.

Hang Yu, Bryan Ng, and Winston K. G. Seah. Forwarding schemes for EM-based wireless nanosensor networks in the terahertz band. In *2nd Annual International Conference on Nanoscale Computing and Communication*, NANOCOM' 15, pages 17:1–17:6, New York, NY, USA, 2015. ACM. ISBN 978-1-4503-3674-1.

Hao Yu, Wentian Tang, Guanyu Mu, Haocheng Wang, Xiaocong Chang, Huijuan Dong, Liqun Qi, Guangyu Zhang, and Tianlong Li. Micro-/nanorobots propelled by oscillating magnetic fields. *Micromachines*, 9(11):540, 2018.

Kaan Yuksel, Jens-Peter Kaps, and Berk Sunar. Universal hash functions for emerging ultra-low-power networks. In *Proceedings of the Communications Networks and Distributed Systems Modeling and Simulation Conference*, 2004.

Sidra Zafar, Mohsin Nazir, Taimur Bakhshi, Hasan Ali Khattak, Muhammad Bilal, Kim-Kwang Raymond Choo, Kyung Kwak, and Aneeqa Sabah. A systematic review of bio-cyber interface technologies and security issues for internet of bio-nano things. *IEEE Access*, 9:93529–93566, June 2021. doi: 10.1109/ACCESS.2021.3093442.

John X. J. Zhang and Kazunori Hoshino. *Molecular Sensors and Nanodevices: Principles, Designs and Applications in Biomedical Engineering*. Academic Press, 2018.

Baozhong Zhang, Roger Wepf, Karl Fischer, Manfred Schmidt, Sébastien Besse, Peter Lindner, Benjamin T. King, Reinhard Sigel, Peter Schurtenberger, Yeshayahu Talmon, et al. The largest synthetic structure with molecular precision: Towards a molecular object. *Angewandte Chemie*, 123(3):763–766, 2011.

Wenhui Zhang, Lijuan Du, Zongren Chen, Juan Hong, and Lu Yue. ZnO nanocrystals as anode electrodes for lithium-ion batteries. *Journal of Nanomaterials*, 2016:8056302, 2016.

Mengmeng Zhang, Wenting Cai, Zhong Wang, Shaoli Fang, Runyu Zhang, Hongbing Lu, Ali E. Aliev, Anvar A. Zakhidov, Chi Huynh, Enlai Gao, et al. Mechanical energy harvesters with tensile efficiency of 17.4% and torsional efficiency of 22.4% based on homochirally plied carbon nanotube yarns. *Nature Energy*, 8(2):203–213, 2023.

Victor Zhirnov and Daniel Herr. New frontiers: Self-assembly and nanoelectronics. *Computer*, 34(1):34–43, January 2001. ISSN 0018-9162. doi: 10.1109/2.895116.

Huaijuan Zhou, Carmen C. Mayorga-Martinez, Salvador Pané, Li Zhang, and Martin Pumera. Magnetically driven micro and nanorobots. *Chemical Reviews*, 121(8):4999–5041, 2021.

Index

Nanonetworks: The Future of Communication and Computation, First Edition. Florian-Lennert A. Lau.

Printed and bound by CPI Group (UK) Ltd, Croydon, CR0 4YY
17/09/2026
14964315-0003